MODERN
THERMODYNAMICS

MODERN THERMODYNAMICS
From Heat Engines to Dissipative Structures

Dilip Kondepudi
Professor of Chemistry, Wake Forest University

Ilya Prigogine
Director, International Solvay Institutes
and
Regental Professor, University of Texas at Austin

JOHN WILEY & SONS
Chichester · New York · Weinheim · Brisbane · Toronto · Singapore

Copyright © 1998 Dilip Kondepudi and Ilya Prigogine
Published in 1998 by John Wiley & Sons Ltd,
Baffins Lane, Chichester,
West Sussex PO19 1UD, England

National   01243 779777
International (+44) 1243 779777

e-mail (for orders and customer service enquiries): cs-books@wiley.co.uk.
Visit our Home Page on http://www.wiley.co.uk
or   http://www.wiley.com

Reprinted with corrections October 1999

Other Wiley Editorial Offices

John Wiley & Sons, Inc., 605 Third Avenue,
New York, NY 10158-0012, USA

WILEY-VCH Verlag GmbH, Pappelallee 3,
D-69469 Weinheim, Germany

Jacaranda Wiley Ltd, 33 Park Road, Milton,
Queensland 4064, Australia

John Wiley & Sons (Asia) Pte Ltd, 2 Clementi Loop #02-01,
Jin Xing Distripark, Singapore 129809

John Wiley & Sons (Canada) Ltd, 22 Worcester Road,
Rexdale, Ontario M9W 1L1, Canada

Library of Congress Cataloging-in-Publication Data

Kondepudi, D. K. (Dilip K.), 1952–
    Modern thermodynamics : from heat engines to dissipative
structures / Dilip Kondepudi, Ilya Prigogine.
        p.   cm.
    Includes bibliographical references and index
    ISBN 0 471 97393 9 (hardbound : alk. paper). — ISBN 0 471 97394 7
(pbk. : alk. paper)
    1. Thermodynamics.   I. Prigogine, I. (Ilya)   II. Title
QC311.K66   1998
536′.7—dc21   97-48745
                                                            CIP

British Library Cataloguing in Publication Data

A catalogue record for this book is available from the British Library

ISBN 0 471 97393 9 (hbk)
    0 471 97394 7 (pbk)

Typeset in 11/13pt Times by Thomson Press (India) Ltd, New Delhi
Printed and bound in Great Britain by Bookcraft (Bath) Ltd
This book is printed on acid-free paper responsibly manufactured from sustainable forestation,
for which at least two trees are planted for each one used for paper production.

CONTENTS

PREFACE: WHY THERMODYNAMICS?

Within the past 50 years our view of Nature has changed drastically. Classical science emphasized equilibrium and stability. Now we see fluctuations, instability, evolutionary processes on all levels from chemistry and biology to cosmology. Everywhere we observe irreversible processes in which time symmetry is broken. The distinction between reversible and irreversible processes was first introduced in thermodynamics through the concept of "entropy", the *arrow of time* as Arthur Eddington called it. Therefore our new view of Nature leads to an increased interest in thermodynamics. Unfortunately, most introductory texts are limited to the study of equilibrium states, restricting thermodynamics to idealized, infinitely slow reversible processes. The student does not see the relationship between irreversible processes that naturally occur, such as chemical reactions and heat conduction, and the rate of increase of entropy. In this text, we present a modern formulation of thermodynamics in which the relation between rate of increase of entropy and irreversible processes is made clear from the very outset. Equilibrium remains an interesting field of inquiry but in the present state of science, it appears essential to include irreversible processes as well.

It is the aim of this book to give a readable introduction to present-day thermodynamics starting with its historical roots as associated with heat engines but including also the thermodynamic description of far-from-equilibrium situations. As is well known today, far-from-equilibrium situations lead to new space-time structures. For this reason the restriction to equilibrium situations hides, in our opinion, some essential features of the behaviour of matter and energy. An example is the role of fluctuations. The atomic structure of matter leads to fluctuations. But at equilibrium or near equilibrium, these fluctuations are inconsequential.

Indeed a characteristic feature of equilibrium thermodynamics is the existence of *extremum principles*. For isolated systems entropy increases and is therefore maximum at equilibrium. In other situations (such as constant temperature) there exist functions called thermodynamic potentials which are also extrema (maxima or minima) at equilibrium. This has important consequences. A fluctuation which leads to a deviation from equilibrium is followed by a response which brings back the system to the extremum of the thermodynamic potential. *The equilibrium world is also a stable world.* This is no longer so in far-from-equilibrium situations. Here fluctuations may be amplified by irreversible dissipative processes and lead to new space-time structures which one of us (Prigogine) has called "dissipative structures" to distinguish them from "equilibrium" structures such as crystals. Therefore distance from equilibrium becomes a parameter somewhat similar to temperature. When we lower the temperature, we go from the gaseous state to a liquid and then a solid. As we shall see, here the variety is even greater. Take the example of chemical reactions. Increasing the distance from equilibrium we may obtain in succession oscillatory reactions, new spatial periodic structures and chaotic situations in which the time behavior becomes so irregular that initially close trajectories diverge exponentially.

One aspect is common to all these nonequilibrium situations, the appearance of long-range coherence. Macroscopically distinct parts become correlated. This is in contrast to equilibrium situations where the range of correlations is determined by short-range intermolecular forces. As a result, situations which are impossible to realize at equilibrium become possible in far-from-equilibrium situations. This leads to important applications in a variety of fields. We can produce new materials in nonequilibrium situations where we escape from the restrictions imposed by the phase rule. Also, nonequilibrium structures appear at all levels in biology. We give some simple examples in Chapters 19 and 20. It is now generally admitted that biological evolution is the combined result of Darwin's natural selection as well as of self-organization which results from irreversible processes.

Since Ludwig Boltzmann (1844–1906) introduced a statistical definition of entropy in 1872, entropy is associated with *disorder*. The increase of entropy is then described as an increase of disorder, as the destruction of any coherence which may be present in the initial state. This has unfortunately led to the view that the consequences of the Second Law are self-evident or trivial. This is, however, not true even for equilibrium thermodynamics, which leads to highly nontrivial predictions. Anyway, equilibrium thermodynamics covers only a small fraction of our everyday experience. We now understand that we cannot describe Nature around us without an appeal to nonequilibrium situations. The biosphere is maintained in nonequilibrium through the flow of energy coming from the sun, and this flow is itself the result of the nonequilibrium situation of our present state in the universe.

It is true that the information obtained from thermodynamics both for equilibrium and nonequilibrium situations is limited to a few general statements. We have to supplement them by the equation of state at equilibrium or the rate laws such as chemical reaction rates. Still, the information we obtain from thermodynamics is quite valuable precisely because of its generality.

II

Our book is subdivided into five parts. Chapters 1–4 deal with the basic principles. The systems considered in thermodynamics are *large* systems (the number of particles is typically Avogadro's number $\sim 10^{23}$). Such systems are described by two types of variables: variables such as pressure or temperature which are independent of the size of the system and are called "intensive" variables, and variables such as the total energy which are proportional to the number of particles (extensive variables). Historically thermodynamics started with empirical observations concerning the relation between these variables (e.g. the relation between pressure and volume). This is the main subject of Chapter 1. But the two conceptual innovations of thermodynamics are the formulation of the "First Law" expressing conservation of energy (Chapter 2), and of the Second Law, introducing entropy (Chapter 3).

Ignis mutat res, fire transforms matter. Fire leads to chemical reactions, to processes such as melting and evaporation. Fire makes fuel burn and release heat. Out of all this common knowledge, nineteenth-century science concentrated on the single fact that combustion produces heat and that heat may lead to an increase in volume; as a result, combustion produces work. Fire leads, therefore, to a new kind of machine, the heat engine, the technological innovation on which industrial society has been founded.

What is then the link between "heat" and "work"? This question was at the origin of the formulation of the principle of energy conservation. Heat is of the same nature as energy. In the heat engine, heat is transferred into work but energy is conserved.

But there was more. In 1811 Baron-Joseph Fourier, the perfect of Isère, won the prize of the French Academy of Sciences for his mathematical description of the propagation of heat in solids. The result stated by Fourier was surprisingly simple and elegant: heat flow is proportional to the gradient of temperature. It is remarkable that this simple law applies to matter, whether its state is solid, liquid or gaseous. Moreover, it remains valid whatever the chemical composition of the body, whether it is iron or gold. It is only the coefficient of proportionality between the heat flow and the gradient of temperature that is specific to each substance.

Fourier's law was the first example describing an *irreversible* process. There is a *privileged direction of time* as heat flows according to Fourier's law, from higher to lower temperature. This is in contrast with the laws of Newtonian dynamics in which past and future play the same role (time enters in Newton's law only through a second derivative, so Newton's law is invariant with respect to time inversion $t \rightarrow -t$). It is the Second Law of thermodynamics which expresses the difference between "reversible" and irreversible processes through the introduction of entropy. Irreversible processes *produce entropy*.

The history of the two principles of thermodynamics is a most curious one. Born in the middle of technological questions, they rapidly acquired a cosmological status. Let us indeed state the two principles as formulated by Rudolf Clausius (1822–1888) in the year 1865:

"The energy of the universe is constant.
The entropy of the universe approaches a maximum."

It was the first evolutionary formulation of cosmology. This was a revolutionary statement as the existence of irreversible processes (and therefore of entropy) conflicts with the time-reversible view of dynamics. Of course, classical dynamics has been superseded by quantum theory and relatively. But this conflict remains because in both quantum theory and relativity the basic dynamical laws are time-reversible.

The traditional answer to this question is to emphasize that the systems considered in thermodynamics are so complex (they contain a large number of interacting particles) that we are obliged to introduce approximations. The Second Law of thermodynamics would have its roots in these approximations! Some authors go so far as to state that entropy is only the expression of our ignorance!

Here again the recent extension of thermodynamics to situations far from equilibrium is essential. Irreversible processes lead then to new space-time structures. They play therefore a basic *constructive* role. No life would be possible without irreversible processes (Chapter 19). It seems absurd to suggest that life would be the result of our approximations! We therefore cannot deny the reality of entropy, the very essence of an arrow of time in nature. *We are the children of evolution and not its progenitors*.

Questions regarding the relation between entropy and dynamics have received great attention recently but they are far from simple. Not all dynamical processes require the concept of entropy. The motion of the earth around the sun is an example in which irreversibility (such as friction due to tides) can be ignored and the motion may be described by time-symmetric equations. But recent developments in nonlinear dynamics have shown that such systems are exceptions. Most systems exhibit chaos and irreversible behaviour. We now begin to be able to characterize the dynamical systems for which irreversibility is an essential feature leading to an increase in entropy.

Let us go back to our book. Entropy production plays a central role in our presentation. As we show in Chapter 15, the entropy production can be expressed in terms of thermodynamic flows J_i and thermodynamic forces X_i. An example is heat conduction where J_i is the flow of heat and X_i the gradient of temperature. We can now distinguish three stages. At equilibrium both the flows and the forces vanish. This is the domain of traditional thermodynamics. It is covered in Chapters 5–11. The reader will find many results familiar from all textbooks on thermodynamics.

However, some subjects neglected in most textbooks are treated here. An example is thermodynamic stability theory, which plays an important role both at equilibrium and away from equilibrium. Thermodynamic theory of stability and fluctuation, which originated in the work of Gibbs, is the subject of Chapters 12–14. We begin with the classical theory of stability, as Gibbs formulated it, which depends on thermodynamic potentials. We then discuss the theory of stability in terms of the modern theory of entropy production, which is more general than the classical theory. This gives us the foundation for the study of stability of nonequilibrium systems discussed in a later part of the book. We then turn to the thermodynamic theory of fluctuations that has its origin in Einstein's famous formula that relates the probability of a fluctuation to decrease in entropy. This theory also gives us the basic results that will later lead us to the Onsager reciprocal relations discussed in Chapter 16.

Chapters 15–17 are devoted to the neighborhood of equilibrium which is defined by linear relations between flows and forces (such as realized in Fourier's law). This is a well-explored field dominated by Onsager's reciprocity relation. Indeed in 1931, Lars Onsager discovered the first general relations in nonequilibrium thermodynamics for the linear, near-equilibrium region. These are the famous "reciprocal relations." In qualitative terms, they state that if a force—say, "one" (corresponding, for example, to a temperature gradient)—may influence a flux "two" (for example, a diffusion process), them force "two" (a concentration gradient) will also influence the flux "one" (the heat flow).

The general nature of Onsager's relations has to be emphasized. It is immaterial, for instance, whether the irreversible processes take place in a gaseous, liquid or solid medium. The reciprocity expressions are valid independently of any microscopic assumptions.

Reciprocal relations have been the first results in the thermodynamics of irreversible processes to indicate that this was not some ill-defined no-man's-land but a worthwhile subject of study whose fertility could be compared with that of equilibrium thermodynamics. Equilibrium thermodynamics was an achievement of the nineteenth century, nonequilibrium thermodynamics was developed in the twentieth century, and Onsager's relations mark a crucial point in the shift of interest away from equilibrium toward nonequilibrium.

It is interesting to notice that due to the flow of entropy, even close to equilibrium, irreversibility can no more be identified with the tendency to disorder. We shall give numerous examples in the text, but let us illustrate it here in a simple situation corresponding to *thermal diffusion*. We take two boxes connected by a cylinder, we heat one box and cool the other. Inside the box there is a mixture of two gases, say hydrogen and nitrogen. We then observe that, at the steady state, the hydrogen concentration is higher in one box and the nitrogen concentration is higher in the other. Irreversible processes, here the flow of heat, produce both *disorder* (thermal motion) and *order* (separation of the two components). We see that a nonequilibrium system may evolve spontaneously to a state of increased complexity. This *constructive* role of irreversibility becomes ever more striking in far-from-equilibrium situations to which we now turn.

The main novelty is that in far-from-equilibrium situations which correspond to the third stages of thermodynamics, an extremum principle seldom exists (Chapters 18–19). As a result, any fluctuations may no longer be damped. Stability is no longer the consequence of the general laws of physics. Fluctuations may grow and invade the whole system. We have called them *dissipative structures*, these new spatiotemporal organizations which may emerge in far-from-equilibrium situations. They correspond to a form of supramolecular coherence involving an immense number of molecules. In far-from-equilibrium situations we begin to observe new properties of matter which are hidden at equilibrium.

We have already mentioned the constructive role of irreversibility and the appearance of long-range correlations in far-from-equilibrium systems. Let us add "unpredictability" because the new nonequilibrium states of matter appear at so-called bifurcation points where the system may in general "choose" between various states. We are far from the classical description of Nature as an automaton.

One often speaks of *self-organization*. Indeed, as there are generally a multitude of dissipative structures available, molecular fluctuations determine which one will be chosen. We begin to understand the enormous variety of structures we observe in the natural world. Today the notion of dissipative structures and of self-organization appear in a wide range of fields from astrophysics up to human sciences and economy. We want to quote a recent report to the European Communities due to C. K. Biebricher, G. Nicolis and P. Schuster:

The maintenance of the organization in nature is not — and cannot be — achieved by central management; order can only be maintained by self-organization. Self-organizing systems allow adaptation to the prevailing environment, i.e., they react to changes in the environment with a thermodynamic response which makes the systems extraordinarily flexible and robust against perturbations of the outer conditions. We want to point out the superiority of self-organizing systems over conventional human technology which carefully avoids complexity and hierarchically manages nearly all technical processes. For instance, in synthetic chemistry, different reaction steps are usually carefully separated from each other and contributions from the diffusion of the reactants are avoided by stirring reactors. An entirely new technology will have to be developed to tap the high guidance and regulation potential of self-organizing systems for technical processes. The superiority of self-organizing systems is illustrated by biological systems where complex products can be formed with unsurpassed accuracy, efficiency and speed. (C. F. Biebricher, G. Nicolis and P. Schuster, *Self-organization in the Physico-Chemical and Life Sciences*, 1994, Report on review studies, PSS 0396, Commission of the European Communities, Director General for Science, Research and Development)

III

This introductory text assumes no familiarity with thermodynamics. For this reason we have excluded a number of interesting problems often associated with "extended thermodynamics." Interested readers are invited to consult more specialized texts in the bibliographies. These are the questions which deal with strong gradients or with very long timescales when memory effects have to be included. Every theory is based on idealizations which have a limited domain of validity. In our presentation the assumption is that, at least locally, quantities such as temperature and pressure take well-defined values. More precisely this is called the "local equilibrium assumption" which is a reasonable approximation for the phenomena studied in this book.

Science has no final formulation. And it is moving away from a static geometrical picture towards a description in which evolution and history play essential roles. For this new description of nature, thermodynamics is basic. This is our message to the reader.

ACKNOWLEDGMENTS

This book is the outcome of decades of work. The senior author was a student of Théophile De Donder (1870–1957), the founder of the Brussels School of Thermodynamics. Contrary to the opinion prevalent at that time, De Donder considered that thermodynamics should not be limited to equilibrium situations. He created an active school in Belgium. But his approach remained isolated. Today the situation has drastically changed. There is a major effort going on in the study of nonequilibrium processes, be it in hydrodynamics, chemistry, optics or biology. The need for an extension of thermodynamics is now universally accepted.

This book is intended to present an introductory text which covers both the traditional aspects of thermodynamics as well as its more recent developments. It the literature there are many texts devoted to classical thermodynamics and others, generally more advanced, specialized in nonequilibrium processes. We believe, however, that from the start the student has to be familiar with both aspects in order to grasp the contrasting behavior of matter at equilibrium and far from equilibrium.

De Donder (seated at the center) with his colleagues on the occasion of his 70th birthday.

Our gratitude goes to Professor Paul Glansdorff who coauthored with the senior author, *Thermodynamic Theory of Stability and Fluctuations*.* His outstanding knowledge of the early history of thermodynamics was most helpful to us in writing Chapters 1–4. We are also indebted to our colleagues in Brussels: P. Borckmans, G. Dewel, A. Goldbeter, R. Lefever, M. Malek-Mansour, G. Nicolis and D. Walgraef for numerous suggestions, especially in the field of nonequilibrium thermodynamics. We thank Maarten Donkersloot for many helpful comments and suggestions for improving the presentation of some topics. Thanks also to John Pojman for figures used in Chapter 19 and to Harry Swinney for the photographs of Turing structures used on the cover and in Chapter 19. We also thank Isabelle Stengers, the coauthor of our book "Order out of Chaos", to which we refer the reader for a deeper discussion of the historical and philosophic aspects of thermodynamics.[†]

This book could not have been realized without the continuous support of many institutions. We especially want to thank the Belgian Communauté Française, the International Solvay Institutes in Brussels, the Department of Energy (DOE) of the United States, The Welch Foundation of Texas, the University of Texas at Austin. We would also like to thank Wake Forest University for granting Reynolds leave for one of us (Kondepudi) to facilitate the writing of this book.

We believe that we are today at a decisive moment in the history of science. At all levels of observation, we see an evolutionary universe, we see fluctuations and instabilities. Thermodynamics is the first science which brought an evolutionary view of Nature. This is in contrast with classical or even quantum mechanics which presents us with the picture of time-reversible and deterministic laws. Although there is no arrow of time and no place for instabilities in equilibrium thermodynamics, this is not true in nonequilibrium thermodynamics, where fluctuations and instabilities play an essential role. Therefore it is important that students already at an early stage become familiar with the role of nonequilibirum processes and learn how thermodynamics describes the transition between the idealized static world of classical physics and the evolving, fluctuating world in which we live.

---

* P. Glansdorff and I. Prigogine, *Thermodynamic Theory of Stability and Fluctuations*, Wiley, New York, 1971.
[†] I. Prigogine and I. Stengers, *Order Out of Chaos*, Bantam Books, New York, 1980; see also *The End of Certitudes*, Free Press, New York, 1997.

NOTES FOR INSTRUCTORS

Chapters 1–11 are intended for a one-semester introductory undergraduate course on modern thermodynamics for students in physics, chemistry and engineering. Not all chapters are meant for all three branches; the instructor may drop a few chapters to emphasize others. Chapters 1–11 include examples of worked problems at the end. The exercises in each chapter are chosen to illustrate applications of the subject in many areas. Interdisciplinary research is becoming increasingly important, so it is necessary to make the student aware of a wide variety of applications for thermodynamics at an early stage.

Chapters 12–19 are meant for an advanced undergraduate or a graduate course in thermodynamics; a good knowledge of vector calculus is assumed. Chapters 12–19 do not include worked examples. The exercises are designed to give the student a deeper understanding of the discussed topics and their applications.

Throughout the text, the student is encouraged to use Mathematica* or Maple† to do to tedious calculations or look at complex physicochemical situations. Appendix 1.2 introduces the reader to the use of Mathematica in solving problems in thermodynamics.

Professor Dilip Kondepudi can be contacted by e-mail at dilip@wfu.edu.

For information related to this book, visit the web site at http://www.wfu.edu/~dilip/. This site contains answers to exercises in Chapters 1–11, links to web sites providing thermodynamic data, answers to frequently asked questions, and comments from instructors and students.

Full solutions to the exercises for Chapters 1–11 are available to instructors using this text. Instructors should contact Professor Dilip Kondepudi for the solutions at dilip@wfu.edu.

---

* Mathematica is a registered trademark of Wolfram Research Inc.
† Maple is a registered trademark of Waterloo Maple Inc.

PART I

HISTORICAL ROOTS:
FROM HEAT ENGINES TO COSMOLOGY

1 THE BASIC CONCEPTS

Introduction

Adam Smith's *Wealth of Nations* was published in the year 1776, seven years after James Watt (1736–1819) obtained a patent for his version of the steam engine. Both men worked at the University of Glasgow. Yet, in Adam Smith's great work the only use for coal was in providing heat for workers [1]. The machines of the eighteenth century were driven by wind, water and animals. Nearly 2000 years had passed since Hero of Alexandria made a sphere spin with the force of steam; but fire's power to generate motion and drive machines remained hidden. Adam Smith (1723–1790) did not see in coal this hidden wealth of nations.

But the steam engine revealed new possibilities. The invention that converted heat to mechanical motion not only heralded the Industrial Revolution, it also gave birth to the science of thermodynamics. Unlike the science of Newtonian mechanics—which had its origins in theories of motion of heavenly bodies—thermodynamics was born out of a more practical interest: heat's possibilities to generate motion.

With time, thermodynamics evolved into a theory that describes transformations of states of matter in general, motion generated by heat being a consequence of certain transformations. It is founded on essentially two fundamental laws, one concerning *energy* and the other concerning *entropy*. A precise definition of energy and entropy, as measurable physical quantities, will be presented in Chapters 2 and 3 respectively. In the following two sections we will give an overview of thermodynamics and familiarize the reader with the terminology and concepts that will be developed in the rest of the book.

Every system is associated with an energy and an entropy. When matter undergoes transformation from one state to another, the total energy remains unchanged or is conserved; the total entropy, however, can only increase or, in idealized cases, remain unchanged. These two simple-sounding statements have far-reaching consequences. Max Planck (1858–1947) was deeply influenced by the breadth of the conclusions that can be drawn from them and made masterly use of thermodynamics in his work. In reading this book, we hope the reader will come to appreciate the significance of this often quoted opinion of Albert Einstein (1879–1955):

"A theory is more impressive the greater the simplicity of its premises is, the more different kinds of things it relates, and the more extended its area of applicability.

Therefore the deep impression which classical thermodynamics made upon me. It is the only physical theory of universal content concerning which I am convinced that, within the framework of the applicability of its basic concepts, it will never be overthrown."

1.1 Thermodynamic Systems

Thermodynamic description of natural processes usually begins by dividing the world into a "system" and its "exterior," which is the rest of the world. This cannot be done, of course, when one is considering the thermodynamic nature of the entire universe. The definition of a thermodynamic system often depends on the existence of "boundaries," boundaries that separate the system of interest from the rest of the world. In understanding the thermodynamic behavior of physical systems, the nature of the interaction between the system and the exterior is important. Accordingly, thermodynamic systems are classified into three types, isolated, closed and open systems according to the way they interact with their exterior (Fig. 1.1).

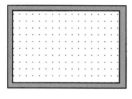

Isolated System

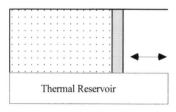

Thermal Reservoir

Closed System

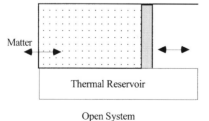

Matter

Thermal Reservoir

Open System

Figure 1.1 Isolated, closed and open systems. Isolated systems exchange neither energy nor matter with the exterior. Closed systems exchange heat and mechanical energy but not matter with the exterior. Open systems exchange both energy and matter with the exterior.

- *Isolated systems* do not exchange energy or matter with the exterior.
- *Closed systems* exchange energy with the exterior but not matter.
- *Open systems* exchange both energy and matter with the exterior.

In thermodynamics, the **state** of a system is specified in terms of macroscopic **state variables** such as volume V, pressure p, temperature T, mole numbers of the chemical constituents N_k, which are self-evident. The two laws of thermodynamics are founded on the concepts of energy U, and entropy S, which, as we shall see, are **functions of state variables**. Since the fundamental quantities in thermodynamics are functions of many variables, thermodynamics makes extensive use of calculus of many variables. A brief summary of some basic identities used in the calculus of many variables is given in Appendix 1.1 (at the end of this chapter). Functions of state variables, such as U and S, are called **state functions**.

It is convenient to classify thermodynamic variables into two categories: variables such as volume and mole number, which are proportional to the size of the system, are called **extensive variables**. Variables such as temperature T and pressure p, that specify a local property, which are independent of the size of the system, are called **intensive variables**.

If the temperature is not uniform, heat will flow until the entire system reaches a state of uniform temperature, the state of **thermal equilibrium**. The state of thermal equilibrium is a special state towards which all isolated systems will inexorably evolve. A precise description of this state will be given later in this book. In the state of thermal equilibrium, the values of total internal energy U and entropy S are completely specified by the temperature T, the volume V and the mole numbers of the chemical constituents N_k:

$$U = U(T, V, N_k) \quad \text{or} \quad S = S(T, V, N_k) \qquad (1.1.1)$$

The values of an extensive variable such as total internal energy U, or entropy S, can also be specified by other extensive variables:

$$U = U(S, V, N_k) \quad \text{or} \quad S = S(U, V, N_k) \qquad (1.1.2)$$

As we shall see in the following chapters, intensive variables can be expressed as derivatives of one extensive variable with respect to another. For example, we shall see that the temperature $T = (\partial U/\partial S)_{V, N_k}$.

1.2 Equilibrium and Nonequilibrium Systems

It is our experience that if a physical system is isolated, its state—specified by macroscopic variables such as pressure, temperature and chemical composition

—evolves *irreversibly* towards a time-invariant state in which we see no further physical or chemical change in the system. It is a state characterized by a uniform temperature throughout the system. This is the state of **thermodynamic equilibrium.** The state of equilibrium is also characterized by several other physical features that we will describe in the following chapters.

The evolution of a state towards the state of equilibrium is due to irreversible processes. At equilibrium, these processes vanish. Thus, a nonequilibrium state can be characterized as a state in which irreversible processes drive the system to the state of equilibrium. In some situations, especially chemical systems, the rate at which the state is transforming due to irreversible processes may be extremely slow, and the isolated system might appear as if it had reached its state of equilibrium. Nevertheless, with appropriate specification of the chemical reactions, the nonequilibrium nature of the state can be identified.

Two or more systems that interact and exchange energy and/or matter will eventually reach the state of thermal equilibrium in which the temperatures of all systems are equal. If a system A is in equilibrium with system B and if B is in equilibrium with system C, then it follows that A is in equilibrium with C. This "transitivity" of the state of equilibrium is some times called the **zeroth law**. Thus, equilibrium systems have one uniform temperature and for these systems there exist state functions of energy and entropy.

Uniformity of temperature, however, is not a requirement for the entropy or energy of a system to be well defined. For **nonequilibrium systems** in which the temperature is not uniform but is well defined locally, we can define densities of

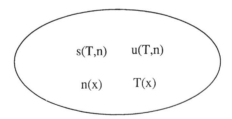

Figure 1.2 In a nonequilibrium system the temperature and number density may vary with position. The entropy and energy of such a system may be described by an entropy density $s(T,n)$ and an energy density $u(T,n)$. The total entropy $S = \int_V s[T(x),n_k(x)]dV$, the total energy $U = \int_V u[T(x),n_k(x)]dV$ and the total mole number $N = \int_V n_k(x)dV$. For such a nonequilibrium system, the total entropy S is not a function of U, N and the total volume V

thermodynamic quantities such as energy and entropy (Fig. 1.2). Thus an energy density

$$u(T, n_k(x)) = \text{internal energy per unit volume} \qquad (1.2.1)$$

can be defined in terms of the local temperature T and the concentration

$$n_k(x) = \text{moles per unit volume} \qquad (1.2.2)$$

Similarly an entropy density, $s(T, n_k)$, can be defined. The total energy U, the total entropy S and the total number of moles of the system then are:

$$S = \int_V s[T(x), n_k(x)]dV \qquad (1.2.3)$$

$$U = \int_V u[T(x), n_k(x)]dV \qquad (1.2.4)$$

$$N = \int_V n_k(x)dV \qquad (1.2.5)$$

In nonuniform systems, the total energy U is no longer a function of other extensive variables such as S, V and N, as in (1.1.2), and obviously one cannot define a single temperature for the entire system because it is not uniform. In general, each of the variables, the total energy U, entropy S, the mole number N and the volume V is no longer a function of the other three variables, as in (1.1.2). But this does not restrict in any way our ability to assign an entropy to a system that is not in thermodynamic equilibrium, as long as the temperature is locally well defined.

In texts on classical thermodynamics, when it is sometimes stated that the entropy of a nonequilibrium system is not defined, it is meant that S is not a function of the variables U, V and N. If the temperature of the system is locally well defined, then indeed the entropy of a nonequilibrium system can be defined in terms of an entropy density, as in (1.2.3).

1.3 Temperature, Heat and Quantitative Laws of Gases

During the seventeenth and eighteenth centuries, a fundamental change occurred in man's conception of Nature. Nature slowly but surely ceased to be solely a vehicle of God's will, comprehensible only through theology. The new "scientific" conception of Nature based on experimentation gave us a different worldview and distanced nature from religion. In the new view, God's creation though it might be, Nature obeyed simple and universal laws, laws that man can know and express in the precise language of mathematics. The key that unlocked this secret was experimentation and quantitative study of physical quantities.

It was during this time of great change that a scientific study of the nature of heat began. This was primarily due to the development of the thermometer, which was constructed and used in scientific investigations since the time of Galileo Galilei (1564–1642) [2, 3]. The impact of this simple instrument was considerable. In the words of Sir Humphry Davy (1778–1829), "Nothing tends to the advancement of knowledge as the application of a new instrument."

The most insightful use of the thermometer was made by Joseph Black (1728–1799), a professor of medicine and chemistry at Glasgow. Black drew a clear distinction between temperature or degree of hotness, and the quantity of heat. His experiments using the newly developed thermometers established the fundamental fact that, *in thermal equilibrium, the temperatures of all the*

Joseph Black (1728–1799) (Courtesy the E. F. Smith Collection, Van Pelt-Dietrich Library, University of Pennsylvania)

substances are the same. This idea was not easily accepted by his contemporaries because it seems to contradict the ordinary experience of touch: a piece of metal feels colder than a piece of wood, even after they have been in contact for a very long time. But the thermometer proved this point beyond doubt. With the thermometer, Black discovered **specific heat** of substances, laying to rest the general belief at his time that the amount of heat required to increase the temperature of a substance by a given amount depended solely on its mass, not on its makeup. He also discovered latent heats of fusion and evaporation of water — the latter with the enthusiastic help from his pupil James Watt (1736–1819) [3, 4].

Though the work of Joseph Black and others established clearly the distinction between heat and temperature, the nature of heat remained an enigma for a long time. Whether heat was an indestructible substance, called the "caloric," that moved from one substance to another or whether it was a form of microscopic motion, continued to be debated as late as the nineteenth century. Finally it became clear that heat was a form of energy that could be transformed to other forms, and the caloric theory was laid to rest — though we still measure the amount of heat in calories.

Temperature can be measured by noting the change of a physical property, such as the volume of a fluid (e.g. mercury) or the pressure of a gas, with degree of hotness. This is an *empirical* definition of temperature. In this case the uniformity of the unit of temperature depends on the uniformity with which the measured property changes as the substance gets hotter. The familiar Celsius scale, which was introduced in the eighteenth century, has largely replaced the Fahrenheit scale, which was introduced in the seventeenth century. As we shall see in the following chapters, the development of the Second Law of thermodynamics during the middle of the nineteenth century gave rise to the concept of *an absolute scale of temperature* that is independent of material properties. Thermodynamics is formulated in terms of the absolute temperature. We shall denote this absolute temperature by *T*.

THE LAWS OF GASES

In the rest of this section we will present an overview of the laws of gases without going into much details. We assume the reader is familiar with the laws of ideal gases.

Box 1.1 Basic Definitions
Pressure is defined as the force per unit area. In SI units the unit of pressure is the

$$\mathbf{Pascal(Pa)} = 1\,\mathrm{N\,m}^{-2}.$$

The pressure due to a column of fluid of uniform density ρ and height h equals

($h\rho g$) where g is the acceleration due to gravity ($9.806\,\mathrm{m\,s^{-2}}$). The pressure due to the earth's atmosphere changes with location and time, but it is often close to $10^5\,\mathrm{Pa}$. For this reason a unit called the **bar** is defined:

$$\mathbf{1\,bar = 10^5\,Pa = 100\,kPa}$$

The atmospheric pressure is also nearly equal to the pressure due to a 760 mm column of mercury. For this reason, the following units are defined:

$$\mathbf{torr = pressure\ due\ to\ 1mm\ column\ of\ mercury}$$

$$\mathbf{atmosphere(atm) = 760\,torr}$$

$$\mathbf{1\,atm = 101.325\,kPa}$$

Temperature is usually measured in Celsius (°C), Fahrenheit (°F) or Kelvin (K). The Celsius and the Fahrenheit scales are empirical but (as we shall see in Chapter 3) the Kelvin scale is the absolute scale based on the second law of thermodynamics. Zero degrees Kelvin is absolute zero, the lowest possible temperature. Temperatures measured in these scales are related as follows:

$$(T/^\circ\mathrm{C}) = \tfrac{5}{9}[(T/^\circ\mathrm{F}) - 32] \qquad (T/\mathrm{K}) = (T/^\circ\mathrm{C}) + 273.15$$

Heat was initially thought to be an indestructible substance called the **caloric**. The theory was that when this caloric passed from one body to another (like some kind of a fluid) it caused changes in temperature. However, in the nineteenth century it was established that heat was not an indestructible caloric but a form of energy that can convert to other forms (Chapter 2). Hence heat is measured in the units of energy. In this text we shall mostly use SI units, in which heat is measured in joules, though the calorie is also common. *A calorie is the amount of heat required to increase the temperature of one gram of water from 14.5 °C to 15.5 °C.* One calorie equals 4.184 Joules.

The **specific heat** of a substance is the amount of heat required to increase a unit mass (usually 1 g or 1 kg) of the substance through 1 °C.

One of the earliest quantitative laws describing the behavior of gases was due to Robert Boyle (1627–1691), an Englishman and a contemporary of Isaac Newton (1642–1727). The same law was also discovered by Edme Mariotte (~1620–1684) in France. In 1660 Boyle published his conclusion in his "New Experiments Physico-Mechanical Touching the Spring of the Air and its Effects": at a fixed temperature T, the volume V, of a gas was inversely proportional to the pressure p, i.e.

$$V = f(T)/p, \quad \text{where } f \text{ is some function} \tag{1.3.1}$$

(Though the temperature that Boyle knew and used was the empirical temperature, as we shall see in Chapter 3, it is appropriate to use the absolute

Robert Boyle (1627–1691) (Courtesy the E. F. Smith Collection, Van Pelt-Dietrich Library, University of Pennsylvania)

temperature T in the formulation of the law of ideal gases. To avoid excessive notation we shall use T whenever it is appropriate.) Boyle also advocated the view that heat was not an indestructible substance (caloric) that passed from one object to another but was "intense commotion of the parts" [5, p. 188].

At constant pressure the variation of volume with temperature was studied by Jacques Charles (1746–1823), who established that

$$\frac{V}{T} = f'(p) \quad \text{where } f' \text{ is some function} \tag{1.3.2}$$

In 1811 Amedeo Avogadro (1776–1856) announced his hypothesis that under conditions of the same temperature and pressure, equal volumes of all gases contained equal number of molecules. This hypothesis greatly helped in

Jacques Charles (1776–1856) (Courtesy the E. F. Smith Collection, Van Pelt-Dietrich Library, University of Pennsylvania)

explaining the changes in pressure due to chemical reactions in which the reactants and products were gases. It implied that at constant pressure and temperature, the volume of a gas is proportional to the number of moles of the gas. Hence, for N moles of a gas:

$$pV = Nf(T) \tag{1.3.3}$$

A comparison of (1.3.1), (1.3.2) and (1.3.3) leads to the well-known equation

$$\boxed{pV = NRT} \tag{1.3.4}$$

in which R is the gas constant ($R = 8.31441\,\mathrm{J\,K^{-1}\,mol^{-1}} = 0.08\,206\,\mathrm{L.atm.}$ $\mathrm{K^{-1}.\,mol^{-1}}$.), known as the **law of ideal gases**.

As more gases were identified and isolated by the chemists during the eighteenth and nineteenth centuries, their properties were studied. It was found that many obeyed Boyle's law approximately. For most gases, this law describes the experimentally observed behavior fairly well for pressures up to a few atmospheres. As we shall see in the next section, the behavior of gases under a wider range of pressures can be described by modifications of the ideal gas law that take into consideration the molecular size and intermolecular forces.

For a *mixture* of ideal gases we have the **Dalton law of partial pressures**, according to which the pressure exerted by each component of the mixture is

Joseph-Louis Gay-Lussac (1778–1850) (Courtesy the E. F. Smith Collection, Van Pelt-Dietrich Library, University of Pennsylvania)

independent of the other components of the mixture, and each component obeys the ideal gas equation. Thus, if p_k is the partial pressure due to component k, we have

$$\boxed{p_k V = N_k R T} \tag{1.3.5}$$

Joseph-Louis Gay-Lussac (1778–1850), who made important contributions to the laws of gases, discovered that a dilute gas expanding into vacuum did so without change in temperature. James Prescott Joule (1818–1889) also verified this fact in his series of experiments that established the equivalence between mechanical energy and heat. In Chapter 2 we will discuss the law of conservation of energy in detail. When the concept of energy and its conservation was established, the implication of this observation became clear. Since a gas expanding into vacuum does not do any work during the processes of expansion, its energy does not change. The fact that temperature does not change during this expansion into vacuum while the volume and pressure do change, implies that the energy of a given amount of ideal gas depends only on its temperature, not on volume or pressure. Furthermore, since the amount of energy (heat) needed to increase the temperature of an ideal gas is proportional to the number of moles of the gas, the energy is proportional to the number of moles N. Thus the energy of the ideal gas, $U(T,N)$, is a function only of the temperature T and the number of moles N. It can be written as

$$U(T,N) = NU_m(T) \tag{1.3.6}$$

in which U_m is the total internal energy per mole. For a mixture of ideal gases the total energy is the sum of the energies of the components:

$$U(T,N) = \sum_k U_k(T,N_k) = \sum_k N_k U_{mk}(T) \tag{1.3.7}$$

in which the components are indexed by k. Later developments established that

$$\boxed{U_m = cRT + U_0} \tag{1.3.8}$$

to a good approximation, in which U_0 is a constant. For monatomic gases such as He and Ar, $c = 3/2$; for diatomic gases such as N_2 and O_2, $c = 5/2$.

The experiments of Gay-Lussac also showed that, at constant pressure, the coefficient of expansion of all dilute gases had nearly the same value of $1/273$ per degree Celsius. Thus a gas thermometer in which the volume of a gas at constant pressure was the indicator of temperature t, had the quantitative relation

$$V = V_0(1 + \alpha t) \tag{1.3.9}$$

in which $\alpha = 1/273$ was the coefficient of expansion at constant pressure. In Chapter 3 we will establish the relation between the temperature t, measured by the gas thermometer, and the absolute temperature T.

These empirical laws of gases played an important part in the development of thermodynamics. They are the testing ground for any general principle and are often used to illustrate these principles.

For most gases, such as CO_2, N_2, O_2, the ideal gas law proved an excellent description of the experimentally observed relation between p, V and T only for pressures up to a few atmospheres. Significant improvements in the laws of gases did not come till the molecular nature of gases was understood. In 1873, more than 200 years after Boyle published his famous results, van der Waals

Johannes van der Waals (1837–1923) (Courtesy the E. F. Smith Collection, Van Pelt-Dietrich Library, University of Pennsylvania)

(1837–1923), proposed an equation in which he incorporated the effects of attractive forces between molecules and molecular size on the pressure and volume of gas. We present van der Waals equation in detail in the next section but here we familiarize the reader with its basic form so it may be compared with the ideal gas equation. According to van der Waals, p, V, N and T are related by the equation

$$\boxed{(p + aN^2/V^2)(V - Nb) = NRT} \tag{1.3.10}$$

In this equation the constant a is a measure of the attractive forces between the molecules and b is proportional to the size of the molecules. For example, the values of a and b for helium are smaller than the corresponding values for CO_2.

Table 1.1 The van der Waals constants a and b for some gases

Gas	$a(\text{L}^2\,\text{atm}\,\text{mol}^{-2})$	$b(\text{L}\,\text{mol}^{-1})$
Acetylene (C_2H_2)	4.40	0.0514
Ammonia (NH_3)	4.18	0.0371
Argon (Ar)	1.34	0.0322
Carbon dioxide (CO_2)	3.59	0.0427
Carbon monoxide (CO)	1.49	0.0399
Chlorine (Cl_2)	6.51	0.0562
Ethyl ether $((CH_3)_2O)$	17.42	0.1344
Helium (He)	0.034	0.0237
Hydrogen (H_2)	0.244	0.0266
Hydrogen chloride (HCl)	3.67	0.0408
Methane (CH_4)	2.25	0.0428
Nitric oxide (NO)	1.34	0.0279
Nitrogen (N_2)	1.39	0.0391
Nitrogen dioxide (NO_2)	5.28	0.0442
Oxygen (O_2)	1.36	0.0318
Sulfur dioxide (SO_2)	6.71	0.0564
Water (H_2O)	5.45	0.0305

Source: These values can be obtained from the critical constants in data source [B]. A more extensive listing of van der Waals constants can be found in data source [F].

The values of the constants a and b for some of the common gases are given in Table 1.1. Unlike the ideal gas equation, this equation explicitly contains molecular parameters and it tells us how the ideal gas pressure and volume are to be "corrected" because of the molecular size and intermolecular forces. We shall see how van der Waals arrived at this equation in the next section. At this point we encourage students to pause and try deriving it on their own before proceeding to the next section.

As one might expect, the energy of the gas is also altered due to forces between molecules. In Chapter 6 we will see that the energy U_{vw} of a van der Waals gas can be written as

$$U_{vw} = U_{ideal} - a\left(\frac{N}{V}\right)^2 V \qquad (1.3.11)$$

The van der Waals equation was a great improvement over the ideal gas law in that it described the observed liquefaction of gases and the fact that, above a certain temperature, called the critical temperature, gases could not be liquefied, regardless of the pressure, as we will see in the following section. But still, it was found that van der Waals' equation failed at very high pressures (exc 1.9). Various improvements suggested by Clausius, Berthelot and others are discussed in Chapter 6.

1.4 States of Matter and the van der Waals Equation

One of the simplest transformations of matter is the melting of solids or the vaporization of liquids. In thermodynamics the various states of matter—solid, liquid, gas—are often referred to as **phases**. At a given pressure every compound has a definite temperature, T_{melt}, at which it melts and a definite temperature, T_{boil}, at which it boils. In fact, these properties can be used to identify a compound or separate the constituents of a mixture. With the development of the thermometer, these properties could be studied with precision. Joseph Black and James Watt discovered another interesting phenomenon associated with the changes of phase: at the melting or the boiling temperature, the heat supplied to a system does not produce an increase in temperature; it only converts the substance from one phase to another. This heat that lays "latent" or hidden without increasing the temperature was called the **latent heat**. When a liquid solidifies or a vapor solidifies, this heat is given out to the surroundings (Fig. 1.3).

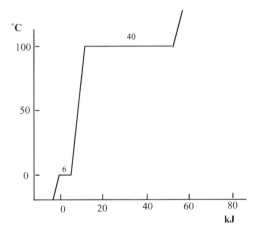

Figure 1.3 The change in temperature of one mole of H_2O versus amount of heat, at a pressure of 1 atm. At the melting point, due to the absorption of heat, the temperature does not change until all the ice melts. It takes about 6 kJ to melt 1 mole of ice, the "latent heat" discovered by Joseph Black. Then the temperature increases until the boiling point is reached, where it remains constant till all the water turns to steam. It takes about 40 kJ to convert 1 mole of water to steam

Clearly the ideal gas equation, good as it is in describing many properties of gases, does not help us to understand the liquefaction of gases. An ideal gas remains a gas at all temperatures and its volume can be compressed without limit. During the eighteenth and nineteenth centuries it became clear that matter is made of atoms and molecules and that there are forces between molecules. It was in this context that Johannes Diederik van der Waals (1837–1923) addressed the problem (in his PhD dissertation) of taking the molecular forces into account in describing the behavior of gases.

van der Waals realized that two main factors were to be added to the ideal gas equation: the effect of molecular attraction and the effect of molecular size. The intermolecular forces would add a correction to the ideal gas pressure whereas the molecular size would decrease the effective volume. In the case of the ideal gas there is no intermolecular attraction. As illustrated in Fig. 1.4, the intermolecular attraction decreases the pressure from its ideal value. If p_{real} is the pressure of a real gas and p_{ideal} is the corresponding pressure of the ideal gas, i.e. the pressure in the absence of intermolecular forces, then $p_{ideal} = p_{real} + \delta p$, where δp is the correction. Since the pressure is proportional to the number density (N/V) (as can be seen from the ideal gas equation), δp should be proportional to (N/V). In addition, the total force on each molecule close to the

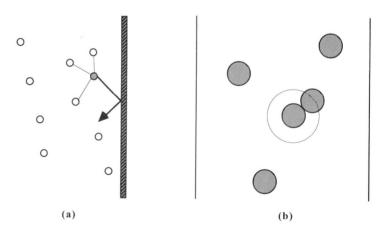

(a)   (b)

Figure 1.4   van der Waals considered molecular interaction and molecular size to improve the ideal gas equation. (a) The pressure of a real gas is less than the ideal gas pressure because intermolecular attraction decreases the speed of the molecules approaching the wall. Therefore $p_{real} = p_{ideal} - \delta p$. (b) The volume available to molecules is less than the volume of the container due to the finite size of the molecules. This "excluded" volume depends on the total number of molecules. Therefore $V_{ideal} = V - bN$

wall of the container is also proportional to the number density (N/V); hence δp should be proportional to two factors of (N/V) so that one may write: $\delta p = a(N/V)^2$. The correction to the volume due to the molecular size, i.e., the "excluded volume," is simply proportional to the number of molecules. Hence $V_{ideal} = V - bN$, in which b is the correction for one mole. Substituting these values in the ideal gas equation $p_{ideal} V_{ideal} = NRT$, we obtain the van der Waals equation

$$(p + aN^2/V^2)(V - Nb) = NRT \qquad (1.4.1)$$

Typical p–V curves at a fixed T, the **p–V isotherms**, for a van der Waals gas are shown in Fig. 1.5. These curves show a transition from a gas to a liquid in the region where the p–V curve is multivalued, i.e. a given pressure does not correspond with a unique volume. This region represents a state in which the liquid and the gas phases are in thermal equilibrium. As we shall see in Chapter 13, the van der Waals curve in this region represents unstable states. The actual state of the gas follows the straight line ACB shown in Fig. 1.5.

The van der Waals equation also exhibits a **critical temperature** T_c: if the temperature T is greater than T_c the p–V curve is always single-valued,

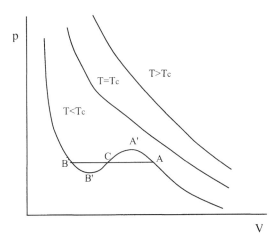

Figure 1.5   The van der Waals isotherms. For $T < T_c$, there is a region AA'CB'B in which, for a given valus of p, the volume V is not uniquely specified by the van der Waals equation. In this region the gas transforms to a liquid. The states on A'B' are unstable. The observed state follows the path ACB. A detailed discussion of this region is given in Chapter 7

indicating there is no transition to the liquid state. Being a cubic, the van der Waals equation has two extrema for $T < T_c$. Below T_c, as T increases, these two extrema move closer and finally coalesce at $T = T_c$. Above the critical temperature there is no phase transition from a gas to a liquid; the distinction between gas and liquid disappears. (This does not happen for a transition between a solid and a liquid because a solid is more ordered than a liquid; the two states are always distinct.) The pressure and volume at which this happens are called the **critical pressure** p_c and **critical volume** V_c. Experimentally, the **critical constants** p_c, V_c and T_c can be measured and they are tabulated. We can relate these critical parameters to the van der Waals parameters a and b by the following means. We note that if we regard $p(V, T)$ as function of V, then for $T < T_c$, the derivative $(\partial p / \partial V)_T = 0$ at the two extrema. As T increases, at the point where the two extrema coincide, i.e. at the critical point $T = T_c$, $p = p_c$ and $V = V_c$, we have an *inflection point*. At an inflection point, the first and second derivatives of a function vanish. Thus, at the critical point

$$\left(\frac{\partial p}{\partial V}\right)_T = 0 \qquad \left(\frac{\partial^2 p}{\partial V^2}\right)_T = 0 \qquad (1.4.2)$$

Using these equations one can obtain the following relations between the critical constants and the constants a and b (exc 1.11):

$$\boxed{a = \frac{9}{8} R T_c V_{mc}} \qquad \boxed{b = \frac{V_{mc}}{3}} \qquad (1.4.3)$$

in which V_{mc} is the molar critical volume. Conversely we can write the critical constants in terms of the van der Waals constants a and b (exc 1.11):

$$\boxed{T_c = \frac{8a}{27Rb}} \qquad \boxed{p_c = \frac{a}{27b^2}} \qquad \boxed{V_{mc} = 3b} \qquad (1.4.4)$$

Table 1.1 contains the values of a and b for some gases.

Thus every gas has a characteristic temperature T_c, pressure p_c, and volume V_{mc}. In view of this, one can introduce dimensionless **reduced variables** defined by

$$T_r = T/T_c \qquad V_r = V_m/V_{mc} \qquad p_r = p/p_c \qquad (1.4.5)$$

If the van der Waals equation is rewritten in terms of the reduced variables, we obtain the following universal equation valid for all gases (exc 1.13):

$$\boxed{p_r = \frac{8T_r}{3V_r - 1} - \frac{3}{V_r^2}} \qquad (1.4.6)$$

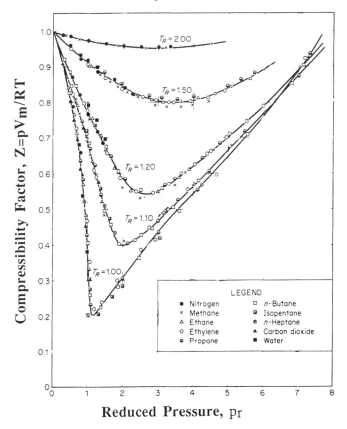

Figure 1.6 Compressibility factor Z as a function of reduced van der Waals variables (Reprinted with permission from Goug-Jen Su, *Industrial and Engineering Chemistry*, **38** (1946), 803. Copyright 1946, American Chemical Society)

This equation has found excellent experimental confirmation. Thus the reduced pressures of all gases have the same value at a given value of reduced volume and reduced temperature. This is called the **law of corresponding states**.

The deviation from the ideal gas behavior is indicated by the compressibility factor Z defined as $Z = (pV_m/RT)$, which equals 1 for ideal gas. Fig. 1.6 shows the agreement between experimental data and (1.4.6) for a plot of Z versus the reduced pressure p_r for various gases.

Sample Mathematica codes for doing numerical and algebraic calculations using the van der Waals equation are provided in Appendix 1.2.

Appendix 1.1: Partial Derivatives

DERIVATIVES OF MANY VARIABLES

When a variable such as energy $U(T, V, N_k)$ is a function of many variables V, T and N_k, its *partial derivative* with respect to each variable is defined by holding all other variables constant. Thus, if $U(T, V, N) = (5/2)NRT - aN^2/V$ then the partial derivatives are

$$\left(\frac{\partial U}{\partial T}\right)_{V,N} = \frac{5}{2}NR \tag{A1.1.1}$$

$$\left(\frac{\partial U}{\partial N}\right)_{V,T} = \frac{5}{2}RT - a\frac{2N}{V} \tag{A1.1.2}$$

$$\left(\frac{\partial U}{\partial V}\right)_{N,T} = a\frac{N^2}{V^2} \tag{A1.1.3}$$

The subscripts indicate the variables that are held constant during the differentiation. In cases where the variables being held constant are understood, the subscripts are often dropped. The change in U, i.e. the differential dU, due to changes in N, V and T is given by

$$dU = \left(\frac{\partial U}{\partial T}\right)_{V,N} dT + \left(\frac{\partial U}{\partial V}\right)_{T,N} dV + \left(\frac{\partial U}{\partial N}\right)_{V,T} dN \tag{A1.1.4}$$

For functions of many variables, there is a second derivative corresponding to every pair of variables: $\partial^2 U/\partial T\,\partial V$, $\partial^2 U/\partial N\,\partial V$, $\partial^2 U/\partial T^2$ etc. For the "cross derivatives" such as $\partial^2 U/\partial T\,\partial V$ that are derivatives with respect to two different variables, the order of differentiation does not matter. That is

$$\frac{\partial^2 U}{\partial T\,\partial V} = \frac{\partial^2 U}{\partial V\,\partial T} \tag{A1.1.5}$$

The same is valid for all higher derivatives such as $\partial^3 U/\partial T^2\,\partial V$, i.e. the order of differentiation does not matter.

BASIC IDENTITIES

Consider three variables x, y and z, each of which can be expressed as a function of the other two variables, $x = x(y, z)$, $y = y(z, x)$ and $z = z(x, y)$. (p, V and T in the ideal gas equation $pV = NRT$ is an example). Then the following identities are valid:

$$\left(\frac{\partial x}{\partial y}\right)_z = \frac{1}{\left(\dfrac{\partial y}{\partial x}\right)_z} \tag{A1.1.6}$$

$$\left(\frac{\partial x}{\partial y}\right)_z \left(\frac{\partial y}{\partial z}\right)_x \left(\frac{\partial z}{\partial x}\right)_y = -1 \tag{A1.1.7}$$

Consider a functions of x and y, $f = f(x, y)$, other than z. Then

$$\left(\frac{\partial f}{\partial x}\right)_z = \left(\frac{\partial f}{\partial x}\right)_y + \left(\frac{\partial f}{\partial y}\right)_x \left(\frac{\partial y}{\partial x}\right)_z \tag{A1.1.8}$$

Appendix 1.2: Mathematica Codes

CODE A: MATHEMATICA CODE FOR EVALUATING THE VAN DER WAALS PRESSURE

```
(* Van der Waals Equation for N = 1 *)
(* values of a and b for CO2 *)
a = 3.59;(* L^2.atm.mol^-2 *)
b = 0.0427;(* L.mol^-1 *)
R = 0.0821;(* L.atm.K^-1.mol^-1 *)
PVW[V_,T_] := (R*T/(V-b))-(a/(V^2));
PID[V_,T_] := R*T/V;
PID[1.5,300]
PVW[1.5,300]
TC = (8/27)*(a/(R*b))
```

Output:
  16.42
  15.3056
  303.424

The van der Waals isotherms can be plotted using a command such as

```
Plot[{PVW[V, 320],PVW[V, 128],PVW[V, 150]},{V, 0.05, 0.3}]
```

CODE B: MATHEMATICA CODE FOR OBTAINING THE CRITICAL CONSTANTS FOR THE VAN DER WAALS EQUATION

```
p[V_,T_] := (R*T/(V-b))-(a/V^2);
(* At the critical point the first and second derivatives of p
with respect to V are zero *)
```

```
(* First derivative *)
D[p[V,T],V]
```

$$\frac{2a}{V^3} - \frac{RT}{(-b+V)^2}$$

```
(* Second derivative *)
D[p[V,T],V,V]
```

$$\frac{-6a}{V^4} + \frac{2RT}{(-b+V)^3}$$

```
(* Solve for T and V when the first and second derivatives
are zero *)
Solve[{(-6*a)/V^]4 + (2*R*T)/(-b+V)^3 == 0,
(2*a)/V^3 - (R*T)/(-b+V)^2 == 0}, {T,V}]
```

$$\{\{T- > \frac{8a}{27bR}, V- > 3b\}\}$$

```
(* Now we can substitute these values in the equation for p
and obtain pc *)
T = (8*a)/(27*b*R); V = 3*b; p[V, T]
```

$$\frac{a}{27b^2}$$

```
(* Thus we have all the critical variables pc = a/(27*b^2);
Tc = (8*a)/(27*b*R); Vc = 3*b; *)
```


CODE C: MATHEMATICA CODE FOR OBTAINING THE LAW OF
CORRESPONDING STATES

```
p[V_,T_] := (R*T/(V-b)) - (a/V^2);
T = Tr*(8*a)/(27*b*R); V = Vr*3*b; pc = a/(27*b^2);
(* In terms of these variables the reduced pressure pr = p/pc.
This can now be calculated *)
p[V,T]/pc
```

$$\frac{27b^2\left(\frac{-a}{9b^2\,Vr^2} + \frac{8a\,Tr}{27b(-b+3b\,Vr)}\right)}{a}$$

```
(* This form of pr is clumsy, so let us simplify it *)
```

```
Simplify [(27*b^2*(-a/(9*b^2*Vr^2)+(8*a*Tr)/(27*b*
(-b+3*b*Vr))))/a]
```

$$\frac{-3}{Vr^2} + \frac{8Tr}{-1+3Vr}$$

```
(* Hence we have the following relation for the reduced
variables, which is the law of corresponding states! *)
pr = (8*Tr)/(3*Vr - 1)) - 3/Vr^2
```

References

1. Prigogine, I. and Stengers, I. *Order Out of Chaos*. 1984, New York: Bantam.
2. Mach, E., *Principles of the Theory of Heat*. 1986, Boston: D. Reidel (original German edition published in 1896).
3. Conant, J. B. (ed.), *Harvard Case Histories in Experimental Science Vol. 1*. 1957, Cambridge MA: Harvard University Press.
4. Mason, S. F., *A History of the Sciences*. 1962, New York: Collier Books.
5. Segrè, E., *From Falling Bodies to Radio Waves*. 1984, New York: W. H. Freeman.
6. Planck, M., *Treatise on Thermodynamics*, 3d ed. 1945, New York: Dover.

Data Sources

[A] NBS table of chemical and thermodynamic properties. *J. Phys. Chem. Reference Data*, **11**, suppl. 2 (1982).
[B] G. W. C. Kaye and T. H. Laby (eds), *Tables of Physical and Chemical Constants*. 1986, London: Longman.
[C] I. Prigogine and R. Defay, *Chemical Thermodynamics*. 1967, London: Longman.
[D] J. Emsley, *The Elements*, 1989, Oxford: Oxford University Press.
[E] L. Pauling, *The Nature of the Chemical Bond*. 1960, Ithaca NY: Cornell University Press.
[F] D. R. Lide (ed.) *CRC Handbook of Chemistry and Physics*, 75th ed. 1994, Ann Arbor MI: CRC Press.

Examples

Example 1.1 The atmosphere consists of 78.08% by volume of N_2, and 20.95% of O_2. Calculate the partial pressures due to the two gases.

Solution: The specification "% by volume" may be interpreted as follows. If the components of the atmosphere were to be separated, at the pressure of 1 atm, the volume occupied by each component would be specified by the volume %. Thus, if we isolate the N_2 in 1.000 L of dry air, at a pressure of 1 atm its volume will be 0.781 L. According to the ideal gas law, at a fixed pressure and temperature, the number of moles $N = V(p/RT)$, i.e. the mole number is proportional to the volume. Hence % by volume is the same as % by mole number, i.e. 1.000 mole of air consists of 0.781 moles of N_2. According to the Dalton law (1.3.5) the partial pressure is proportional to the number of moles, the partial pressure of N_2 is 0.781 atm and that of O_2 is 0.209 atm.

Example 1.2   Using the ideal gas approximation, estimate the change in the total internal energy of 1.00 L of N_2 at $p = 2.00$ atm and $T = 298.15$ K, if its temperature is increased by 10.0 K. What is the energy required to heat 1.00 mole of N_2 from 0.0 K to 298 K?

Solution: The energy of an ideal gas depends only number of moles and the temperature. For a diatomic gas such as N_2 the energy per mole equals $(5/2)RT + U_0$. Hence for N moles of N_2 the change in energy ΔU for a change in temperature from T_1 to T_2 is

$$\Delta U = N(5/2)R(T_2 - T_1)$$

In the above case

$$N = pV/RT = \frac{2.00\,\text{atm} \times 1.00\,\text{L}}{0.0821\,\text{L atm mol}^{-1}\,\text{K}^{-1}(298.15)} = 8.17 \times 10^{-2}\,\text{mol}$$

Hence

$$\Delta U = (8.17 \times 10^{-2}\,\text{mol})\frac{5}{2}\left(8.314\,\text{J mol}^{-1}\,\text{K}^{-1}\right)(10.0\,K)$$

$$= 17.0\,\text{J}$$

(Note the different units of R used in this calculation.)

The energy required to heat 1.00 mole of N_2 from 0 K to 298 K is

$$(5/2)RT = (5/2)(8.314\,\text{J K}^{-1})298\,\text{K} = 6.10\,\text{kJ}$$

Example 1.3   At $T = 300$ K, 1.00 mole of CO_2 occupies a volume of 1.50 L. Calculate the pressures given by the ideal gas equation and the van der Waals equation. (The van der Waals constants a and b can be obtained from Table 1.1.)

Solution: The ideal gas pressure is

$$p = 1.00\frac{\text{mol} \times 0.0821\,\text{atm L mol}^{-1}\,\text{K}^{-1} \times 300\,\text{K}}{1.50\,L} = 16.4\,\text{atm}$$

The pressure according to the van der Waals equation is

$$p = \frac{NRT}{V - Nb} - a\frac{N^2}{V^2}$$

Since the van der Waals constants a and b given in Table 1.1 are in units of $L^2\,atm\,mol^{-2}$ and $L\,mol^{-2}$ respectively, we will use the value $R = 0.0821\,atm\,L\,mol^{-1}\,K^{-1}$. This will give the pressure in atm:

$$p = \frac{1.00(0.0821)300}{1.50 - 1.00(0.0421)} - 3.59\frac{1.00}{1.50^2} = 15.3\,atm$$

Exercises

1.1 Consider two identical vessels, one containing $CO_2(g)$ and the other He(g). Assume that both contain the same number of molecules and are maintained at the same temperatures. The pressure exerted by a gas is a result of molecular collisions against the container walls. Thus, intuitively one might expect that the pressure exerted by the heavier molecules of $CO_2(g)$ will be larger than the pressure due to He(g). Compare this expectation with the pressure predicted by the ideal gas law (or the Avogadro hypothesis). How would you explain the prediction of the ideal gas law which is validated by the experiment?

1.2 Describe an experimental method, based on the ideal gas law, to obtain the molecular mass of a gas.

1.3 (a) Calculate the number of moles of gas per cubic meter of atmosphere at $p = 1\,atm$ and $T = 298\,K$ using the ideal gas equation.
(b) The atmospheric content of CO_2 is about 360 ppmv (parts per million by volume). Assuming a pressure of 1.00 atm, estimate the amount of CO_2 in a 10.0 km layer of the atmosphere at the surface of the earth. The radius of the earth is 6370 km. (The actual amount of CO_2 in the atmosphere is about 6.0×10^{16} moles.)
(c) The atmospheric content of O_2 is 20.946% by volume. Using the result in part (b), estimate the total amount of O_2 in the atmosphere.
(d) Life on earth consumes about 0.47×10^{16} moles of O_2 per year. If the photosynthetic production of O_2 were to suddenly stop, how long would the oxygen in the atmosphere last at the present rate of consumption?

1.4   The production of fertilizers begins with the Haber processes which is the reaction $3H_2 + N_2 \rightarrow 2NH_3$ conducted at about 500 K and a pressure of about 300 atm. Assume this reaction occurs in a vessal of fixed volume and temperature. If the initial pressure due to 300.0 mol H_2 and 100.0 mol N_2 is 300.0 atm, what will the final pressure be? What would the final pressure be if initially the system contained 240.0 mol H_2 and 160.0 mol N_2 at $p = 300.0$ atm?

1.5   The van der Waals constants for N_2 are $a = 1.390\,L^2\,atm\,mol^{-2}$ and $b = 0.0391\,L\,mol^{-1}$. Consider 0.50 moles of $N_2(g)$ is in a vessel of volume 10.0 L. Assuming that the temperature is 300 K, compare the pressures predicted by the ideal gas equation and the van der Waals equation.
(a) What is the percentage difference in using the ideal gas equation instead of the van der Waals equation?
(b) Keeping $V = 10.0\,L$, use Maple or Mathematica to plot p versus N for $N = 1$ to 100, using the ideal gas and van der Waals equations. What do you notice about the difference between the pressure predicted by the two equations?

1.6   For 1.00 mol of Cl_2 in a volume of 2.50 L, calculate the difference in the energy between U_{ideal} and U_{vw}. What is the percentage difference when compared to $U_{ideal} = \frac{5}{2}\,NRT$? (Use Table 1.1.) Assume $T = 298$ K.

1.7   (a) Using the ideal gas equation, calculate the volume of one mole of gas at a temperature of $25\,°C$ and a pressure of 1 atm. This volume is called the *Avogadro volume*.
(b) The atmosphere of Venus is 98% $CO_2(g)$. The surface temperature is about 750 K and the pressure is about 90 atm. Using the ideal gas equation, calculate the volume of 1 mole of $CO_2(g)$ under these conditions (Avogadro volume on Venus).
(c) Use Maple or Mathematica and the van der Waals equation to obtain the Avogadro volume on Venus and compare it (find the percentage difference) with the result obtained using the ideal gas equation.

1.8   The van der Waals parameter b is a measure of the volume excluded due to the finite size of the molecules. Estimate the size of a single molecule from the data in Table 1.1.

1.9   Though the van der Waals equation was a big improvement over the ideal gas equation, its validity is also limited. Compare the following experimental data with the predictions of the van der Waals equation for **1 mole** of CO_2 at $T = 40\,°C$.

p(atm)	V_m (L mol^{-1})
1	25.574
10	2.4490
25	0.9000
50	0.3800
80	0.1187
100	0.0693
200	0.0525
500	0.0440
1000	0.0400

Source: I. Prigogine and R. Defay, *Chemical Thermodynamics*, 1967, Longman, London.

1.10 Use Mathematica or Maple to plot the van der Waals p–V curves for some of the gases listed in Table 1.1; see Appendix 1.2 for sample programs. In particular, compare the van der Waals curves for CO_2 and He with the ideal gas equation.

1.11 From the van der Waals equation, using (1.4.2) obtain (1.4.3) and (1.4.4). (These calculations may also be done using Mathematica or Maple.)

1.12 Using Table 1.1 and the relations (1.4.4) obtain the critical temperature T_c, critical pressure p_c and critical molar volume V_{mc} for CO_2, H_2 and CH_4. Write a Maple or Mathematica code to calculate the van der Waals constants a and b given T_c, p_c and V_{mc} for any gas.

1.13 Using Mathematica or Maple obtain equation (1.4.6) from equation (1.4.5).

2 THE FIRST LAW OF THERMODYNAMICS

The Idea of Energy Conservation Amidst New Discoveries

The concepts of kinetic energy, associated with motion, and of potential energy, associated with conservative forces such as gravitation, were well known at the beginning of the nineteenth century. For a body in motion, the conservation of the sum of kinetic and potential energy is a direct consequence of Newton's laws (exc. 2.1). But this concept had no bearing on the multitude of thermal, chemical and electrical phenomena that were being investigated at that time. And, during the final decades of the eighteenth century and the beginning decades of the nineteenth century, new phenomena were being discovered at a rapid pace.

The Italian physician Luigi Galvani (1737–1798) discovered that a piece of charged metal could make the leg of a dead frog twitch! The amazed public was captivated by the idea that electricity could generate life, as dramatized by Mary Shelley (1797–1851) in her *Frankenstein*. Summarizing the results of his investigations in a paper published in 1791, Galvani attributed the source of electricity to animal tissue. But it was the physicist Alessandro Volta (1745–1827) who recognized that the "Galvanic effect" is due to the passage of electric current. In 1800 Volta went on to construct the so-called voltaic pile, the first "chemical battery": electricity could now be generated from chemical reactions. The inverse effect, the driving of chemical reactions by electricity, was demonstrated by Michael Faraday (1791–1867) in the 1830s. The newly discovered electric current could also produce heat and light. To this growing list of interrelated phenomena, in 1819 the Danish physicist Hans Christian Ørsted (1777–1851) added the generation of a magnetic field by an electrical current. In Germany, in 1822, Thomas Seebeck (1770–1831) (who helped Goethe in his scientific investigations) demonstrated the "thermoelectric effect," the generation of electricity by heat. The well-known Faraday's law of induction, the generation of an electrical current by a changing magnetic field, came in 1831. All these discoveries presented a great web of interrelated phenomena in heat, electricity, magnetism and chemistry to the nineteenth-century scientists.

Soon, within the scientific community that investigated this multitude of new phenomena, the idea that all these effects really represented the transformation of one indestructible quantity, "the energy," began to take shape (see the article "Energy Conservation as an Example of Simultaneous Discovery" in [1]). This law of conservation of energy is the First Law of thermodynamics. We will see details of its formulation in the following sections.

The mechanical view of nature holds that all energy is ultimately reducible to kinetic and potential energy of interacting particles. Thus, the law of conservation of energy may be thought of as essentially the law of conservation of the sum of kinetic and potential energies of all the constituent particles. Cornerstones for the formulation of the First Law are the decisive experiments of James Prescott Joule (1818–1889) of Manchester, a brewer and an amateur scientist. Here is how Joule expressed his view of conservation of energy [2, 3]:

"Indeed the phenomena of nature, whether mechanical, chemical or vital, consist almost entirely in a continual conversion of attraction through space*, living force† and heat into one another. Thus it is that order is maintained in the universe—nothing is deranged, nothing ever lost, but the entire machinery, complicated as it is, works smoothly and harmoniously. And though, as in the awful vision of Ezekiel, 'wheel may be in the middle of wheel', and everything may appear complicated and involved in the apparent confusion and intricacy of an almost endless variety of causes, effects, conversion, and arrangements, yet is the most perfect regularity preserved—the whole being governed by the sovereign will of God."

In practice, however, we measure energy in terms of heat and changes in macroscopic variables. Energy can take many forms: mechanical work, heat, chemical energy and energy associated with electric and magnetic fields. For each of these forms we can specify the energy in terms of macroscopic variables.

2.1 The Nature of Heat

Though the distinction between temperature and heat was recognized in the seventeenth century as a result of the work of Joseph Black and others, the nature of heat was not clearly understood till the middle of the nineteenth century. Robert Boyle, Isaac Newton and others held the view that heat was microscopic agitative motion of particles. An opposing view, which prevailed in France, was that heat was an indestructible fluid-like substance that was exchanged between material bodies. This indestructible substance was called **caloric** and it was measured in "calories" (Box 2.1). In fact, such figures as Antoine-Laurent Lavoisier (1743–1794), Jean-Baptiste-Joseph Fourier (1768–1830), Pierre-Simon de Laplace (1749–1827), Siméon-Denis Poisson (1781–1840) all supported the caloric theory of heat. Even Sadi Carnot (1796–1832), in whose insights the Second Law originated, initially used the concept of caloric, though he later rejected it.

The true nature of heat, a form of energy that can interconvert to other forms of energy, was established after much debate. One of the most dramatic demonstrations of the conversion of mechanical energy to heat was done by

---

* Potential energy.
† Kinetic energy.

Benjamin Thompson, an American born in Woburn, Massachusetts, whose adventurous life took him to Bavaria, where he became Count von Rumford (1753–1814) [4]. Rumford immersed metal cylinders in water and drilled holes in them. The heat produced due to mechanical friction could bring the water to a boil! He even estimated that the production of 1 calorie of heat requires about 5.5 joules [5].

It was the results of the careful experiments of James Prescott Joule, reported in 1847, which established beyond doubt that heat was not an indestructible substance, that in fact it can be transformed to mechanical energy and vice versa [5,6]. Furthermore, Joule showed there is an equivalence between heat and mechanical energy in the following sense: a certain amount of mechanical energy, regardless of the particular means of conversion, always produces the

---

Box 2.1 Basic Definitions

Heat can be measured by the increase in temperature it causes in a body. In this text we shall mostly use SI units, in which heat is measured in joules, though the calorie is also commonly used.

The Calorie. The calorie was originally defined as the amount of heat required to increase the temperature of 1 g of water by 1 °C. When it was realized that this amount depended on the initial temperature of water, the following definition was adopted: *A calorie is the amount of heat required to increase the temperature of one gram of water from* 14.5 °C *to* 15.5 °C.

Work and Heat. In classical mechanics, when a force **F** displaces a body by an amount $d\mathbf{s}$, the work done $dW = \mathbf{F} \cdot d\mathbf{s}$. Work is measured in joules. Dissipative forces, such as friction between solids in contact, or viscous forces in liquids, can generate heat from work. Joule's experiments demonstrated that a certain amount of work, regardless of the manner in which it is performed, always produces the same amount of heat. Thus, the following equivalence between work and heat is established*:

$$1 \text{ calorie} = 4.184 \text{ Joules}$$

Heat capacity. The heat capacity C of a body is the ratio of the heat absorbed dQ to the resulting increase in temperature dT:

$$\frac{dQ}{dT} = C$$

The change in temperature depends on whether the substance is maintained at constant volume or constant pressure. The corresponding heat capacities are denoted by C_V and C_p respectively.
Molar heat capacity. The molar heat capacity is the heat capacity of one mole of the substance.

---

* The current practice is to define a calorie as 4.184 Joules.

same amount of heat. (4.184 joules produce 1 calorie of heat.) This meant heat and mechanical energy could be thought of as different manifestations of the same physical quantity, the "energy".

But still, what is heat? In the classical picture of particle motion, it is a disordered form of kinetic energy. Molecules in incessant motion, collide and randomize their kinetic energy. When a body is heated or cooled, the average kinetic energy of its molecules changes. In fact, we will see in later chapters that the average kinetic energy of the molecules $(mv_{avg}^2/2) = 3k_BT/2$, in which v_{avg} is the average speed of the molecule, $k_B = 1.381 \times 10^{-23}$ J K^{-1} is the Boltzmann constant and T is the temperature in Kelvin. In the special situations of phase transformations, heat does not change the temperature of the body, but causes a transformation of the state.

That is not all we can say about heat. In addition to matter, we also have fields. The interaction between the particles is described by fields, such as electromagnetic fields. Classical physics had established that electromagnetic radiation was a physical quantity that can carry energy and momentum. So when particles gain or lose energy, some of it can transform into the energy of the field. The energy associated with electromagnetic radiation is an example. The interaction between matter and radiation also leads to a state of thermal equilibrium in which a temperature can be associated with radiation. Radiation in thermal equilibrium with matter is called "heat radiation" or "thermal radiation". We shall study the thermodynamics of thermal radiation in some detail in Chapter 11.

During the twentieth century, our view of particles and fields has been unified by modern quantum field theory. According to quantum field theory, all particles are excitations of quantum fields. We now know that electromagnetic fields are associated with particles we call photons, though they also have a wave nature.

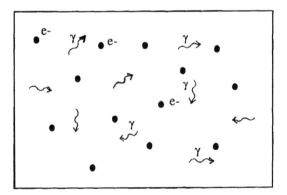

Figure 2.1 Classical picture of a gas of electrons at low temperatures in equilibrium with radiation

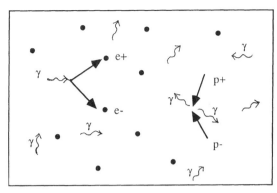

Figure 2.2   A gas of electrons and positrons in equilibrium with radiation at very high temperatures. At temperatures over 10^{10} K, particle–antiparticle pair creation and annihilation begins to occur and the total number of particles is no longer a constant. At these temperatures, electrons, positrons and photons are in the state called thermal radiation. The energy density of thermal radiation depends only on the temperature

Other fields, such as those associated with nuclear forces, have corresponding particles. Just as photons are emitted or absorbed by molecules undergoing transition from one state to another (Fig. 2.1)—which in the classical picture corresponded to emission or absorption of radiation—during high-energy particle interactions, particles such as electrons, mesons and protons can be spontaneously emitted or absorbed. One of the most remarkable discoveries of modern physics is that every particle has an antiparticle. When a particle encounters its antiparticle, they may annihilate each other, converting their energy into other forms, such as photons. All this has expanded our knowledge of the possible states of matter. At temperatures that we normally experience, collisions between molecules produce emission of photons but not other particles. At sufficiently high temperatures (greater than 10^{10} K), other particles can also be similarly created as a result of collisions. Particle creation is often in the form of particle–antiparticle pairs (Fig. 2.2). Thus, there are states of matter in which there is incessant creation and annihilation of particle-antiparticle pairs, a state in which the number of particles does not remain constant. This state of matter is a highly excited state of a field. The notions of thermodynamic equilibrium and thermodynamic temperature should also apply to this state.

Fields in thermal equilibrium can be more generally referred to as **thermal radiation**. One of the characteristic properties of thermal radiation is that its energy density is only a function of temperature; unlike the ideal gas, the number of particles of each kind itself depends on the temperature. ''Blackbody radiation," the study of which led Max Planck (1858–1947) to the quantum hypothesis, is thermal radiation associated with the electromagnetic field. At high enough temperatures, all particles—electrons and positrons, protons and antiprotons—can exist in the form of thermal radiation just as electromagnetic radiation does at temperatures we experience. Immediately after the Big Bang, when the temperature of the universe was extremely high, the state of matter in the universe was in the form of thermal radiation. As the universe expanded and cooled, the photons remained in the state of thermal radiation which can be associated with a temperature, but the protons, electrons and neutrons are no longer in that state. In its present state, the radiation that fills the universe is in an equilibrium state of temperature about 3 K, but the observed abundance of elements in the universe is not that expected in a state of thermodynamic equilibrium. (A non-technical description of the thermal radiation in the early universe may be found in Steven Weinberg's well-known book, *The First Three Minutes* [7].)

2.2   The First Law of Thermodynamics: Conservation of Energy

Though mechanical energy (kinetic energy + potential energy) and its conservation was known from the time of Newton and Leibniz, energy was not thought of as a general and universal quantity until the nineteenth century [5, 8].

With the establishment of the mechanical equivalence of heat by Joule, heat became a form of energy that could be converted to work and vice versa. It was in the second half of the nineteenth century that the concept of *conservation* of energy was clearly formulated. Many contributed to this idea which was very much "in the air" at that time. For example, the law of "constant summation of heats of reaction," formulated by the Russian chemist Germain Henri Hess (1802–1850), was the law of energy conservation in chemical reactions. It can be said that the most important contributions to the idea of conservation of energy as a universal law of Nature came from Julius Robert von Mayer (1814–1878), James Prescott Joule (1818–1889) and Hermann von Helmholtz (1821–1894). Two landmarks in the formulation of the law of conservation of energy are a paper by Robert Mayer entitled "Remarks on the Forces of Inanimate Nature" (Bermerkungen über die Kräfte der unbelebten Natur), published in 1842, and a paper by Helmholtz "On the Conservation of Force" (Uber die Erhaltung der Kraft), which appeared in 1847 [5,6].

In the year 1840, seven years before Helmholtz's paper appeared, the Russian chemist Germain Henri Hess (1802–1850) published his "law of constant summation," essentially the law of conservation of energy applied to chemical reactions. This law, now called Hess's law, is routinely used to calculate heats of chemical reactions.

James Prescott Joule (1818–1889) (Reproduced, by permission, from the Emilio Segré Visual Archives of the American Institute of Physics)

The conservation of energy can be stated and utilized entirely in terms of macroscopic variables. A transformation of state may occur due to exchange of heat, performance of work and change in chemical composition. Each of them is associated with a change in energy. The First Law of thermodynamics states that:

"When a system undergoes a transformation of state, the algebraic sum of the different energy changes, heat exchanged, work done, etc., is independent of the manner of the transformation. It depends only on the initial and final states of the transformation."

Figure 2.3 shows how a transformation of volume and temperature of a gas from the state O to the state X may occur in two different ways, each following different intermediate volumes and temperatures. In each case the total amount of heat exchanged and the mechanical work done will be different. But, as the

Hermann von Helmholtz (1821–1894) (Courtesy the E. F. Smith Collection, Van Pelt-Dietrich Library, University of Pennsylvania)

First Law states, the sum of the two will be the same, independent of the path. Since the total change in energy is independent of the path, the infinitesimal change dU associated with any infinitesimal transformation is solely a function of the initial and final states. An alternative way of stating that the change in the energy U depends only on the initial and final states is that in a *cyclic process* i.e., a process in which the system returns to its initial state, the integral of the energy change is zero:

$$\oint dU = 0 \qquad (2.2.1)$$

Equation (2.2.1) may also be considered a statement of the First Law. Since changes in U are independent of the transformation path, the change from a fixed

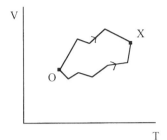

Figure 2.3 The change of energy U during a transformation from normal or reference state O to the state X is independent of the manner of transformation. (The state of this system is defined by its volume V and temperature T)

state O to any final state X (Fig. 2.3) is entirely specified by X. If the value of U at the state O is arbitrarily defined as U_0, then U is a *state function* that is specified by the state X:

$$U = U(T, V, N_k) + U_0 \qquad (2.2.2)$$

The energy U can only be defined up to an arbitrary additive constant U_0.

Yet another way of stating the First Law is as an "impossibility," a restriction Nature imposes on physical processes. For example, in Max Planck's treatise [18], the First Law is stated thus:

"It is in no way possible, either by mechanical, thermal, chemical, or other devices, to obtain perpetual motion, i.e. it is impossible to construct an engine which will work in a cycle and produce continuous work, or kinetic energy, from nothing." (author's italics)

It is easy to see the equivalence of this statement to the above formulation summarized in equation (2.2.1). And note how this statement is entirely in macroscopic, operational terms; it has no reference whatsoever to the microscopic structure of matter. The process described above is called *perpetual motion of the first kind*.

For a closed system, the energy exchanged by a system with the exterior in a time dt may be divided into two parts: dQ, the amount of heat, and dW, the amount of mechanical energy. Unlike the total internal energy dU, the quantities dQ and dW *are not independent of the manner of transformation*; we cannot specify dQ or dW simply by knowing the initial and final states. Hence it is not possible to define a function Q that depends only on the initial and final states, i.e. heat is not a state function. Although every system can be said to possess a certain amount of energy U, the same cannot be said of heat Q or work W. But there is no difficulty in specifying the amount of heat exchanged in a particular transformation. If the rate process that results in the exchange of heat is specified, then dQ is the heat exchanged in a time interval dt.

Most introductory texts on thermodynamics do not include irreversible processes but describe all transformations as idealized, infinitely slow, reversible processes. In this case, dQ cannot be defined in terms of a time interval dt

because the transformation does not occur in finite time, and one has to use the initial and final states to specify dQ. This poses a problem because Q is not a state function, so dQ cannot be uniquely specified by the initial and final states. To overcome this difficulty, an "imperfect differential" $đQ$ is defined to represent the heat exchanged in a transformation, a quantity that depends on the initial and final states *and* the manner of transformation. In our approach we will avoid the use of imperfect differentials. The heat flow is described by processes that occur in a finite time and, with the assumption that the rate of heat flow is known, the heat exchanged dQ in a time dt is well defined. The same is true for the work dW. Idealized, infinitely slow reversible processes still remain useful for some conceptual reasons and we will use them occasionally, but we will not restrict our presentation to reversible processes as many texts do.

The total change in energy dU of a closed system in a time dt is

$$\boxed{dU = dQ + dW}$$
(2.2.3)

The quantities dQ and dW can be specified in terms of the rate laws for heat transfer and the forces that do the work. For example, the heat supplied in a time dt by a heating coil of resistance R carrying a current I is given by $dQ = (I^2 R)dt = VI\,dt$, in which V is the voltage drop across the coil.

For open systems, there is an additional contribution due to the flow of matter dU_{matter} (Fig. 2.4):

$$dU = dQ + dW + dU_{\text{matter}}$$
(2.2.4)

Also, for open systems, we define the volume not as the volume occupied by a fixed number of moles but by the boundary of the system, perhaps a membrane. Since the flow of matter into and out of the system can be associated with mechanical work (e.g. the flow of molecules into the system through a semipermeable membrane due to excess external pressure), dW is not necessarily associated with changes in the system volume. The calculation of changes in energy dU in open systems does not pose a fundamental difficulty. In any process, if changes in T, V and N_k can be computed, the change in energy can be calculated. The total change in the energy can then be obtained by integrating $U(T, V, N_k)$ from the initial state A to the final state B:

$$\int_A^B dU = U_B - U_A$$
(2.2.5)

Because U is a state function, this integral is independent of the path.

Let us now consider some specific examples of exchange of energy in forms other than heat.

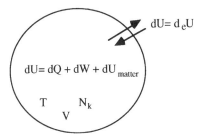

Figure 2.4 Conservation of energy means the total energy U of an isolated system remains a constant. The change in the energy dU of a system, in a time dt, can only be due to exchange of energy d_eU with the exterior through exchange of heat, mechanical processes that are associated with work dW, and through the exchange of matter dU_{matter}. The energy change of the system is equal and opposite to that of the exterior

- For closed systems, if dW is the mechanical work due to a volume change then we may write

$$dW_{mech} = -p\,dV \qquad (2.2.6)$$

in which p is the pressure at the moving surface and dV is the change in volume (Box 2.2).
- For transfer of charge dq across a potential difference ϕ,

$$dU_q = \phi\,dq \qquad (2.2.7)$$

- For dielectric systems, the change of electric dipole moment dP in the presence of an electric field E is associated with a change of energy

$$dU_{elect} = -E\,dP \qquad (2.2.8)$$

- For magnetic systems, the change of magnetic dipole moment dM in the presence of a magnetic field B is associated with a change of energy

$$dU_{mag} = -B\,dM \qquad (2.2.9)$$

- For a change of surface area dA with an associated surface tension γ,

$$dU_{surf} = \gamma\,dA \qquad (2.2.10)$$

Box 2.2 Mechanical Work due to Change in Volume

Mechanical work: $dW = \mathbf{F} \cdot \mathbf{ds}$

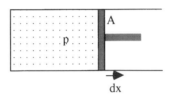
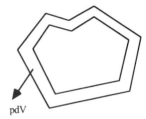

The force on the piston of area A due to the pressure p is pA. The work done by the expanding gas (which causes a decrease in the energy of the gas) in moving the piston by an amount dx is

$$dW = -pA\,dx = -p\,dV$$

in which dV is the change in volume of the gas. The negative sign is because the energy is transferred to the exterior when a gas expands. By considering small displacements of the surface of a body at pressure p, the above expression for the work done can be shown to be generally valid.

Isothermal expansion:
By keeping the gas in contact with a reservoir at temperature T, the volume of gas can be made to expand or contract with no change in temperature. In such an isothermal expansion

$$\text{Work} = \int_{V_i}^{V_f} -p\,dV = \int_{V_i}^{V_f} -\frac{NRT}{V}\,dV = -NRT \ln\left(\frac{V_f}{V_i}\right)$$

in which the negative sign indicates that as the gas expands, energy is transferred to the exterior. In this process, to keep T constant, heat must flow into the gas from a reservoir.

In general, the quantity dW is a sum of all the different forms of "work," *each term being a product of an intensive variable and a differential of an extensive variable*.

Thus, in general, the change in the total internal energy may be written as

$$dU = dQ - p\,dV + \phi\,dq - E\,dP + \dots \tag{2.2.11}$$

This change of energy of system is a function of state variables such as T, V and N_k.

For systems undergoing chemical transformations, the total energy may be considered as a function of T, V and the mole numbers N_k, $U = U(T, V, N_k)$.

The total differential of U can be written as:

$$dU = \left(\frac{\partial U}{\partial T}\right)_{V,N_k} dT + \left(\frac{\partial U}{\partial V}\right)_{T,N_k} dV + \sum_k \left(\frac{\partial U}{\partial N_k}\right)_{V,T,N_{i\neq k}} dN_k$$

$$= dQ + dW + dU_{\text{matter}} \tag{2.2.12}$$

Box 2.3 Calorimetry

Calorimeter. Heat evolved or absorbed during a transformation, such as a chemical reaction, is measured using a calorimeter. The transformation of interest is made to occur inside a chamber which is well insulated from the environment to keep heat loss to a minimum. To measure the heat generated by a process, first the heat capacity of the calorimeter should be determined. This is done by noting the increase in the temperature of the calorimeter due to a process for which the heat evolved is known. The heat produced by a current-carrying resistor, for example, is known to be I^2R joules per second in which I is the current in amps and R the resistance in ohms. (Using Ohm's law, $V = IR$, in which V is the voltage across the resistor in volts, the heat generated per second may also be written as VI.) If the heat capacity C_{cal} of the calorimeter is known, then one only needs to note the change in the temperature of the calorimeter to determine the heat generated by a process.

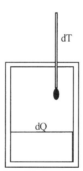

Bomb calorimeter. The heat of combustion of a compound is determined in a bomb calorimeter. In a bomb calorimeter the combustion is made to occur in a chamber pressurized to about 20 atm with pure oxygen to ensure that the combustion is complete.

The exact form of the function $U(T, V, N_k)$ for a particular system is obtained empirically. One way of obtaining the temperature dependence of U is to measure the *molar heat capacity, C_V, at constant volume*. (Box 2.1 gives basic definitions of heat capacity.) At constant volume, since no work is performed, $dU = dQ$. Hence

$$C_V(T, V) \equiv \left(\frac{dQ}{dT}\right)_{V=\text{const}} = \left(\frac{\partial U}{\partial T}\right)_{V,N=1} \tag{2.2.13}$$

If C_V is determined experimentally, the internal energy $U(T, V)$ is then obtained through integration of C_V:

$$U(T, V, N) - U(T_0, V, N) = N \int_{T_0}^{T} C_V(T, V)dT \qquad (2.2.14)$$

in which T_0 is the temperature of a reference state. If C_V is independent of temperature and volume, as is the case for an ideal gas, we have

$$\boxed{U_{\text{ideal}} = C_V N T + U_0} \qquad (2.2.15)$$

in which U_0 is an arbitrary additive constant. As noted earlier, U can only be defined up to an additive constant. For ideal monatomic gases $C_V = (3/2)R$ and for diatomic gases $C_V = (5/2)R$.

The notion of total internal energy is not restricted to homogeneous systems in which quantities such as temperature are uniform. For many systems, temperature is locally well defined but may vary with the position x and time t. In addition, the equations of state may remain valid in every elemental volume δV (i.e. in a small volume element defined appropriately at every point x) in which all the state variables are specified as densities. For example, corresponding to the energy $U(T, V, N_k)$ we may define the energy destiny $u(x, t) =$ energy per unit volume at the point x at time t, which can be expressed as a function of the local temperature $T(x, t)$ and the molar density $n_k(x, t)$ (= number of moles per unit volume) which in general are functions of both position x and time t:

$$u(x, t) = u(T(x, t), n_k(x, t)) \qquad (2.2.16)$$

The law of conservation of energy is a *local conservation law*: the change in energy in a small volume can only be due to a flow of energy into or out of the volume (Fig. 2.4). Two spatially separated regions cannot exchange energy unless the energy passes through the region connecting the two parts[*].

---

[*] One might wonder why energy conservation does not take place by its disappearance at one location and simultaneous appearance at another. Such conservation, it turns out, is not compatible with the theory of relativity. According to relativity, events that are simultaneous but occurring at different locations to one observer, may not be simultaneous to another. Hence the simultaneous disappearance and appearance of energy as seen by one observer will not be simultaneous for all. For some observers, energy would have disappeared at one location first and only some time later would it reappear at the other location, thus violating the law of conservation of energy during the time interval separating the two events.

2.3 Elementary Applications of the First Law

RELATION BETWEEN C_p AND C_V

The first law of thermodynamics leads to many simple and useful conclusions. It leads to a relation between the *molar heat capacities*, C_p, at constant pressure, and C_V, at constant volume (Fig. 2.5 and Table 2.1). Consider a one-component substance with fixed mole number $N = 1$. Then, using (2.2.3) and (2.2.6), since the energy U is a function of the volume and temperature, the change in the energy dU can be written as

$$dU = dQ - p\,dV = \left(\frac{\partial U}{\partial T}\right)_V dT + \left(\frac{\partial U}{\partial V}\right)_T dV \tag{2.3.1}$$

From this it follows that the heat supplied can be written as

$$dQ = \left(\frac{\partial U}{\partial T}\right)_V dT + \left[p + \left(\frac{\partial U}{\partial V}\right)_T\right] dV \tag{2.3.2}$$

If the system is heated at a constant volume, since no work is done, the change in the energy of the system is entirely due to the heat supplied. Therefore

$$C_V \equiv \left(\frac{dQ}{dT}\right)_V = \left(\frac{\partial U}{\partial T}\right)_V \tag{2.3.3}$$

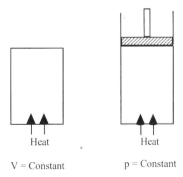

Figure 2.5 Molar heat capacity at constant pressure is larger than molar heat capacity at constant volume

Table 2.1 Molar heat capacities C_V and C_p for some compounds at $T = 298.15\,\mathrm{K}$ and $p = 1$ atm

Compound	$C_p(\mathrm{J\,mol^{-1}})$	$C_V(\mathrm{J\,mol^{-1}})$
Ideal monatomic gas	$(5/2)R$	$(3/2)R$
Ideal diatomic gas	$(7/2)R$	$(5/2)R$
Noble gases	20.79	12.47
(He, Ne, Ar, Kr, Xe)		
$N_2(g)$	29.12	20.74
$O_2(g)$	29.36	20.95
$CO_2(g)$	37.11	28.46
$H_2(g)$	28.82	20.44
$H_2O(g)$	75.29	
$C_2H_5OH(l)$	111.5	
$C_6H_6(l)$	136.1	
$Cu(s)$	244.4	
$Fe(s)$	25.1	

On the other hand, if the system is heated at constant pressure, we have from (2.3.2) that

$$C_p \equiv \left(\frac{dQ}{dT}\right)_p = \left(\frac{\partial U}{\partial T}\right)_V + \left[p + \left(\frac{\partial U}{\partial V}\right)_T\right]\left(\frac{\partial V}{\partial T}\right)_p \qquad (2.3.4)$$

Comparing (2.3.3) and (2.3.4) we see that C_V and C_p are related by

$$\boxed{C_p - C_V = \left[p + \left(\frac{\partial U}{\partial V}\right)_T\right]\left(\frac{\partial V}{\partial T}\right)_p} \qquad (2.3.5)$$

The right-hand side of (2.3.5) is equal to the additional amount of heat required in a constant-pressure, or "isobaric," process to compensate for the energy expended due to expansion of volume.

Relation (2.3.5) is general. For an ideal gas it reduces to a simple form. Following equations (1.3.6) and (1.3.8), the energy U is only a function of the temperature and is independent of the volume. Hence in (2.3.5), $(\partial U/\partial V)_T = 0$; for one mole of an ideal gas, since $pV = RT$, the remaining term $p(\partial V/\partial T)_p = R$. Therefore (2.3.5) reduces to the simple relation

$$\boxed{C_p - C_V = R} \qquad (2.3.6)$$

between the molar heat capacities of an ideal gas.

ADIABATIC PROCESSES IN AN IDEAL GAS

In an *adiabatic process* the state of a gas changes without any exchange of heat. Using the equation $dU = dQ - p\,dV$, we can write

$$dQ = dU + p\,dV = \left(\frac{\partial U}{\partial T}\right)_V dT + \left(\frac{\partial U}{\partial V}\right)_T dV + p\,dV = 0 \qquad (2.3.7)$$

For an ideal gas, since U is a function of temperature but not of volume, this reduces to

$$C_V N\,dT + p\,dV = 0 \qquad (2.3.8)$$

Since the change in volume occurs such that we may use the ideal gas equation to relate the pressure to the volume and temperature, we have

$$C_V\,dT + \frac{RT}{V}\,dV = 0 \qquad (2.3.9)$$

(For very rapid changes in volume, the relation between pressure, volume and temperature will deviate from the ideal gas law.) But since $R = C_p - C_V$ for an ideal gas, we can write (2.3.9) as

$$\frac{dT}{T} + \frac{(C_p - C_V)}{C_V V}\,dV = 0 \qquad (2.3.10)$$

Integration of (2.3.10) gives

$$\boxed{TV^{(\gamma-1)} = \text{const}} \quad \text{where} \quad \boxed{\gamma = \frac{C_p}{C_V}} \qquad (2.3.11)$$

Table 2.2   Ratios of heat capacities and velocity of sound for some gases at 1 atm[*]

Gas	Value of γ at 15 °C, 1 atm[*]	Velocity of sound at 0 °C (m/s)
Ar(g)	1.667	308
CO_2(g)	1.304	259
H_2(g)	1.410	1284
He(g)	1.667	965
N_2(g)	1.404	334
O_2(g)	1.401	316

[*] More extensive data may be found in sources [B] and [F].

Using $pV = nRT$, the above relation can be transformed into

$$\boxed{pV^\gamma = \text{const}} \quad \text{or} \quad \boxed{T^\gamma p^{1-\gamma} = \text{const}} \tag{2.3.12}$$

Thus the first law gives us (2.3.11) and (2.3.12), which characterize adiabatic processes in an ideal gas. Values of γ are given in Table 2.2.

SOUND PROPAGATION

An adiabatic process in Nature is the rapid variation of pressure during the propagation of sound. Pressure variations, which are a measure of the sound intensity, are small. A measure of these pressure variations is p_{rms}, the root mean square value of the sound pressure with respect to the atmospheric pressure, i.e. p_{rms} is the square root of the average value of $(p - p_{\text{atms}})^2$. The intensity of sound I is measured in units of *decibels* (dB). The decibel is a logarithmic measure of the pressure variations defined by

$$I = 10 \log_{10} (p_{\text{rms}}^2 / p_0^2) \tag{2.3.13}$$

in which the reference pressure $p_0 = 2 \times 10^{-8}\,\text{kPa}(= 2 \times 10^{-10}\,\text{bar} = 1.974 \times 10^{-10}\,\text{atm})^*$. This logarithmic scale is used because it roughly corresponds to the sensitivity of the ear. We normally encounter sound whose intensity is in the range 10–100 dB, corresponding to a pressure variations in the range 6×10^{-10} to 2×10^{-5} bar. These small variations of pressure for audible sound occur in the frequency range 20 Hz to 20 kHz (music is in the range 40 Hz to 4 kHz).

Due to the rapidity of pressure variations, hardly any heat is exchanged between the surroundings and the volume of air that is undergoing the pressure variations. It is essentially an adiabatic process. As a first approximation, we may assume the ideal gas law is valid for these rapid changes. In introductory physics texts it is shown that the velocity of sound, C_{sound}, in a medium depends on the bulk modulus B and the density ρ according to the relation

$$C_{\text{sound}} = \sqrt{\frac{B}{\rho}} \tag{2.3.14}$$

The bulk modulus $B = -\delta p / (\delta V / V)$ relates the relative change in the volume of a medium $(\delta V / V)$ due to a change in pressure δp. (The negative sign indicates that for positive δp the change δV is negative.) If the propagation of sound is an adiabatic process, in the ideal gas approximation, the changes in volume and pressure are such that $pV^\gamma = $ constant. By differentiating this

---

$^*p_0$ is the threshold of audibility

relation, one can easily see that the bulk modulus B for an adiabatic process is

$$B = -V\frac{dp}{dV} = \gamma p \tag{2.3.15}$$

For an ideal gas of density ρ and molar mass M,

$$p = \frac{NRT}{V} = \frac{NM}{V}\frac{RT}{M} = \frac{\rho RT}{M}$$

Hence

$$B = \frac{\gamma \rho RT}{M} \tag{2.3.16}$$

Using this expression in (2.3.14) we arrive at the conclusion that if the propagation of sound is an adiabatic process, then the velocity C_{sound} is given by

$$\boxed{C_{\text{sound}} = \sqrt{\frac{\gamma RT}{M}}} \tag{2.3.17}$$

Experimental measurements of sound velocity confirm this conclusion to a good approximation.

2.4 Thermochemistry: Conservation of Energy in Chemical Reactions

During the first half of the nineteenth century, chemists mostly concerned themselves with the analysis of compounds and chemical reactions and they paid little attention to the heat evolved or absorbed in a chemical reaction. Though the early work of Lavoisier (1743–1794) and Laplace (1749–1827) established that heat absorbed in a chemical reaction was equal to the heat released in the reverse reaction, the relation between heat and chemical reactions was not investigated very much. The Russian chemist Germain Henri Hess (1802–1850) was rather an exception among the chemists of his time in regard to his interest in the heat released or absorbed by chemical reactions [9]. Hess conducted a series of studies in neutralizing acids and measuring the heats released (Box 2.4). This and several other experiments on the heats of chemical reactions lead Hess to his "law of constant summation," which he published in 1840, two years before the appearance of Robert Mayer's paper on the conservation of energy:

"The amount of heat evolved during the formation of a given compound is constant, independent of whether the compound is formed directly or indirectly in one or in a series of steps." [10]

Hess's work was not very well known for many decades after its publication. The fundamental contribution of Hess to thermochemistry was made known to the chemists largely through Wilhelm Ostwald's (1853–1932) *Textbook of General Chemistry*, published in 1887. The above statement, known as **Hess's law**, was amply confirmed in the detailed work of Marcellin Berthelot (1827–

Germain Henri Hess (1802–1850) (Courtesy the E. F. Smith Collection, Van Pelt-Dietrich Library, University of Pennsylvania)

1907) and Julius Thomsen (1826–1909) [11]. As we shall see below, Hess's law is a consequence of the law of conservation of energy and is most conveniently formulated in terms of a function of state called *enthalpy*.

Hess's law refers to the heat evolved in a chemical reaction under constant (atmospheric) pressure. Under such conditions, a part of the energy released during the reaction may be converted to work $W = -\int_{V_1}^{V_2} p\,dV$, if there is a change in volume from V_1 to V_2. Following the First Law, the heat evolved ΔQ_p when a reaction occurs at a constant pressure can be written as

$$\Delta Q_p = \int_{U_1}^{U_2} dU + \int_{V_1}^{V_2} p\,dV = (U_2 - U_1) + p(V_2 - V_1) \qquad (2.4.1)$$

Box 2.4 The Experiments of Germain Henri Hess

Hess conducted a series of studies in which he first diluted sulfuric acid by adding different amounts of water and then neutralized the acid by adding a solution of ammonia. Heat was released in both steps. Hess found that, depending on the amount of water added during the dilution, different amounts of heat were released during this first step and the subsequent step of neutralization with ammonia, but the sum of the heats released in the two processes remained the same [10]. The following example, in which ΔH is the heat released, illustrates Hess's experiments:

$$1L \text{ of } 2M \, H_2SO_4 \xrightarrow[\Delta H_1]{\text{Dilution}} 1.5M \, H_2SO_4 \xrightarrow[\Delta H_2]{NH_3 \text{ solution}} 3L \text{ neutral solution}$$

$$1L \text{ of } 2M \, H_2SO_4 \xrightarrow[\Delta H'_1]{\text{Dilution}} 1.0M \, H_2SO_4 \xrightarrow[\Delta H'_2]{NH_3 \text{ solution}} 3L \text{ neutral solution}$$

Hess found that $(\Delta H_1 + \Delta H_2) = (\Delta H'_1 + \Delta H'_2)$ to a good approximation.

From this we see that the heat released can be written as a difference between two terms, one referring to the initial state (U_1, V_1) and the other to the final state (V_2, U_2):

$$\Delta Q_p = (U_2 + pV_2) - (U_1 + pV_1) \tag{2.4.2}$$

Since U, p and V are specified by the state of the system and are independent of the manner in which that state was reached, the quantity $U + pV$ is a state function, specified by the state variables. According to (2.4.2), the heat evolved ΔQ_p is the difference between the values of the function $(U + pV)$ at the initial and final states. The state function $(U + pV)$ is called the **enthalpy** H:

$$\boxed{H \equiv U + pV} \tag{2.4.3}$$

The heat released by a chemical reaction at constant pressure $\Delta Q_p = H_2 - H_1$. Since ΔQ_p depends only on the values of enthalpy at the initial and final states, it is independent of the "path" of the chemical transformation, in particular if the transformation occurs "directly or indirectly in one or in a series of steps," as Hess concluded.

As a simple example, consider the reaction

$$2P(s) + 5Cl_2(g) \rightarrow 2PCl_5(s) \tag{2.A}$$

in which 2 moles of P react with 5 moles of Cl_2 to produce 2 moles of PCl_5. This reaction releases 886 kJ of heat. It can occur directly when an adequate amount of Cl is present or it can be made to occur in two steps:

$$2P(s) + 3Cl_2(g) \rightarrow 2PCl_3(l) \tag{2.B}$$

$$2PCl_3(g) + 2Cl_2(g) \rightarrow 2PCl_5(s) \tag{2.C}$$

In reactions (2.B) and (2.C), for the molar quantities shown in the reaction, the heats evolved are 640 kJ and 246 kJ respectively. If the enthalpy changes between the initial and final states of reactions (A), (B) and (C) are denoted by $\Delta H_{rxn(A)}$, $\Delta H_{rxn(B)}$ and $\Delta H_{rxn(C)}$ respectively, then we have

$$\Delta H_{rxn(A)} = \Delta H_{rxn(B)} + \Delta H_{rxn(C)} \tag{2.4.4}$$

Box 2.5 Basic Definitions used in Thermochemistry

Like the energy U, the enthalpy can also be specified with reference to a **standard state**.

The standard state of a pure substance at particular temperature is its state, gas, liquid or solid, at a pressure of one bar.

Notation used to indicate the standard state: g = gas; l = liquid; s = pure crystalline solid.

For solutions, the standard state is a hypothetical solution of concentration (1 mol/kg of solvent) at a pressure of 1 bar. Notation used to indicate the standard state: ai = completely dissociated electrolyte in water; ao = undissociated compound in water.

Standard reaction enthalpies at a specified temperature are reaction enthalpies in which the reactants and products are in their standard states.

Standard molar enthalpy of formation $\Delta H_f^0[X]$ of a compound X, at a specified temperature T, is the enthalpy of formation of the compound X from its constituent elements in their standard state. Here is an example:

$$X = CO_2(g): \quad C(s) + O_2(g) \xrightarrow{\Delta H_f^0[CO_2(g)]} CO_2(g)$$

The enthalpies of formation of elements in their standard state are defined to be zero at any temperature.

The consistency of this definition is based on the fact that in chemical reactions the elements do not transform among themselves, i.e. reactions between elements resulting in the formation of other elements do not occur (though energy is conserved in such a reaction).

Thus, the enthalpies of formation $\Delta H_f^0[H_2]$, $\Delta H_f^0[O_2]$, $\Delta H_f^0[Fe]$ are defined to be zero at all temperatures.

The heat evolved or enthalpy change in a chemical reaction under constant pressure is usually denoted by ΔH_{rxn} and called the enthalpy of reaction. *The enthalpy of reaction is negative for exothermic reactions and positive for endothermic reactions.*

The First Law of thermodynamics applied to chemical reactions in the form of Hess's law gives us a very useful way of predicting the heat evolved or absorbed in a chemical reaction if we can formally express it as a sum of chemical reactions for which the enthalpies of reaction are known. In fact, if we can assign a value to the enthalpy of one mole of each compound, then the heats of reaction can be expressed as the difference between sums of enthalpies of the initial reactants and the final products. In reaction (2.C), if we can assign enthalpies for a mole of $PCl_3(g)$, $Cl_2(g)$ and $PCl_5(s)$, then the enthalpy of this reaction will be the difference between the enthalpy of the product $PCl_5(s)$ and the sum of the enthalpies of the reactants $PCl_3(g)$ and $Cl_2(g)$. But from the definition of enthalpy (2.4.3) it is clear that it could only be specified with respect to a reference or normal state because U can only be defined in this way.

A convenient way to obtain the standard reaction enthalpies of a reaction at a specified temperature has been developed by defining a standard molar enthalpy of formation $\Delta H_f^0[X]$ for each compound X, as described in Box 2.5.

Standard enthalpies of formation of compounds can be found in tables of thermodynamic data [12]. Using these tables and Hess's law, the standard enthalpies of reactions can be computed by viewing the reaction as a "dismantling" of the reactants to their constituent elements followed by a recombination to form the products. Since the enthalpy for the dismantling step is the negative of the enthalpy of formation, the enthalpy of the reaction

$$aX + bY \rightarrow cW + dZ \tag{2.4.5a}$$

can be written as

$$\Delta H_{\text{rxn}}^0 = -a\Delta H_f^0[X] - b\Delta H_f^0[Y] + c\Delta H_f^0[W] + d\Delta H_f^0[Z] \tag{2.4.5b}$$

Enthalpies of various chemical processes are discussed in detail in texts on physical chemistry [13].

Though it is most useful in thermochemistry, enthalpy as defined in (2.4.3) is a function of state which can be related to quantities other than heats of reaction. Since the change in enthalpy corresponds to the heat exchanged in a process at constant pressure,

$$dQ_p = dU + p\,dV = dH_p \tag{2.4.6}$$

in which the subscripts denote a constant pressure process. If the system consists of *one mole* of a substance, and if the corresponding change in temperature due to the exchange of heat is dT, it follows that

$$C_p \equiv \left(\frac{dQ}{dT}\right)_p = \left(\frac{\partial H}{\partial T}\right)_p \qquad (2.4.7)$$

Also, in general, the change in enthalpy in a chemical reaction (not necessarily occurring at constant pressure) can be written as

$$\begin{aligned} \Delta H_{rxn} = H_2 - H_1 &= (U_2 - U_1) + (p_2 V_2 - p_1 V_1) \\ &= \Delta U_{rxn} + (p_2 V_2 - p_1 V_1) \end{aligned} \qquad (2.4.8)$$

in which the subscripts 1 and 2 denote the initial and final states. In an isothermal process occurring at temperature T, if all the gaseous components in the reaction can be approximated to ideal gases and if the change in volume of the nongaseous components can be neglected, then the changes of enthalpy and energy are related by

$$\Delta H_{rxn} = \Delta U_{rxn} + \Delta N_{rxn} RT \qquad (2.4.9)$$

in which ΔN_{rxn} is the change in the total number of moles of the gaseous reactants, a relation used in obtaining enthalpies of combustion using a bomb calorimeter.

VARIATION OF ENTHALPY WITH TEMPERATURE

Using relation (2.4.7) the enthalpy change due to temperature can be expressed in terms of the specific heat C_p:

$$H(T, p, N) - H(T_0, p, N) = N \int_{T_0}^{T} C_p(T) dT \qquad (2.4.10)$$

Though the variation of C_p with temperature is generally small, an empirical form of

$$\boxed{C_p = \alpha + \beta T + \gamma T^2} \qquad (2.4.11)$$

is often used. Some typical values of α, β and γ are shown in Table 2.3.

From (2.2.14) and (2.4.10) it is clear that the temperature dependence of the total internal energy U and enthalpy H of any particular gas can be obtained if

Table 2.3 Values of constants α, β, and γ in (2.4.11) for some gases*

	$\alpha(\mathrm{J\,K^{-1}\,mol^{-1}})$	$\beta(\mathrm{J\,K^{-2}\,mol^{-1}})$	$\gamma(\mathrm{J\,K^{-3}\,mol^{-1}})$
$O_2(g)$	25.503	13.612×10^{-3}	-42.553×10^{-7}
$N_2(g)$	26.984	5.910×10^{-3}	-3.376×10^{-7}
$CO_2(g)$	26.648	42.262×10^{-3}	-142.4×10^{-7}
$HCl(g)$	28.166	1.809×10^{-3}	15.465×10^{-7}
$H_2O(g)$	30.206	9.936×10^{-3}	11.14×10^{-7}

* The range of validity is 300–1500 K ($p = 1$ atm).

the molar heat capacities are known as a function of temperature. Sensitive calorimetric methods are available to measure specific heats experimentally.

Using relation (2.4.10), if the reaction enthalpy in the standard state ($p_0 = 1$ bar) is known at one temperature T_0, then the reaction enthalpy at any other temperature T can be obtained if C_p is known for the reactants and the products. The enthalpies of the reactants and products at the desired temperature T, satisfy

$$H_x(T, p_0, N) - H_x(T_0, p_0, N) = N \int_{T_0}^{T} C_{p,x}(T)dT \qquad (2.4.12)$$

in which the subscript identifies the reactants and products. Then by subtracting the sum of the enthalpies of reactants from the sum of the enthalpies of the products, as shown in (2.4.5b), the variable $\Delta H_{\mathrm{rxn}}(T, p_0)$ can be written as

$$\Delta H_{\mathrm{rxn}}(T, p_0) - \Delta H_{\mathrm{rxn}}(T_0, p_0) = \int_{T_0}^{T} \Delta C_p(T)dT \qquad (2.4.13)$$

in which ΔC_p is the difference in the heat capacities of the products and the reactants. Thus $\Delta H_{\mathrm{rxn}}(T, p_0)$ at any arbitrary temperature T can be obtained knowing $\Delta H_{\mathrm{rxn}}(T_0, p_0)$ at a standard temperature T_0. Relation (2.4.13) was first noted by Gustav Kirchhoff (1824–1887) and is sometimes called Kirchhoff's law. The change in reaction enthalpy with temperature is generally small.

VARIATION OF ENTHALPY WITH PRESSURE

Finally the variation of H with pressure, at a fixed temperature, can be obtained from the definition $H = U + pV$. Generally, H and U can be expressed as functions of p, T and N. For changes in H we have

$$\Delta H = \Delta U + \Delta(pV) \qquad (2.4.14)$$

At constant T, and N, in the ideal gas approximation, $\Delta H = 0$ for gases. This is because the product pV and the energy U are functions only of temperature

(Chapter 1), they are independent of pressure, hence H is independent of pressure. The change in H due to a change in p is mainly due to intermolecular forces, and it becomes significant only for large densities. These changes in H may be calculated using the van der Waals equation, for example.

For most solids and liquids, at a constant temperature, the total energy U does not change much with pressure. Since the change in volume is rather small unless the changes in pressure are very large, the change in enthalpy ΔH due a change in pressure Δp can be approximated by

$$\Delta H \approx V \Delta p \qquad (2.4.15)$$

A more accurate estimate can be made from a knowledge of the compressibility of the compound.

The First Law thus provides a powerful means of understanding the heats of chemical reactions. The above scheme, the measurement and tabulation of the heats of formation of compounds at a standard temperature and pressure, enables us to compute the heats of reaction of an enormous number of reactions. The data table at the end of the book lists the standard heats of formation of some compounds. An extensive list can be found in the literature [12]. In addition, with a knowledge of heat capacities and compressibilities, heats of reaction can be calculated at any temperature and pressure.

COMPUTATION OF ΔH_{rxn} USING BOND ENTHALPIES

The concept of a chemical bond gives us a better understanding of the nature of a chemical reaction: it is essentially the breaking and making of bonds between atoms. The heat evolved or absorbed in a chemical reaction can be obtained by adding the heat absorbed in the breaking of bonds and the heat evolved in the making of bonds. The heat or enthalpy needed to break a bond is called the **bond enthalpy**.

For a particular bond, such as a C–H bond, the bond enthalpy may change a little from compound to compound, but one can meaningfully use an average bond enthalpy to estimate the enthalpy of a reaction. For example, the reaction $2H_2(g) + O_2(g) \rightarrow 2H_2O(g)$ can be written explicitly, indicating the bonds as

$$2(H–H) + O–O \rightarrow 2(H–O–H)$$

This shows that the reaction involves the breaking of two H–H bonds and one O–O bond followed by the making of four O–H bonds. If the bond enthalpy of the H–H bond is denoted by $\Delta H[H–H]$, etc., the reaction enthalpy ΔH_{rxn} may be written as

$$\Delta H_{rxn} = 2\Delta H[H–H] + \Delta H[O–O] - 4\Delta H[O–H]$$

Table 2.4 Average bond enthalpies for some common bonds (kJ mol^{-1})

	H	C	N	O	F	Cl	Br	I	S	P	Si
H	436										
C (single)	412	348									
C (double)		612									
C (triple)		811									
C (aromatic)		518									
N (single)	388	305	163								
N (double)		613	409								
N (triple)		890	945								
O (single)	463	360	157	146							
O (double)		743		497							
F	565	484	270	185	155						
Cl	431	338	200	203	254						
Br	366	276				219	193				
I	299	238				210	178	151			
S	338	259				250	212		264		
P	322									172	
Si	318		374								176

Source: L. Pauling, *The Nature of the Chemical Bond*, 1960, Cornell University Press, Ithaca NY.

This is a good way of estimating the reactions enthalpy of a large number of reaction using a relatively small table of average bond enthalpies. Table 2.4 lists some average bond enthalpies. With this table the reaction enthalpies of a large number of reactions can be estimated.

2.5 Extent of Reaction: A State Variable for Chemical Systems

In every chemical reaction the changes in the mole numbers dN_k are related through the stoichiometry of the reaction. In fact, *only one parameter is required to specify the changes in mole numbers resulting from a particular chemical reaction*. Consider the following elementary chemical reaction:

$$H_2(g) + I_2(g) \rightleftharpoons 2HI(g) \tag{2.5.1}$$

which is of the form

$$A + B \rightleftharpoons 2C \tag{2.5.2}$$

In this case the changes in the mole numbers dN_A, dN_B and dN_C of the components A, B and C are related by the stoichiometry. We can express this relation as

$$\frac{dN_A}{-1} = \frac{dN_B}{-1} = \frac{dN_C}{2} \equiv d\xi \qquad (2.5.3)$$

in which we have introduced a single variable $d\xi$ that expresses all the changes in the mole numbers due to the chemical reaction. This variable ξ, introduced by Théophile de Donder [14,15] is basic for the thermodynamic description of chemical reactions and is called the **extent of reaction** or the **degree of advancement**. The velocity of reaction is the rate at which the extent of reaction changes with time:

$$\text{Velocity of reaction} = \frac{d\xi}{dt} \qquad (2.5.4)$$

If the initial values of N_k are written as N_{k0} then the values of all N_k during the reactions can be specified by the extent of reaction ξ:

$$N_k = N_{k0} + \nu_k \xi \qquad (2.5.5)$$

in which ν_k is the stoichiometric coefficient of the reacting component N_k. Note that ν_k is negative for the reactants and positive for the products. In this definition $\xi = 0$ at the initial state.

   If the changes in the mole numbers in a system are due to chemical reactions, the total internal energy U of such a system can be expressed in terms of the initial mole numbers N_{k0}, which are constants, and the extents of reaction ξ_i defined for each of the reactions. For example, consider a chemical system consisting of three substances A, B and C undergoing a single reaction (2.5.2). Then the mole numbers can be expressed as $N_A = N_{A0} - \xi$, $N_B = N_{B0} - \xi$ and $N_C = N_{C0} + 2\xi$. A given value of ξ completely specifies all the mole numbers. Hence the total energy U may be regarded as a function $U(T, V, \xi)$ with the understanding that the initial mole numbers N_{A0}, N_{B0} and N_{C0} are constants in the function U. If more than one chemical reaction is involved, then one extent of reaction ξ_i is defined for each independent reaction i and each mole number is specified in terms of the extents of reaction of all the chemical reactions in which it takes part. Clearly the ξ_i are state variables and internal energy can be expressed as a function of T, V and ξ_i, i.e. $U(T, V, \xi_i)$.

   In terms of the state variables T, V, and ξ_i, the total differential of U becomes

$$dU = \left(\frac{\partial U}{\partial T}\right)_{V, \xi_k} dT + \left(\frac{\partial U}{\partial V}\right)_{T, \xi_k} dV + \sum_k \left(\frac{\partial U}{\partial \xi_k}\right)_{V, T, \xi_{i \neq k}} d\xi_k \qquad (2.5.6)$$

   Using the First Law, the partial derivatives of U can be related to "thermal coefficients" which characterize the system's response to heat under various conditions. Consider a system with one chemical reaction. We have one extent of reaction ξ. Then, by using the First Law,

$$dU = dQ - p\,dV = \left(\frac{\partial U}{\partial T}\right)_{V,\xi} dT + \left(\frac{\partial U}{\partial V}\right)_{T,\xi} dV + \left(\frac{\partial U}{\partial \xi}\right)_{T,V} d\xi \quad (2.5.7)$$

which can be written as

$$dQ = \left(\frac{\partial U}{\partial T}\right)_{V,\xi} dT + \left[p + \left(\frac{\partial U}{\partial V}\right)_{T,\xi}\right] dV + \left(\frac{\partial U}{\partial \xi}\right)_{T,V} d\xi \quad (2.5.8)$$

Just as the partial derivative $(\partial U/\partial T)_V$ has the physical meaning of being the heat capacity at constant volume NC_V, the other derivatives, called *thermal coefficients*, can be related to experimentally measurable quantities. The derivative $r_{T,V} \equiv (\partial U/\partial \xi)_{V,T}$ for example has the physical meaning of being the amount of *heat evolved* per unit change in the extent of reaction (one equivalent of reaction) at constant V and T. *If $r_{T,V}$ is negative the reaction is exothermic; if it is positive the reaction is endothermic.* Just as we derived the relation (2.3.6) between the heat capacities C_p and C_V, one can derive several interesting relations between these thermal coefficients as a consequence of the First Law [16].

Also, since the extent of reaction is a state variable, the enthalpy of a reacting system can be expressed as a function of the extent of reaction:

$$H = H(p, V, \xi) \quad (2.5.9)$$

The heat of reaction per unit change of ξ, $h_{p,T}$, is the derivative of H with respect to ξ:

$$\boxed{h_{p,T} = \left(\frac{\partial H}{\partial \xi}\right)_{p,T}} \quad (2.5.10)$$

2.6 Conservation of Energy in Nuclear Reactions

At terrestrial temperatures, transformations of states of matter are mostly chemical, radioactivity being an exception. Just as molecules collide and react at terrestrial temperatures, at very high temperatures that exceed 10^6 K, typical of temperatures attained in the stars, nuclei collide and undergo nuclear reactions. At these temperatures, the electrons and nuclei of atoms are completely torn apart. Matter turns into a state that is unfamiliar to us and the transformations that occur are between nuclei, which is why they are called nuclear chemistry.

All the elements heavier than hydrogen on our planet and on other planets are a result of nuclear reactions, generally known as nucleosynthesis, that occurred in the stars [17]. Just as we have unstable molecules that dissociate into other more stable molecules, some of the nuclei that were synthesized in the stars are unstable and they disintegrate—they are the "radioactive" elements. The energy released by radioactive elements turns into heat; it is a source of heat in the Earth. For example, the natural radioactivity in granite due to ^{238}U, ^{235}U, ^{232}Th and ^{40}K produces about 5 µcal/g per year.

Under special circumstances, nuclear reactions can occur on the Earth, e.g. nuclear fission of uranium and nuclear fusion of hydrogen in special reactors. Nuclear reactions release vastly greater amounts of energy than chemical reactions. The energy released in a nuclear reaction can be calculated from the difference in the rest mass of the reactants and the products using Einstein's famous relation $E^2 = p^2c^2 + m_0^2c^4$, in which p is the momentum, m_0 the rest mass and c is the velocity of light. If the total rest mass of the products is lower than total rest mass of the reactants, the difference in energy due to change in the rest mass turns into the kinetic energy of the products. This excess kinetic energy turns into heat due to collisions. If the difference in the kinetic energy of the reactants and products is negligible, the heat released is $\Delta Q = \Delta m_0 c^2$ where Δm_0 is the difference in the rest mass between the reactants and the products. In nuclear fusion, two deuterium nuclei ^{2}H can combine to form a helium atom, releasing a neutron:

$$^2\text{H} + {}^2\text{H} \rightarrow {}^3\text{He} + \text{n}$$

$$\begin{aligned}
\Delta m_0 &= 2 \times (\text{mass of } {}^2\text{H}) - (\text{mass of } {}^3\text{He} + \text{mass of n}) \\
&= 2(2.0141) \text{ amu} - (3.0160 + 1.0087) \text{ amu} \\
&= 0.0035 \text{ amu}^*
\end{aligned}$$

Since 1 amu $= 1.6605 \times 10^{-27}$ kg, when 2 moles of ^{2}H react to produce 1 mole of ^{3}He and one mole of neutrons, the difference in mass is $\Delta m = 3.5 \times 10^{-6}$ kg. The corresponding heat released is

$$\Delta E = \Delta m_0 c^2 = 3.14 \times 10^8 \text{ kJ mol}^{-1}$$

If a nuclear process occurs at constant pressure, the heat released is equal to the enthalpy, and all the thermodynamic formalism that was applied to the chemical reactions can be applied to nuclear reactions. Needless to say, in accordance with the First Law, Hess's law of additivity of reaction enthalpies is also valid for nuclear reactions.

---

* amu = atomic mass unit.

GENERAL REMARKS

Thermodynamically, energy is only defined up to an additive constant. In physical processes, it is only the change in energy (2.2.11) that can be measured and there is no way to measure the absolute value of energy. With the advent of the theory of relativity, which has given us the relation between rest mass, momentum and energy, $E^2 = p^2 c^2 + m_0^2 c^4$, the definition of energy has become as absolute as the definition of mass and momentum. In later chapters we will use the absolute value of the energy of elementary particles to describe matter in the state of thermal radiation.

The conservation of energy has become the founding principle of physics. During the early days of nuclear physics, studies of β radiation, often called "β decay," showed initially that the energy of the products was not equal to the energy of the initial nucleus. This led to a reexamination of the law of conservation of energy by some physicists, who wondered if it could be violated in certain processes. Based on the validity of the conservation of energy, Wolfgang Pauli (1900–1958) suggested in 1930 that the missing energy was carried by a new particle which interacted extremely weakly with other particles and was therefore difficult to detect. This particle was later given the name *neutrino*. Pauli was proven right 25 years later. Experimental confirmation of the neutrino's existence came in 1956 from the careful experiments conducted by Frederick Reines and the late Clyde Cowan. Since then our faith in the law of conservation of energy has become stronger than ever. Frederick Reines was awarded the Nobel prize for physics in 1995. (*Physics Today*, Dec. 1995, pp. 17–19 gives the interesting history behind the discovery of the neutrino.)

References

1. Kuhn, T., *The Essential Tension*. 1977, Chicago: University of Chicago Press.
2. Steffens, H. J., *James Prescott Joule and the Concept of Energy*. 1979, New York: Science History Publications, p. 134.
3. Prigogine, I. and Stengers, I., *Order Out of Chaos*. 1984, New York: Bantam, p. 108.
4. Dornberg, J., Count Rumford: the most successful Yank abroad, ever. *Smithsonian*, December 1994, p. 102–115.
5. Segrè, E., *From Falling Bodies to Radio Waves*. 1984, New York: W. H. Freeman.
6. Mach, E., *Principles of the Theory of Heat*. 1986, Boston: D. Reidel.
7. Weinberg, S., *The First Three Minutes*. 1980, New York: Bantam.
8. Mason, S. F., *A History of the Sciences*. 1962, New York: Collier Books.
9. Leicester, H. M., *J. Chem. Ed.*, **28** (1951) 581–583.
10. Davis, T. W., *J. Chem. Ed.*, **28** (1951) 584–585.
11. Leicester, H. M., *The Historical Background of Chemistry*. 1971, New York: Dover.
12. *The National Bureau of Standards Tables of Chemical Thermodynamic Properties*. 1982, town: NBS.
13. Atkins, P. W., *Physical Chemistry*, 4th ed. 1990, New York: W. H. Freeman.

14. de Donder, T., *Lecons de Thermodynamique et de Chimie-Physique*. 1920, Paris: Gauthiers-Villars.
15. de Donder, T. and Van Rysselberghe P., *Affinity*. 1936, Menlo Park, CA: Stanford University Press.
16. Prigogine, I. and Defay R., *Chemical Thermodynamics*, 4th ed. 1967, London: Longman, p. 542.
17. Mason, S. F., *Chemical Evolution*. 1991, Clarendon Press, Oxford.
18. Planck, M., *Treatise on Thermodynamics*, 3rd ed. 1945, New York: Dover.

Data Sources

[A] NBS table of chemical and thermodynamic properties. *J. Phys. Chem. Reference Data*, **11**, suppl. 2 (1982).
[B] G. W. C. Kaye and T. H. Laby (eds) *Tables of physical and chemical constants*. 1986, London: Longman.
[C] I. Prigogine and R. Defay, *Chemical Thermodynamics*. 1967, London: Longman.
[D] J. Emsley, *The Elements* 1989, Oxford: Oxford University Press.
[E] L. Pauling, *The Nature of the Chemical Bond* 1960, Ithaca NY: Cornell University Press.
[F] D. R. Lide (ed.) *CRC Handbook of Chemistry and Physics*, 75th ed. 1994, Ann Arbor, MI: CRC Press.

Examples

Example 2.1   A bullet of mass 20.0 g, traveling at a speed of 350.0 m/s is lodged into a block of wood. How many calories of heat are generated in this process?

Solution: In this process, kinetic energy (KE) of the bullet is converted to heat:

$$\text{KE of the bullet} = mv^2/2 = (1/2)20.0 \times 10^{-3} \, \text{kg} \times (350 \, \text{m/s})^2 = 1225 \, \text{J}$$

$$1225 \, \text{J} = 1225 \, \text{J}/(4.184 \, \text{J cal}^{-1}) = 292.8 \, \text{cal}$$

Example 2.2   Calculate the energy ΔU required to increase the temperature of 2.50 mol of an ideal monatomic gas from 15.0 °C to 65.0 °C.

Solution: Since the heat capacity $C_V = (\partial U/\partial T)_V$, we see that

$$\Delta U = \int_{T_i}^{T_f} C_V \, dT = C_V(T_f - T_i)$$

Since C_V for a monatomic ideal gas is $(3/2)R$, we have

$$\Delta U = (3/2)(8.314 \; \text{J mol}^{-1} \, \text{K}^{-1})(2.5 \, \text{mol})(65.0 - 15.0) \, \text{K}$$
$$= 1559 \, \text{J}$$

Example 2.3 The velocity of sound in CH_4 at $41.0\,°C$ was found to be 466 m/s. Calculate the value of γ, the ratio of heat capacities, at this temperature.

Solution: Equation (2.3.17) gives the relation between γ and the velocity of sound:

$$\gamma = \frac{MC^2}{RT} = \frac{16.04 \times 10^{-3}\,\text{kg}((466\,\text{m/s})^2}{8.314(314.15\,\text{K})} = 1.33$$

Example 2.4 One mole of $N_2\,(g)$ at $25.0\,°C$ and a pressure of 1.0 bar undergoes an isothermal expansion to a pressure of 0.132 bar. Calculate the work done.

Solution: For an isothermal expansion:

$$\text{Work} = -NRT\,\ln\left(\frac{V_f}{V_i}\right)$$

For an ideal gas, at constant T, $p_i V_i = p_f V_f$, hence

$$\text{Work} = -NRT\,\ln\left(\frac{V_f}{V_i}\right)$$

$$= -NRT\,\ln\left(\frac{p_i}{p_f}\right) = -1.0(8.314\,\text{JK}^{-1})\,(298.15\,\text{K})\ln\left(\frac{1.0\,\text{bar}}{0.132\,\text{bar}}\right)$$

$$= -5.02\,\text{kJ}$$

Example 2.5 Calculate the heat of combustion of propane in the following reaction at $25\,°C$:

$$C_3H_8(g) + 5O_2(g) \rightarrow 3CO_2(g) + 4H_2O(l)$$

Solution: From the table of heats of formation at 298.15 K we obtain

$$\Delta H_{\text{rxn}}^0 = -\Delta H_f^0[C_3H_8] - 5\Delta H_f^0[O_2] + 3\Delta H_f^0[CO_2] + 4\Delta H_f^0[H_2O]$$

$$= -(-103.85\,\text{kJ}) - (0) + 3(-393.51\,\text{kJ}) + 4(-285.83\,\text{kJ}) = -2220\,\text{kJ}$$

Example 2.6 For the reaction $N_2(g) + 3H_2(g) \rightarrow 2NH_3(g)$ at $T = 298.15\,\text{K}$ the standard enthalpy of reaction is $-46.11\,\text{kJ/mol}$. At *constant volume*, if 1.0 mol of $N_2(g)$ reacts with 3.0 mol of $H_2(g)$, what is the energy released?

Solution: The standard enthalpy of reaction is the heat released at a constant pressure of 1.0 bar. At constant volume, since no mechanical work is done, the heat released equals the change in internal energy ΔU. From equation (2.4.9) we see that

$$\Delta H_{\text{rxn}} = \Delta U_{\text{rxn}} + \Delta N_{\text{rxn}}RT$$

In this reaction, $\Delta N_{rxn} = -2$, hence

$$\Delta U_{rxn} = \Delta H_{rxn} - (-2)RT = -46.11 + 2(8.314\,\mathrm{JK}^{-1})298.15 = -41.15\,\mathrm{kJ}$$

Exercises

2.1 For a conservative force $F = -\partial V(x)/\partial x$, in which $V(x)$ is the potential, using Newton's laws of motion, show that the sum of kinetic and potential energy is a constant.

2.2 How many joules of heat are generated by the brakes of a 1000 kg car when it is brought to rest from a speed of 50 kph? If we use this heat to heat 1.0 L of water from an initial temperature of 30 °C, estimate the final temperature, assuming that the heat capacity of water is about 1 cal per milliliter per °C. (1 cal = 4.184 J)

2.3 The manufacturer of a heater coil specifies that it is a 500 W device.
(a) At a voltage of 110 V what is the current through the coil?
(b) Given that the latent heat of fusion of ice is 6.0 kJ/mol, how long will it take for this heater to melt 1.0 kg of ice at 0 °C.

2.4 Use the relation $dW = -p\,dV$ to show the following:
(a) The work done in an *isothermal* expansion of N moles of an ideal gas from intial volume V_i to final volume V_f is:

$$\text{Work} = -NRT\,\ln\left(\frac{V_f}{V_i}\right)$$

(b) For an ideal gas, calculate the work done in an isothemal expansion of one mole from $V_i = 10.0\,\mathrm{L}$ to $V_f = 20.0\,\mathrm{L}$ at temperature $T = 350\,\mathrm{K}$.
(c) Repeat the calculation of part (a) using the van der Waals equation in place of the ideal gas equation and show that

$$\text{Work} = -NRT\,\ln\left(\frac{V_f - Nb}{V_i - Nb}\right) + aN^2\left(\frac{1}{V_i} - \frac{1}{V_f}\right)$$

2.5 Given that for the gas Ar, the heat capacity $C_V = (3R/2) = 12.47\,\mathrm{J\,K^{-1}\,mol^{-1}}$, calculate the velocity of sound in argon at $T = 298\,\mathrm{K}$ using the ideal gas relation between C_p and C_V. Do the same for N_2, for which $C_V = 20.74\,\mathrm{J\,K^{-1}\,mol^{-1}}$.

2.6 Calculate the sound velocities of He, N, and CO_2 using (2.3.17) and the values of γ at 15 °C in Table 2.2 and compare them to the experimentally measured velocities shown in the same table.

2.7   We have seen in equation (2.3.5) that for any system

$$C_p - C_V = \left[p + \left(\frac{\partial U}{\partial V} \right)_T \right] \left(\frac{\partial V}{\partial T} \right)_p$$

For the van der Waals gas, the energy $U_{vw} = U_{ideal} - a(N/V)^2 \, V$, in which $U_{ideal} = C_V NT + U_0$ (2.2.15). Use these two expressions and the van der Waals equation to obtain an explicit expression for the difference between C_p and C_V.

2.8   For nitrogen at $p = 1$ atm, and $T = 298$ K, calculate the change in temperature when it undergoes an adiabatic compression to a pressure of 1.5 atm. ($\gamma = 1.404$ for nitrogen)

2.9   As shown in (2.4.11), for many gases the empirical value of molar heat capacity satisfies $C_p = \alpha + \beta T + \gamma T^2$. The values of α, β and γ for some gases are given in Table 2.3. Use this table to calculate the change in the enthalpy of one mole of $CO_2(g)$ when it is heated from 350.0 K to 450.0 K at $p = 1$ atm.

2.10   Using the table of thermodynamic data that contains heats of formation of compounds at $T = 298.15$ K, calculate the standard heats of reaction for the following reactions:
(a) $H_2(g) + F_2(g) \rightarrow 2HF(g)$
(b) $C_7H_{16}(g) + 11O_2(g) \rightarrow 7\,CO_2(g) + 8\,H_2O(l)$
(c) $2\,NH_3(g) + 6\,NO(g) \rightarrow 3\,H_2O_2(l) + 4\,N_2(g)$

2.11   Gasoline used as motor fuel consists of a mixture of the hydrocarbons heptane (C_7H_{16}), octane (C_8H_{18}) and nonane (C_9H_{20}). Using Table 2.4, estimate the enthalpy of combustion of one gram of each of these fluids. In a combustion reaction, an organic compound reacts with $O_2(g)$ to produce $CO_2(g)$ and $H_2O(g)$.

2.12   Calculate the amount of energy released in the combustion of 1 g of sucrose and compare it with the mechanical energy needed to lift 100 kg through 1 m. The equation for the reaction is

$$C_{12}H_{22}O_{11}(s) + 12\,O_2(g) \rightarrow 11\,H_2O(l) + 12\,CO_2(g)$$

2.13   Consider the reaction

$$CH_4(g) + 2\,O_2(g) \rightarrow CO_2(g) + 2\,H_2O(l)$$

Assume the initial quantities are 3.0 mol CH_4 and 2.0 mol O_2 and the initial extent of reaction is $\xi = 0$. When the extent of reaction is $\xi = 0.25$ mol, what are the concentrations of the reactants and the products? How much heat is released at this point? What is the value of ξ when all the O_2 has reacted?

2.14 The sun radiates energy approximately at a rate of 3.9×10^{26} J/s. What will be the change in the mass of the sun in one million years, if it radiates at this rate? (Sun's mass $= 2 \times 10^{30}$ kg)

2.15 Calculate the energy released in the reaction

$$2\,^1H + 2n \rightarrow\,^4He$$

given the following masses: mass of $^1H = 1.0078$ amu, mass of n $= 1.0087$ amu, mass of $^4He = 4.0026$ amu. (1 amu $= 1.6605 \times 10^{-27}$ kg)

3 THE SECOND LAW OF THERMODYNAMICS AND THE ARROW OF TIME

3.1 The Birth of the Second Law

James Watt (1736–1819), the most famous of Joseph Black's pupils, obtained a patent for his modifications of Thomas Newcomen's steam engine in the year

James Watt (1736–1819) (Courtesy the E. F. Smith Collection, Van-Pelt-Dietrich Library, University of Pennsylvania)

1769. Soon this invention brought unimagined power and speed to everything: mining of coal, transportation, agriculture and industry. This revolutionary generation of motion from heat that began on the British Isles quickly crossed the Channel and spread throughout Europe.

Nicolas-Léonard-Sadi Carnot (1796–1832), a brilliant French military engineer, lived in this rapidly industrializing Europe. "Every one knows," he wrote in his memoirs, "that heat can produce motion. That it possesses vast motive-power no one can doubt, in these days when the steam-engine is everywhere so well known" [1, p. 3]. Sadi Carnot's father, Lazare Carnot (1753–1823), held many high positions during and after the French Revolution and was known for his contributions to mechanics and mathematics. Lazare Carnot had a strong influence on his son Sadi. Both had their scientific roots in engineering, and both had a deep interest in general principles in the tradition of the French

Sadi Carnot (1796–1832) (Courtesy the E. F. Smith Collection, Van Pelt-Dietrich Library, University of Pennsylvania)

Encyclopedists. It was his interest in general principles that led Sadi Carnot to his abstract analysis of steam engines. Carnot pondered over the principles that governed the working of the steam engine and identified the *flow of heat* as the fundamental process required for the generation of "motive-power"— "work" in today's terminology. He analyzed the amount of work generated by such **heat engines**, engines that performed mechanical work through the flow of heat, and realized that there must be a fundamental limit to the work that can be obtained from the flow of a given amount of heat. Carnot's great insight was that this limit was *independent* of the machine and the manner in which work was obtained; it depended only on the temperatures that caused the flow of heat. Further development of this principle led to the Second Law of thermodynamics.

Carnot described his general analysis of heat engines in his only scientific publication, *Réflexions sur la Puissance Motrice du Feu, et sur les Machines Propres a Développer cette Puissance* (*Reflections on the Motive Force of Fire and on the Machines Fitted to Develop that Power*) [1]. Six hundred copies of this work were published in 1824 at Carnot's own expense. At that time, the name Carnot was well known to the scientific community in France due to the fame of Sadi's father, Lazare Carnot. Still, Sadi Carnot's book did not attract much attention from the scientific community at the time of its publication. Eight years after the publication of his *Réflexions*, Sadi Carnot died of cholera. A year later, Émile Clapeyron (1799–1864) came across Carnot's book, realized its fundamental importance and made it known to the scientific community.

CARNOT'S THEOREM

Carnot's analysis proceeded as follows. First, Carnot observed, *"Wherever there exists a difference of temperature, motive force can be produced"* [1, p. 9]. Every heat engine that produced work from the flow of heat operated between two heat reservoirs of unequal temperatures. In the processes of transferring heat from a hot reservoir to a cold reservoir, the engine performed mechanical work (Fig. 3.1). Carnot then specified the following condition for the production of maximum work [1, p. 13]:

"The necessary condition of the maximum (work) is, *that in the bodies employed to realize the motive power of heat there should not occur any change of temperature which may not be due to a change of volume*. Reciprocally, every time that this condition is fulfilled the maximum will be attained. This principle should never be lost sight of in the construction of a heat engine; it is its fundamental basis. If it cannot be strictly observed, it should at least be departed from as little as possible."

Thus, for maximum work generation, all changes in volume—such as the expansion of a gas (steam) that pushes a piston—should occur with minimal temperature gradients so that changes in temperature are almost all due to volume expansion and not due to flow of heat caused by temperature gradients.

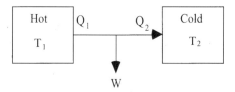

Figure 3.1 Illustration of the fundamental observation made ' by Sadi Carnot: "Wherever there exists a difference of temperature, motive force can be produced" [1, p. 9]. The engine absorbs heat Q_1 from the hot reservoir, converts part of it to work W, and delivers the rest of the heat to the cold reservoir. The efficiency η is given by $W = \eta Q_1$. (According to the caloric theory of heat used by Carnot, $Q_1 = Q_2$)

This is achieved in heat engines that absorb and discard heat during very slow changes in volume, keeping their internal temperature as uniform as possible.

Furthermore, in the limit of infinitely slow transfer of heat during changes of volume, with infinitesimal temperature difference between the source of heat (the heat reservoir) and the engine, the operation of a heat engine is a *reversible process*. In a reversible process the series of states that the engine goes through could be retraced in the exact opposite order. A reversible engine can perform mechanical work, W, by transferring heat from a hot reservoir to a cold reservoir; it can do the exact opposite by transferring the *same* amount of heat from a cold reservoir to a hot reservoir using the *same* amount of work, W. The next idea Carnot introduces is that of a *cycle*: during its operation, the engine went through a cycle of states so that, after producing work from the flow of heat, it returned to its initial state, ready to go through the cycle once again. A modern version of Carnot's reversible cycle will be discussed later in this section.

Carnot argued that the reversible cyclic engine must produce the maximum work (motive force), but he used the caloric theory of heat, according to which heat was an indestructible substance. Carnot's argument proceeded as follows: If any engine could produce a greater amount of work than that produced by a reversible cyclic engine, then it would be possible to produce work endlessly by the following means. Begin by moving heat from the hot reservoir to a cold reservoir using the more efficient engine. Then move the same amount of heat back to the hot reservoir using the reversible engine. Because the forward process does more work than is needed to perform the reverse process, there is a net gain in work. In this cycle of operations, a certain amount of heat was simply moved from the hot reservoir to the cold reservoir and back to the hot reservoir, with a net gain of work. By repeating this cycle, an unlimited amount of work can be obtained simply by moving a certain amount of heat back and forth between a hot reservoir and a cold reservoir. This, Carnot asserted, was impossible [1, p. 12]:

"This would be not only perpetual motion, but an unlimited creation of motive power without consumption either of caloric or of any other agent whatever. Such a creation is entirely contrary to ideas now accepted, to laws of mechanics and of sound physics. It is inadmissible."

Hence, reversible cyclic engines must produce the maximum amount of work. A corollary of this conclusion is that *all* reversible cyclic engines must produce the same amount of work regardless of their construction. Furthermore, and most important, since all reversible engines produce the same amount of work from a given amount of heat, the amount of work generated by a reversible heat engine is independent of the material properties of the engine; it can depend only on the temperatures of the hot and cold reservoirs. This brings us to the most important conclusion in Sadi Carnot's book [1, p. 20]:

"The motive power of heat is independent of the agents employed to realize it; its quantity is fixed solely by the temperatures of the bodies between which is effected, finally, the transfer of caloric."

Carnot did not derive a mathematical expression for the maximum efficiency attained by a reversible heat engine in terms of the temperatures between which it operated. This was done later by others who realized the importance of his conclusion. Carnot did, however, find a way of calculating the maximum work that can be generated. (For example, he concluded that "1000 units of heat passing from a body maintained at the temperature of 1 degree to another body maintained at zero would produce, in acting upon the air, 1.395 units of motive power" [1, p. 42].)

Though Sadi Carnot used the caloric theory of heat to reach his conclusions, his later scientific notes reveal his realization that the caloric theory was not supported by experiments. In fact, Carnot understood the mechanical equivalence of heat and even estimated the conversion factor to be approximately 3.7 joules per calorie (the more accurate value being 4.18 J/cal) [1–3]. Unfortunately, Sadi Carnot's brother, Hippolyte Carnot, who was in possession of Sadi's scientific notes from the time of his death in 1832, did not make them known to the scientific community until 1878 [3]. That was the year in which Joule published his last paper. By then the equivalence between heat and work and the law of conservation of energy were well known through the work of Joule, Helmholtz, Mayer and others. (It was also in 1878 that Gibbs published his famous work *On the Equilibrium of Heterogeneous Substances*).

Sadi Carnot's brilliant insight went unnoticed until Émile Clapeyron (1799–1864) came across Carnot's book in 1833. Realizing its importance, he reproduced the main ideas in an article that was published in the *Journal de l'Ecole Polytechnique* in 1834. Clapeyron represented Carnot's example of a reversible engine in terms of a p–V diagram (which is used today) and described it with mathematical detail. Clapeyron's article was later read by Lord Kelvin and others who realized the fundamental nature of Carnot's conclusions and investigated its consequences. These developments led to the formulation of the Second Law of thermodynamics as we know it today.

To obtain the efficiency of a reversible heat engine, we shall not follow Carnot's original reasoning because it considered heat as an indestructible substance. Instead, we shall modify it by incorporating the First Law. For the heat engine represented in Fig. 3.1, the law of conservation of energy gives $W = Q_1 - Q_2$. This means, a fraction η of the heat Q_1 absorbed from the hot reservoir is converted into work W, i.e. $\eta = W/Q_1$. The fraction η is called *the efficiency of the heat engine*. Since $W = (Q_1 - Q_2)$ in accordance with the First Law, $\eta = (Q_1 - Q_2)/Q_1 = (1 - Q_2/Q_1)$. Carnot's discovery that the reversible engine produces maximum work amounts to the statement that its efficiency is maximum. This efficiency is independent of the properties of the engine and is a function only of the temperatures of the hot and the cold reservoirs:

$$\eta = 1 - \frac{Q_2}{Q_1} = 1 - f(t_1, t_2) \tag{3.1.1}$$

in which $f(t_1, t_2)$ is a function only of the temperatures t_1 and t_2 of the hot and cold reservoirs. The scale of the temperatures (Celsius or other) t_1 and t_2 is not specified here. Equation (3.1.1) is **Carnot's theorem**. In fact, Carnot's observation enables us to define an absolute scale of temperature that is independent of the material property used to measure it.

EFFICIENCY OF A REVERSIBLE HEAT ENGINE

Now we turn to the task of obtaining the efficiency of reversible heat engines. Since the efficiency of a reversible heat engine is the maximum, all of them must have the same efficiency. Hence obtaining the efficiency of one particular reversible engine will suffice. The following derivation also makes it explicit that the efficiency of Carnot's engine is only a function of temperature.

Carnot's reversible engine consists of an ideal gas that operates between a hot reservoir and a cold reservoir, at temperatures θ_1 and θ_2 respectively. Until their identity has been is established, we shall use θ for the temperature that appears in the ideal gas equation and T for the absolute temperature (which, as we shall see in the next section, is defined by the efficiency of a reversible cycle). Thus, the ideal gas equation is written as $pV = NR\theta$, in which θ is the temperature measured by noting the change of some quantity such as volume or pressure. (Note that measuring temperature by volume expansion is purely empirical; each

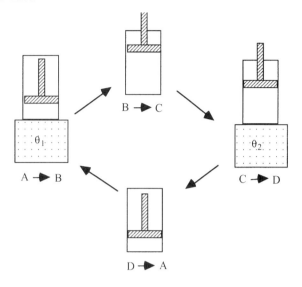

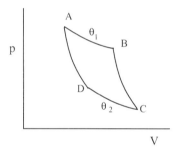

Figure 3.2 The Carnot cycle. The upper part shows the four steps of the Carnot cycle, during which the engine absorbs heat from the hot reservoir, produces work and returns heat to the cold reservoir. The lower part shows the representation of this process in a $p–V$ diagram used by Clapeyron in his exposition of Carnot's work

unit of temperature is simply correlated with a certain change in volume.) The cycle consists of the following four steps (Fig. 3.2).

STEP 1

The gas has an initial volume V_A and is in contact with the hot reservoir at temperature θ_1. Staying in contact with the reservoir, the gas undergoes an infinitely slow *reversible expansion* (as Carnot specified it) to the state B, of volume V_B. During this process, the work done *by the gas* is

$$W_{AB} = \int_{V_A}^{V_B} p\,dV = \int_{V_A}^{V_B} N\,\frac{R\theta_1}{V}\,dV = NR\theta_1 \ln\left(\frac{V_B}{V_A}\right) \tag{3.1.2}$$

During this isothermal processes, heat is absorbed from the reservoir. Since the internal energy of an ideal gas depends only on the temperature, according to equations (1.3.8) and (2.2.15), there is no change in the energy of the gas; all the work done equals the heat absorbed. Hence the heat absorbed is

$$Q_{AB} = W_{AB} \tag{3.1.3}$$

STEP 2

In the second step, the gas is thermally insulated from the reservoir and the environment; it is made to undergo an adiabatic expansion from state B to a state C, decreasing the temperature from θ_1 to θ_2. During this adiabatic processes, work is done by the gas. Noting that on the adiabat BC we have $pV^\gamma = p_B V_B^\gamma = p_C V_C^\gamma$, we can calculate the work as

$$W_{BC} = \int_{V_B}^{V_C} p\,dV = \int_{V_B}^{V_C} \frac{p_B V_B^\gamma}{V^\gamma}\,dV = \frac{p_C V_C^\gamma V_C^{1-\gamma} - p_B V_B^\gamma V_B^{1-\gamma}}{1-\gamma}$$

$$= \frac{p_C V_C - p_B V_B}{1-\gamma}$$

Using $pV = NR\theta$, the above equation can be further simplified to

$$W_{BC} = \frac{NR(\theta_1 - \theta_2)}{\gamma - 1} \tag{3.1.4}$$

in which θ_1 and θ_2 are the initial and final temperatures during the adiabatic expansion.

STEP 3

In the third step, the gas is in contact with the reservoir of temperature θ_2, and undergoes an isothermal compression to the point D at which the volume V_D is such that an adiabatic compression can return it to the initial state A. (V_D can be specified by finding the point of intersection of the adiabat through the point A and the isotherm at temperature θ_2.) During this processes, work is done *on the gas* and it equals

$$W_{CD} = \int_{V_C}^{V_D} p\, dV = \int_{V_C}^{V_D} N \frac{R\theta_2}{V} dV = NR\theta_2 \ln\left(\frac{V_D}{V_C}\right) = Q_{CD} \qquad (3.1.5)$$

STEP 4

In the final step, an adiabatic compression takes the gas from state D to its initial state A. Since this process is similar to step 2, we can write

$$W_{DA} = \frac{NR(\theta_2 - \theta_1)}{\gamma - 1} \qquad (3.1.6)$$

The total work obtained in this reversible Carnot cycle is

$$W = W_{AB} + W_{BC} + W_{CD} + W_{DA} = Q_{AB} - Q_{CD}$$
$$= NR\theta_1 \ln\left(\frac{V_B}{V_A}\right) - NR\theta_2 \ln\left(\frac{V_C}{V_D}\right) \qquad (3.1.7)$$

The efficiency $\eta = W/Q_{AB}$ can now be written using (3.1.2), (3.1.3) and (3.1.7):

$$\eta = \frac{W}{Q_{AB}} = 1 - \frac{NR\theta_2 \ln(V_C/V_D)}{NR\theta_1 \ln(V_B/V_A)} \qquad (3.1.8)$$

For the isothermal processes, we have $p_A V_A = p_B V_B$; $p_C V_C = p_D V_D$; and for the adiabatic processes, we have $p_B V_B^\gamma = p_C V_C^\gamma$ and $p_D V_D^\gamma = p_A V_A^\gamma$. Using these relations, it can easily be seen that $V_C/V_D = V_B/V_A$. Using this relation in (3.1.8) we arrive at a simple expression for the efficiency:

$$\eta = \frac{W}{Q_{AB}} = 1 - \frac{\theta_2}{\theta_1} \qquad (3.1.9)$$

In this expression for the efficiency, θ is the temperature defined by one particular property (such as volume at a constant pressure) and we assume that it satisfies the ideal gas equation. The temperature t, measured by any other empirical means such as measuring the volume of mercury, is related to θ. We may denote this relation by $\theta(t)$, i.e. t measured by one means is equal to $\theta = \theta(t)$, measured by another means. In terms of any other temperature t, the efficiency may take a more complex form. In terms of the temperature θ that obeys the ideal gas equation, however, the efficiency of the reversible heat engine takes a particularly simple form (3.1.9).

3.2   The Absolute Scale of Temperature

The fact that the efficiency of a reversible heat engine is independent of the physical and chemical nature of the engine has an important consequence which was noted by Lord Kelvin, William Thomson (1824–1907). Following Carnot's work, Lord Kelvin introduced the *absolute scale of temperature*. The efficiency of a reversible heat engine is a function only of the temperatures of the hot and cold reservoirs, independent of the material properties of the engine. Furthermore, the efficiency cannot exceed 1, in accordance with the First Law. These two facts can be used to define an *absolute scale of temperature which is independent of any material properties.*

William Thomson (Lord Kelvin) (1824–1907) (Courtesy the E. F. Smith Collection, Van Pelt-Dietrich Library, University of Pennsylvania)

First, by considering two successive Carnot engines, one operating between t_1 and t', the other operating between t' and t_2, we can see that the function $f(t_2, t_1)$ in equation (3.1.1) is a ratio of functions of t_1 and t_2 as follows. If Q' is the heat exchanged at temperature t', we can write

$$f(t_2, t_1) = \frac{Q_2}{Q_1} = \frac{(Q_2/Q')}{(Q'/Q_1)} = \frac{f(t_2, t')}{f(t', t_1)} \qquad (3.2.1)$$

This relation implies that we can write the function $f(t_2, t_1)$ as a ratio $f(t_2)/f(t_1)$. Hence the efficiency of a reversible Carnot engine can be written as

$$\eta = 1 - \frac{Q_2}{Q_1} = 1 - \frac{f(t_2)}{f(t_1)} \qquad (3.2.2)$$

One can now define a temperature $T \equiv f(t)$, based solely on the efficiencies of reversible heat engines. This is the absolute temperature measured in Kelvin. In terms of this temperature scale, the efficiency of a reversible engine is given by

$$\boxed{\eta = 1 - \frac{Q_2}{Q_1} = 1 - \frac{T_2}{T_1}} \qquad (3.2.3)$$

in which T_1 and T_2 are the respective absolute temperatures of the cold reservoir and the hot reservoir. An efficiency of 1 defines the absolute zero of this scale. **Carnot's theorem** is the statement that reversible engines have the maximum efficiency given by (3.2.3).

Comparing (3.2.3) with (3.1.9), we see that the ideal gas temperature coincides with the absolute temperature, hence we can use the same symbol, T, for both[*].

In summary, for an idealized, *reversible heat engine* that absorbs heat Q_1 from a hot reservoir at absolute temperature T_1 and discards heat Q_2 to a cold reservoir at absolute temperature T_2, from (3.2.3) we have

$$\boxed{\frac{Q_1}{T_1} = \frac{Q_2}{T_2}} \qquad (3.2.4)$$

---

[*] The empirical temperature t, of a gas thermometer is defined through the increase in volume at constant pressure (1.3.9):

$$V = V_0(1 + \alpha t)$$

Gay-Lussac found that α is close to $1/273$ per degree Celsius. From this equation it follows that $dV/V = dt/(1 + \alpha t)$. On the other hand, from the ideal gas equation $pV = NRT$, we have, at constant p, $dV/V = dT/T$. This enables us to relate the absolute temperature T to the empirical temperature t by $T = (1 + \alpha t)$.

All *real* heat engines that go through a cycle in finite time must involve irreversible processes, such as flow of heat due to a temperature gradient; they are less efficient. The efficiency $\eta' = 1 - Q_2/Q_1 < 1 - T_2/T_1$ for such engines. This implies $T_2/T_1 < Q_2/Q_1$ whenever irreversible processes are included. Therefore, while the equality (3.2.4) is valid for a reversible cycle, for the operation of an *irreversible cycle* that we encounter in reality, we have the *inequality*

$$\frac{Q_1}{T_1} < \frac{Q_2}{T_2}$$

(3.2.5)

This inequality plays a fundamental role in what follows.

3.3   The Second Law and the Concept of Entropy

The full import of the concepts originating in Carnot's *Reflexions* was realized in the generalizations made by Rudolf Clausius (1822–1888), who introduced the concept of *entropy*, a new physical quantity as fundamental and universal as energy.

Clausius began by generalizing expression (3.2.4) that follows from Carnot's theorem to an arbitrary cycle. This was done by considering composites of Carnot cycles in which the corresponding isotherms differ by an infinitesimal amount ΔT, as shown in Fig. 3.3(a). Let Q_1 be the heat absorbed during the transformation from A to A$'$, at temperature T_1, and Q'_1 be the heat absorbed during the transformation A$'$B at temperature $(T_1 + \Delta T)$. Similarly we define Q'_2 and Q_2 for the transformations CC$'$ and C$'$D occurring at temperatures $T_2 + \Delta T$ and T_2 respectively. Then the reversible cycle AA$'$BCC$'$DA can be thought of as a sum of the two reversible cycles AA$'$C$'$DA and A$'$BCC$'$A$'$ because the adiabatic work A$'$C$'$ in one cycle cancels that of the second cycle, C$'$A$'$. For the reversible cycle AA$'$BCC$'$D, we can therefore write

$$\frac{Q_1}{T_1} + \frac{Q'_1}{T_1 + \Delta T} - \frac{Q_2}{T_2} - \frac{Q'_2}{T_2 + \Delta T} = 0$$

(3.3.1)

The above composition of cycles can be extended to an arbitrary closed path (Fig. 3.3(b)) by considering it as a combination of an infinite number of Carnot cycles. With notation $dQ > 0$ if heat is absorbed by the system and $dQ < 0$ if it is discarded, the generalization of (3.3.1) of an arbitrary closed path gives

$$\oint \frac{dQ}{T} = 0$$

(3.3.2)

Rudolf Clausius (1822–1888) (Courtesy the E. F. Smith Collection, Van Pelt-Dietrich Library, University of Pennsylvania)

This equation has an important consequence: it means that the integral of the quantity dQ/T along a path representing a reversible process from a state A to a state B depends only on the states A and B and is independent of the path (Fig. 3.4). Thus Clausius saw how one can define a function S *that only depends on the initial and final states of a reversible process*. If S_A and S_B are the values of this function in the states A and B, we can write

$$S_B - S_A = \int_A^B \frac{dQ}{T} \quad \text{or} \quad dS = \frac{dQ}{T} \tag{3.3.3}$$

By defining a reference state O, the new function of state S could now be defined for any state X as the integral of dQ/T for a *reversible process* transforming the state O to the state X.

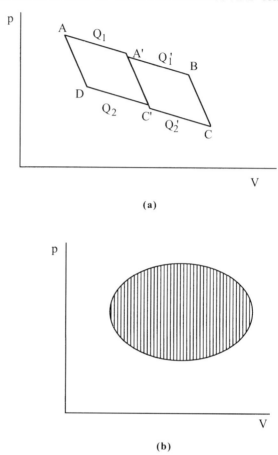

Figure 3.3 Clausius's generalization of the Carnot cycle

Clausius introduced this new quantity S in 1865 saying, "I propose to call the magnitude S the *entropy* of the body, from the Greek word τροπη, *transformation*." [4, p. 357]. The usefulness of this definition depends on the assumption that any two states can be connected by a reversible transformation.

If the temperature remains fixed, it follows from (3.3.3) that for a reversible flow of heat Q, the change in entropy is Q/T. In terms of entropy, Carnot's theorem (3.2.3) becomes the statement that the sum of the *entropy changes in a reversible cycle is zero*:

$$\boxed{\frac{Q_1}{T_1} - \frac{Q_2}{T_2} = 0} \qquad (3.3.4)$$

In a reversible process, since the system and the reservoir have the same

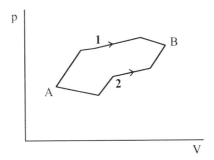

Figure 3.4 Any function, whose integral over an arbitrary closed path is zero, as in (3.3.2), can be used to define a function of state. Since the total integral for a closed path from A to B along 1 and from B to A along 2 is zero, it follows that: $\oint dQ/T = \int_{A,\text{path }1}^{B} dQ/T + \int_{B,\text{path }2}^{A} dQ/T = 0$. Now we note that $\int_{A}^{B} dQ/T = -\int_{B}^{A} dQ/T$ along paths 1 or 2. Hence $\int_{A,\text{path }1}^{B} dQ/T = \int_{A,\text{path }2}^{B} dQ/T$, i.e. the integral of dQ/T from point A to point B is independent of the path. It depends only on the points A and B

temperature when heat is exchanged, the change of entropy of the reservoir in any part of the cyclic process is the negative of the entropy change of the system.

In a less efficient *irreversible cycle* a smaller fraction of Q_1 (the heat absorbed from the hot reservoir) is converted into work. This means the amount of heat delivered to the cold reservoir by an irreversible cycle, Q_2^{irr}, is greater than Q_2. We have therefore

$$\frac{Q_1}{T_1} - \frac{Q_2^{\text{irr}}}{T_2} < 0 \tag{3.3.5}$$

Since the cyclic engine returns to its initial state, whether it is reversible or irreversible, there is no change in its entropy. On the other hand, since the heat transferred to the reservoirs and the irreversible engine have opposite sign, the *total change of entropy of the reservoirs is*

$$\frac{(-Q_1)}{T_1} - \frac{(-Q_2^{\text{irr}})}{T_2} > 0 \tag{3.3.6}$$

if the reservoir temperatures can be assumed to be the same as the temperatures at which the engine operates. In fact, for heat to flow at a finite rate, the reservoir temperatures T_1' and T_2' must be such that $T_1' > T_1$ and $T_2' < T_2$. In this case, the increase in entropy is even larger than (3.3.6).

Generalizing (3.3.5) and (3.3.6) for a system that goes through an arbitrary cycle, with the equalities holding for a reversible process, we have

$$\oint \frac{dQ}{T} \leq 0 \qquad \text{(system)} \tag{3.3.7}$$

For the "exterior" with which the system exchanges heat, since dQ has the opposite sign, we have

$$\oint \frac{dQ}{T} \geq 0 \qquad \text{(exterior)} \tag{3.3.8}$$

At the end of the cycle, be it reversible or irreversible, there is no change in the system's entropy because it has returned to its original state. For irreversible cycles it means that the system expels more heat to the exterior. This is generally a conversion of mechanical energy into heat through irreversible processes. Consequently, the entropy of the exterior increases. This may be summarized as follows:

For a reversible cycle: $\quad dS = \dfrac{dQ}{T} \qquad \oint dS = \oint \dfrac{dQ}{T} = 0 \tag{3.3.9}$

For an irreversible cycle: $\quad dS > \dfrac{dQ}{T} \qquad \oint dS = 0, \qquad \oint \dfrac{dQ}{T} < 0 \tag{3.3.10}$

As we shall see in the following section, this statement can be made more precise by expressing the entropy change dS as a sum of two parts:

$$\boxed{dS = d_e S + d_i S} \tag{3.3.11}$$

Here $d_e S$ is the change of the system's entropy due to exchange of energy and matter and $d_i S$ is the change in entropy due to irreversible processes within the system. For a closed system that does not exchange matter, $d_e S = dQ/T$. The quantity $d_e S$ could be positive or negative, but $d_i S$ can only be greater than or equal to zero. In a cyclic process that returns the system to its initial state, since the net change in entropy must be zero, we have

$$\oint dS = \oint d_e S + \oint d_i S = 0 \tag{3.3.12}$$

Since $d_iS \geq 0$, we must have $\oint d_iS \geq 0$. For a closed system, from (3.3.12) we immediately obtain the previous result (3.3.10):

$$\oint d_eS = \oint \frac{dQ}{T} \leq 0$$

This means that, for the system to return to its initial state, the entropy $\oint d_iS$ generated by the irreversible processes within the system has to be discarded through the expulsion of heat to the exterior. There is no real system in nature that can go through a cycle of operations and return to its initial state without increasing the entropy of the exterior, or more generally the "universe." *The increase of entropy distinguishes the future from the past: there exists an arrow of time.*

STATEMENTS OF THE SECOND LAW

The limitation on the convertibility of heat to work that Carnot discovered is one manifestation of a fundamental limitation in all natural processes: it is the Second Law of thermodynamics. The Second Law can be formulated in many equivalent ways. For example, as a statement about a macroscopic impossibility, without any reference to the microscopic nature of matter:

"It is impossible to construct an engine which will work in a complete cycle, and convert *all* the heat it absorbs from a reservoir into mechanical work."

A statement perfectly comprehensible in macroscopic, operational terms. A cyclic engine that converts all heat to work is shown in Fig. 3.5. Since the reservoir or the "exterior" only loses heat, inequality (3.3.8) is clearly violated. This engine is sometimes called a *perpetual motion machine of the second kind* and the Second Law is the statement that such a machine is impossible. The

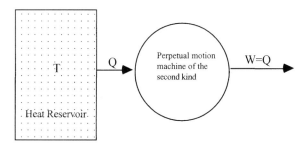

Figure 3.5 Perpetual motion machine of the second kind, which according to the Second Law is impossible. The existence of such a machine would violate inequalities (3.3.7) and (3.3.8)

equivalence between this statement and Carnot's theorem can be seen easily and it is left as an exercise for the reader.

Another way of stating the Second Law is due to Rudolf Clausius (1822–1888):

"Heat cannot by itself pass from a colder to a hotter body."

If heat could pass spontaneously from a colder body to a hotter body, then a perpetual motion machine of the second kind could be realized by simply making the heat Q_2 expelled by a cyclic heat engine to the colder reservoir pass by itself to the hotter reservoir. The result would be the complete conversion of the heat $(Q_1 - Q_2)$ to work.

As we have seen, any real system that goes through a cycle of operations and returns to its initial state does so only by *increasing* the entropy of its exterior with which it is interacting. This also means that in no part of the cycle, the *sum* of entropy changes of the system and its exterior can be negative. If this were so, we could complete the rest of the cycle through a reversible transformation, which would not contribute to the change of entropy. The net result would be a *decrease* of entropy in a cyclic process. Thus the Second Law may also be stated as:

"The sum of the entropy changes of a system and its exterior can never decrease."

Thus the universe as a whole can never return to its initial state. Remarkably, the Carnot analysis of heat engines has led to a formulation of a cosmological principle. The two laws of thermodynamics are best summarized by Rudolf Clausius:

"The energy of the universe is a constant.
The entropy of the universe approaches a maximum."

3.4 Entropy, Reversible and Irreversible Processes

The usefulness of the concept of entropy and the Second Law depends on our ability to define entropy of a physical system in a calculable way. Using (3.3.3), if the entropy S_0 of a *reference or standard state is defined*, then the entropy of an arbitrary state S_X can be obtained through a *reversible process* that transforms the state 0 to the state X (Fig. 3.6).

$$S_X = S_0 + \int_0^X \frac{dQ}{T} \tag{3.4.1}$$

(In practice dQ is measured with the knowledge of the heat capacity using $dQ = CdT$.) In a real system the transformation from state O to state X occurs in a finite time and involves irreversible processes along the path I. *In classical*

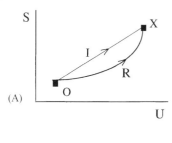

(A)

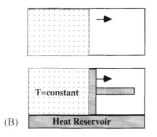

(B)

Figure 3.6   Reversible and irreversible processes. (A) The system reaches the state X from the standard state O through a path I involving irreversible processses. It is assumed that the same transformation can be achieved through a reversible transformation R. (B) An example of an irreversible process is the spontaneous expansion of a gas into vacuum. The same change can be achieved reversibly through an isothermal expansion of a gas that occurs infinitely slowly so that the heat absorbed from the reservoir equals the work done on the piston. In a reversible isothermal expansion the change in entropy can be calculated using $dS = dQ/T$.

thermodynamics it is assumed that every irreversible transformation that occurs in nature can also be achieved through a reversible process for which (3.4.1) *is valid.* In other words, it is assumed that every irreversible transformation that results in a certain change in the entropy can be exactly reproduced through a reversible process in which the entropy change is solely due to the exchange of heat. Since the change in entropy depends only on the initial and final states, the change in entropy calculated using a reversible path will be equal to the entropy change produced by the irreversible processes. (Some authors restrict the above assumption to transformations between equilibrium states; this restriction excludes chemical reactions in which the transformations are often from a nonequilibrium state to an equilibrium state).

A process is reversible only in the limit of infinite slowness: as perfect reversibility is approached, the speed of the process tends to zero. As Max Planck notes in his treatise [5, p. 86], "Whether reversible processes exist in nature or not, is not *a priori* evident or demonstrable". But irreversibility, if it exists, has to be universal because a spontaneous decrease of entropy in one system could be utilized to decrease the entropy of any other system through appropriate interaction; a spontaneous decrease of entropy of one system implies a spontaneous decrease of entropy of all systems. Hence, either all systems are irreversible, or none are.

The notion of an idealized reversible path provides a convenient way for calculating the changes of entropy. But it is also lacking in providing the real connection between naturally occurring irreversible processes and entropy. Addressing this issue in his 1943 monograph, *The Nature of Thermodynamics*, P.W. Bridgman wrote [6, p. 133]:

"It is almost always emphasized that thermodynamics is concerned with reversible processes and equilibrium states and that it can have nothing to do with irreversible processes or systems out of equilibrium in which changes are progressing at a finite rate. The reason for the importance of equilibrium states is obvious enough when one reflects that temperature itself is defined in terms of equilibrium states. But the admission of general impotence in the presence of irreversible processes appears on reflection to be a surprising thing. Physics does not usually adopt such an attitude of defeatism."

Today, in most texts on thermodynamics, an irreversible transformation is usually identified by the **Clausius inequality**

$$\boxed{dS \geq \frac{dQ}{T}} \tag{3.4.2}$$

that we have seen in the last section. But the fact that Clausius considered irreversible processes as an integral part of formulating the Second Law is generally not mentioned. In his ninth memoir, Clausius included irreversible processes explicitly into the formalism of entropy and replaced the inequality (3.4.2) by an equality [4, p. 363, eqn 71]:

$$N = S - S_0 - \int \frac{dQ}{T} \tag{3.4.3}$$

in which S is the entropy of the final state and S_0 is the entropy of the initial state. He identified the change in entropy due to exchange of heat (heat gain or loss compensated by equal gain or loss of heat by the exterior) with the exterior by the term dQ/T; "The magnitude N thus determines," wrote Clausius, "the uncompensated *transformation*" (uncompensirte Verwandlung) [4, p. 363]. It is the entropy produced by the irreversible processes within the system. Although dQ can be positive or negative, the Clausius inequality (3.4.2) implies that the change in entropy due to irreversible processes must be positive:

$$N = S - S_0 - \int \frac{dQ}{T} > 0 \tag{3.4.4}$$

Clausius also stated the Second Law like this: "Uncompensated transformations can only be positive." [4, p. 247].

Perhaps Clausius hoped to, but did not, provide a way of computing N associated with irreversible processes. Nineteenth-century thermodynamics remained in the restricted domain of idealized reversible transformation and without a theory that related entropy explicitly to irreversible processes. Some expressed the view that entropy is a physical quantity that is spatially distributed and transported (e.g. Bertrand [7] in his 1887 text), but still no theory relating irreversible processes to entropy was formulated in the nineteenth century.

Noticing the importance of relating entropy to irreversible processes, Pierre Duhem (1861–1916) began to develop a formalism. In his extensive and difficult two-volume work, *Energétique* [8], Duhem explicitly obtained expressions for the entropy produced in processes involving heat conductivity and viscosity [9]. Some of these ideas for calculating the uncompensated heat also appeared in the work of the Polish researcher L. Natanson [10] and the Viennese school led by G. Jaumann [11–13], where the notions of entropy flow and entropy production were developed.

Formulation of a theory of entropy along these lines continued during the twentieth century, and for a large class of systems we now have a theory in which the entropy change can be calculated in terms of the variables that characterize the irreversible processes. For example, the modern theory relates the rate of change of entropy to the rate of heat conduction or the rates of chemical reaction. *To obtain the change in entropy, it is not necessary to use infinitely slow reversible processes.*

With reference to Fig. 3.6, in the classical formulation of entropy it is often stated that, along the irreversible path I, the entropy may not be a function of the total energy and the total volume, hence it is not defined. However, for a large class of systems the notion of *local equilibrium* makes entropy a well-defined quantity, even if it is not a function of the total energy and volume, as was discussed in Chapter 1.

In his pioneering work on the thermodynamics of chemical processes, Théophile De Donder (1872–1957) [14–16] incorporated the "uncompensated transformation" or "uncompensated heat" of Clausius into the formalism of the Second Law through the concept of *affinity*, which is presented in the next chapter. This modern approach incorporates irreversibility into the formalism of the Second Law by providing explicit expressions for the computation of entropy produced by irreversible processes [17–19]. We shall follow this more general approach in which, along with thermodynamic *states*, irreversible *processes* appear explicitly in the formalism.

The basis of our approach is the notion of *local equilibrium*. For a very large class of systems that are not in thermodynamic equilibrium, thermodynamic quantities such as temperature, concentration, pressure and internal energy remain well-defined concepts locally, i.e. one could meaningfully formulate a thermodynamic description of a system in which intensive variables such as temperature and pressure are well defined in each elemental volume, and extensive variables such as entropy and internal energy are replaced by their corresponding *densities*. Thus, thermodynamic variables can be considered as functions of position and time. This is the assumption of *local equilibrium*. There are systems in which this assumption is not a good approximation but they are exceptional. In most hydrodynamic and chemical systems, local equilibrium is an excellent approximation. Modern computer simulations of molecular dynamics have shown that if initially the system is in such a state that temperature is not well defined, in a very short time (a few molecular collisions) the system relaxes to a state in which temperature is a well-defined quantity [20].

We may begin the modern formalism by expressing the changes in entropy as a sum of two parts [17]:

$$\boxed{dS = d_eS + d_iS}$$ (3.4.5)

in which d_eS is the entropy change due to exchange of matter and energy with the exterior and d_iS is the entropy change due to "uncompensated transformation," the entropy produced by the irreversible processes in the interior of the system (Fig. 3.7).

Our task now is to obtain explicit expressions for d_eS and d_iS in terms of experimentally measurable quantities. Irreversible processes can be described in terms of **thermodynamic forces** and **thermodynamic flows**. The thermodynamic flows are a consequence of the thermodynamic forces. Figure 3.8 shows how a difference in temperature between adjacent parts of a system (or temperature gradient) is the thermodynamic force that causes the irreversible flow of heat. Similarly, a concentration difference between two adjacent parts of a system is the thermodynamic force that causes the flow of matter. In general, the irreversible change d_iS is associated with a flow dX of a quantity such as heat or matter that has occurred in a time dt. For the flow of heat, we have $dX = dQ$, the amount of heat that flowed in time dt; for the case of matter, we have $dX = dN$, the number of moles of the substance that flowed in time dt. In each case the change in entropy can be written in the form

$$d_iS = F\,dX$$ (3.4.6)

in which F is the thermodynamic force. In this formalism the thermodynamic forces are expressed as functions of thermodynamic variables such as temperature and concentration. In the following section we shall see that, for the flow of heat shown in Fig. 3.8, the thermodynamic force takes the form

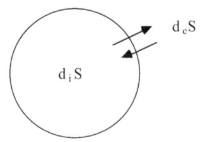

Figure 3.7 Entropy changes in a system consist of two parts: d_iS due to irreversible processes, and d_eS, due to exchange of energy and matter. According to the second law, the change d_iS is always positive. The entropy change d_eS can be positive or negative

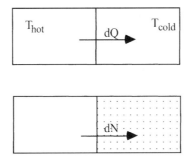

Figure 3.8 Irreversible processes that increase the entropy within a system

$F = (1/T_{\text{cold}} - 1/T_{\text{hot}})$. For the flow of matter, the corresponding thermodynamic force is expressed in terms of **affinity**, a concept developed in Chapter 4. All irreversible processes can be described in terms of thermodynamic forces and thermodynamic flows. The entropy change is the sum of all the changes due to irreversible flows dX_k. We then have the general expression

$$d_{\text{i}}S = \sum_k F_k \, dX_k \geq 0 \quad \text{or} \quad \frac{d_{\text{i}}S}{dt} = \sum_k F_k \frac{dX_k}{dt} \geq 0 \tag{3.4.7}$$

Equation (3.4.7) embodies the Second Law of thermodynamics. The entropy production due to each irreversible process is a product of the corresponding thermodynamic force F_k and the flow $J_k = dX_k/dt$.

The entropy exchange with the exterior, $d_{\text{e}}S$, is expressed in terms of the flow of heat and matter: For **isolated systems**, since there is no exchange of energy or matter, we have

$$d_{\text{e}}S = 0 \quad \text{and} \quad d_{\text{i}}S \geq 0 \tag{3.4.8}$$

For **closed systems** that exchange energy, but not matter, we have

$$d_{\text{e}}S = \frac{dQ}{T} = \frac{dU + p \, dV}{T} \quad \text{and} \quad d_{\text{i}}S \geq 0 \tag{3.4.9}$$

In this expression dQ is the amount of heat exchanged by the system in a time dt. (By defining dQ in this way, we avoid the use of "imperfect differentials.")

For **open systems** that exchange both matter and energy:[*]

$$d_{\text{e}}S = \frac{dU + p \, dV}{T} + (d_{\text{e}}S)_{\text{matter}} \quad \text{and} \quad d_{\text{i}}S \geq 0 \tag{3.4.10}$$

---

[*] In an open system, $dU + p \, dV \neq dQ$.

$(d_e S)_{\text{matter}}$ is the exchange of entropy due to matter flow. It can be written in terms of *chemical potential*, a concept that will be developed in the next chapter.

Whether we consider isolated, closed or open systems, $d_i S \geq 0$. This is the statement of the Second Law in its most general form. There is another important aspect to this statement: it is valid for all subsystems, not just for the entire system. For example, if we assume that the entire system is divided into two subsystems, we not only have

$$d_i S = d_i S^1 + d_i S^2 \geq 0 \qquad (3.4.11)$$

in which $d_i S^1$ and $d_i S^2$ are the entropy productions in each subsystem, but we also have

$$d_i S^1 \geq 0 \qquad d_i S^2 \geq 0 \qquad (3.4.12)$$

We cannot have

$$d_i S^1 > 0, \, d_i S^2 < 0 \quad \text{and} \quad d_i S = d_i S^1 + d_i S^2 \geq 0 \qquad (3.4.13)$$

This statement is stronger and more general than the classical statement that the entropy of an isolated system always increases.

In summary, for closed systems, the First and the Second Laws can be stated as

$$dU = dQ + dW \qquad (3.4.14)$$
$$dS = d_i S + d_e S \quad \text{in which} \quad d_i S \geq 0, \, d_e S = dQ/T \qquad (3.4.15)$$

If the transformation of state is assumed to take place through a reversible process, $d_i S = 0$, and the entropy change is solely due to flow of heat. We then obtain the equation

$$\boxed{dU = T\,dS + dW = T\,dS + p\,dV} \qquad (3.4.16)$$

which is found in texts that confine the formulation of thermodynamics to idealized reversible processes. For open systems, the changes in energy and entropy have an additional contribution due to flow of matter. In this case, though the definitions of heat and work need careful consideration, there is no fundamental difficulty in obtaining dU and $d_e S$.

Finally, we note that the above formulation enables us to calculate only the *changes* of entropy. It does not give us a way to obtain the absolute value of entropy. In this formalism, entropy can be known only up to an additive constant. However, in 1906, Walther Nernst (1864–1941) formulated a law which stated that *at the absolute zero of temperature the entropy of every*

Walther Nernst (1864–1941) (Courtesy the E. F. Smith Collection, Van Pelt-Dietrich Library, University of Pensylvania)

chemically homogeneous solid or liquid body has a zero value [21, p. 85].

$$S \to 0 \quad \text{as} \quad T \to 0\text{K} \tag{3.4.17}$$

This law is often referred to as the **Third Law** of thermodynamics or the **Nernst heat theorem**. Its validity has been well verified by experiment.

The Third Law *enables us to give the absolute value for the entropy*. The physical basis of this law lies in the behavior of matter at low temperature that can only be explained by quantum theory. It is remarkable that the theory of relativity gave us the means to define absolute values of energy, and quantum theory enables us to define absolute values of entropy.

Box 3.1 Statistical Interpretation of Entropy

As we have seen in this chapter, the foundation of the concept of entropy as a state function is entirely macroscopic. The validity of the Second Law is rooted in the reality of irreversible processes. In stark contrast to the macroscopic irreversible processes we see all around us, the laws of both classical and quantum mechanics are time symmetric, i.e., according to the laws of mechanics, a system that can evolve from a state A to a state B can also evolve from the state B to the state A. For example, the spontaneous flow of gas molecules from a part that has a higher density to a part that has a lower density and its reverse (which violates the Second

Law) are both in accord with the laws of mechanics. Processes that are ruled impossible by the Second Law of thermodynamics do not violate the laws of mechanics. Yet all irreversible macroscopic processes, such as the flow of heat, are the consequence of motion of atoms and molecules that are governed by the laws of mechanics; the flow of heat is a consequence of molecular collisions that transfer energy. How can irreversible macroscopic processes emerge from the reversible motion of molecules? To reconcile the reversibility of mechanics with the irreversibility of thermodynamics, Ludwig Boltzmann (1844–1906) proposed the following relation between microscopic states and entropy:

$$S = k_B \ln W$$

in which W is the number of microstates corresponding to the macrostate whose entropy is S. The constant k_B introduced by Boltzmann is now called the Boltzmann constant; $k_B = 1.381 \times 10^{-23}$ J K^{-1}. The gas constant $R = k_B N_A$, in which N_A is the Avogadro number. The following example will illustrate the meaning of W. Consider the macrostate of a box containing a gas with N_1 molecules in one half and N_2 in the other (see the figure below). Each molecule can be in one half or the other. The total number of ways in which the $(N_1 + N_2)$ molecules can be distributed between the two halves such that N_1 molecules are in one and N_2 molecules in the other is equal to W. The number of distinct "microstates" with N_1 molecules in one half and N_2 in the other is:

$$W = \frac{(N_1 + N_2)!}{N_1! N_2!}$$

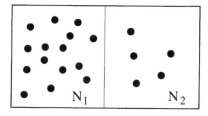

According to Boltzmann, macrostates with larger W are more probable. The irreversible increase of entropy then corresponds to evolution to states of higher probability. Equilibrium states are those for which W is a maximum. In the above example, it can be shown that W reaches its maximum when $N_1 = N_2$.

It should be mentioned that the introduction of "probability" requires a deeper discussion which is outside the scope of this text (for a non-technical discussion, see [22]). In dynamics, initial conditions are arbitrary and the introduction of probability is usually based on some approximation ("worse graining").

3.5   Examples of Entropy Changes due to Irreversible Processes

To illustrate how entropy changes are related to irreversible processes, we shall consider some simple examples. All of them are "discrete systems" that consist

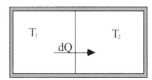

Figure 3.9 Entropy production due to heat flow. The irreversible flow of heat between parts of unequal temperature results in the increase in entropy. The rate at which entropy is produced, $P = d_iS/dt$, is given by (3.5.3)

of two parts that are not mutually in equilibrium. Continuous systems are described by the use of vector calculus in parts IV and V.

HEAT CONDUCTION

Consider an *isolated* system which we assume (for simplicity) consists of two parts, each having a well-defined temperature, i.e. each part is locally in equilibrium. Let the temperatures of the two parts be T_1 and T_2 (Fig. 3.9), $T_1 > T_2$. Let dQ be the amount of heat flow from the hotter part to the colder part in a time dt. Since this isolated system does not exchange entropy with the exterior, $d_eS = 0$. And since the volume of each part is a constant, $dW = 0$. The energy change in each part is solely due to the flow of heat: $dU_i = dQ_i$, $i = 1, 2$. In accordance with the First Law, the heat gained by one part is equal to the heat lost by the other. Therefore, $-dQ_1 = dQ_2 = dQ$. Both parts are locally in equilibrium with a well-defined temperature and entropy. The total change in entropy, d_iS, of the system is the sum of the changes of entropy in each part due to the flow of heat:

$$d_iS = -\frac{dQ}{T_1} + \frac{dQ}{T_2} = \left(\frac{1}{T_2} - \frac{1}{T_1}\right)dQ \qquad (3.5.1)$$

Since the heat flows irreversibly from the hotter part to the colder part, dQ is positive if $T_1 > T_2$. Hence $d_iS > 0$. In expression (3.5.1), dQ and $(1/T_2 - 1/T_1)$ respectively correspond to dX and F in (3.4.6). In terms of the *rate* of flow of heat dQ/dt, the rate of entropy production can be written as

$$\frac{d_iS}{dt} = \left(\frac{1}{T_2} - \frac{1}{T_1}\right)\frac{dQ}{dt} \qquad (3.5.2)$$

Now the rate of heat flow or the heat current $J_Q \equiv dQ/dt$ is given by the laws of heat conduction. For example according to the Fourier law of heat conduction, $J_Q = \alpha(T_1 - T_2)$, in which α is the coefficient of heat conductivity. Note that the "thermodynamic flow" J_Q is driven by the "thermodynamic force" $F = (1/T_2 - 1/T_1)$. For the rate of entropy production we have from (3.5.2) that

$$\frac{d_iS}{dt} = \left(\frac{1}{T_2} - \frac{1}{T_1}\right)\alpha(T_1 - T_2) = \frac{\alpha(T_1 - T_2)^2}{T_1 T_2} \geq 0 \qquad (3.5.3)$$

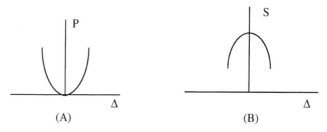

Figure 3.10   Two equivalent principles that characterize the state of equilibrium. (A) The total entropy production $P \equiv d_i S/dt$ as a function of the difference in temperatures $\Delta \equiv (T_1 - T_2)$ of the two parts of the system shown in Fig. 3.6; at equilibrium the entropy production vanishes. (B) At equilibrium the system can also be characterized by the principle that the entropy of the system is a maximum

Due to the flow of heat from the hot part to the cold part, the temperatures eventually become equal, and the entropy production ceases. This is the state of equilibrium. *The entropy production must vanish in the state of equilibrium,* which implies that the force F and the corresponding flux J_Q both vanish. In fact, we can deduce the properties of the equilibrium state by stipulating that all entropy production must vanish in that state.

If we denote the total entropy production by $P \equiv d_i S/dt$, from (3.5.3) we see that it is a quadratic function of the deviation $\Delta \equiv (T_1 - T_2)$. At equilibrium the entropy production takes its minimum value, equal to zero. This is indicated graphically in Fig. 3.10(A).

A nonequilibrium state in which $T_1 \neq T_2$ evolves to the equilibrium state in which $T_1 = T_2 = T_{eq}$ through continuous increase of entropy. Therefore, the entropy of the equilibrium state must be larger than the entropy of any nonequilibrium state. In Chapter 12 we will see explicitly that for a small deviation, $\Delta = (T_1 - T_2)$, from the state of equilibrium, the corresponding change, ΔS, is a quadratic function of Δ attaining a maximum at $\Delta = 0$ (Fig. 3.10(B))

This example illustrates the general assertion that the state of equilibrium can be characterized either by minimum (zero) entropy production, or by maximum entropy.

IRREVERSIBLE EXPANSION OF A GAS

In a reversible expansion of gas the pressure of the gas and the pressure on the piston are assumed to be the same. If we consider an expansion of a gas at constant temperature T by virtue of its contact with a heat reservoir, the change in entropy of the gas $d_e S = dQ/T$, in which dQ is the the heat flow from the reservoir to the gas that is necessary to maintain the temperature constant. This is an ideal situation. In any real expansion of a gas that takes place in a finite time, the pressure of the gas is greater than the pressure on the piston. If p_{gas} is the pressure of the gas, and p_{piston} is the pressure on the piston, the difference $(p_{gas} - p_{piston})$ is the force per unit area that moves the piston. The irreversible increase in entropy in this case is given by

$$d_i S = \frac{(p_{gas} - p_{piston})}{T} dV > 0 \qquad (3.5.4)$$

Here the term $(p_{gas} - p_{piston})/T$ corresponds to the "thermodynamic force" and dV/dt the corresponding "flow." The term $(p_{gas} - p_{piston})dV$ may be identified as the "uncompensated heat" of Clausius. Since the change in the volume and $(p_{gas} - p_{piston})$ have the same sign, $d_i S$ is always positive. The change in entropy $dS = d_e S + d_i S = dQ/T + (p_{gas} - p_{piston})dV/T$. In the case of an ideal gas, since the energy is only a function of T, the initial and final energies of the gas remain the same; the heat absorbed is equal to the work done in moving the piston, $p_{piston} dV$. For a given change in volume, the maximum work is obtained for a reversible process in which $p_{gas} = p_{piston}$.

3.6 Entropy Changes Associated with Phase Transformations

In this section we will consider a simple example of entropy exchange $d_e S$. Changes in the phase of a system, from a solid to a liquid or from a liquid to a vapor (Fig. 1.3), provide a convenient situation because, at the melting or boiling point, the temperature remains constant even though heat is exchanged. Hence, in the expression for the entropy change associated with the heat exchange, $d_e S = dQ/T$, the temperature T remains constant. The total entropy change ΔS due to the exchange of heat ΔQ is now easy to determine. In a solid-to-liquid transition, if the melting temperature is T_{melt}, we have

$$\Delta S = \int_0^{\Delta Q} \frac{dQ}{T_{melt}} = \frac{\Delta Q}{T_{melt}} \qquad (3.6.1)$$

As was discovered by Joseph Black, the heat absorbed, "the latent heat," converts the solid to a liquid at the fixed temperature. Generally this change happens at a fixed pressure, hence we may equate ΔQ to ΔH, the enthalpy change associated with melting. The enthalpy associated with the conversion of one mole of the solid to liquid is called **the molar enthalpy of fusion** ΔH_{fus}. The corresponding change in entropy, **the molar entropy of fusion** ΔS_{fus} can now be written as

$$\boxed{\Delta S_{fus} = \frac{\Delta H_{fus}}{T_{melt}}} \qquad (3.6.2)$$

Table 3.1   Enthalpies of fusion and vaporization at $p = 1$ bar $= 10^5$ Pa $= 0.987$ atm*

Compound	T_{melt} (K)	ΔH_{fus} (kJ/mol)	T_{boil} (K)	ΔH_{vap} (kJ/mol)
H_2O	273.15	6.008	373.15	40.656
CH_3OH	175.2	3.16	337.2	35.27
C_2H_5OH	156	4.60	351.4	38.56
CH_4	90.68	0.941	111.7	8.18
CCl_4	250.3	2.5	350	30.0
NH_3	195.4	5.652	239.7	23.35
CO_2	217.0	8.33	194.6	25.23
CS_2	161.2	4.39	319.4	26.74
N_2	63.15	0.719	77.35	5.586
O_2	54.36	0.444	90.18	6.820

* More extensive data may be found in sources [B] and [F]. It is found that for many compounds ΔS_{vap} is about 88 J K^{-1}. This back-of-the-envelope estimate of ΔS_{vap} is called Trouton's rule.

Water has a heat of fusion of 6.008 kJ/mol and a melting temperature of 273.15 K at a pressure of 1 bar. When one mole of ice turns to water, the entropy change is $\Delta S_{\mathrm{fus}} = 21.99$ J K^{-1} mol^{-1}.

Similarly, if the conversion of a liquid to vapor occurs at a constant pressure at its boiling point T_{boil}, then the **molar entropy of vaporization** ΔS_{vap} and the **molar enthalpy of vaporization** ΔH_{vap} are related by

$$\boxed{\Delta S_{\mathrm{vap}} = \frac{\Delta H_{\mathrm{vap}}}{T_{\mathrm{boil}}}} \tag{3.6.3}$$

The heat of vaporization of water is 40.656 kJ/mol. Since the boiling point is 373.15 K at a pressure of 1 bar ($= 0.987$ atm), from the above equation it follows that the molar entropy change $\Delta S_{\mathrm{vap}} = 108.95$ J K^{-1} mol^{-1}, about five times the entropy change associated with the melting of ice. Since entropy increases with volume, the large increase in volume from about 18 mL (volume of one mole of water) to about 30 L (volume of one mole of steam at $p = 1$ bar) is partly responsible for this larger change. Molar enthalpies of fusion and vaporization of some compounds are given in Table 3.1.

3.7   Entropy of an Ideal Gas

As a last example, we obtain the entropy of an ideal gas as a function of volume, temperature and mole number. For a closed system in which the changes of entropy are only due to flow of heat, if we assume that the changes in volume V and temperature T take place so as to make $d_iS = 0$, then we have seen from (3.4.16) that $dU = T\,dS + dW$. If $dW = -p\,dV$, and if we express dU as a function of V and T, we obtain

$$T\,dS = \left(\frac{\partial U}{\partial V}\right)_T dV + \left(\frac{\partial U}{\partial T}\right)_V dT + p\,dV \qquad (3.7.1)$$

For an ideal gas, $(\partial U/\partial V)_T = 0$, because the energy U is only a function of T— as was demonstrated in the experiments of Joule and Gay-Lussac (section 1.3 and equation (1.3.6)). Also, by definition $(\partial U/\partial T)_V = NC_V$ in which C_V is the molar heat capacity at constant volume, which is found to be a constant. Hence (3.7.1) may be written as

$$dS = \frac{p}{T}dV + NC_V\frac{dT}{T} \qquad (3.7.2)$$

Using the ideal gas law $pV = NRT$, (3.7.2) can be integrated to obtain

$$S(V,\,T) = S_0 + NR\ln V + NC_V\ln T \qquad (3.7.3)$$

In this expression the extensivity of S as a function of V and N is not explicit. Since S is assumed to be extensive in V and N, the entropy of an ideal gas is usually written as

$$\boxed{S(V,\,T,\,N) = N[s_0 + R\ln(V/N) + C_V\ln T]} \qquad (3.7.4)$$

in which s_0 is a constant. Thus we see that the ideal gas entropy has a logarithmic dependence on the number density (number of particles per unit volume) and the temperature.

3.8 Remarks on the Second Law and Irreversible Processes

The Second Law is quite general. However, when intermolecular forces are long range, as in the case of particles interacting through gravitation, there are difficulties because our classification into extensive variables (proportional to volume) and intensive variables (independent of volume) does not apply. The total energy U, for example, is no longer proportional to the volume. Fortunately gravitational forces are very weak as compared to the short range intermolecular forces. It is only on the astrophysical scale that this problem becomes important. We shall not discuss this problem in this text.

The generality of the Second Law gives us a powerful means to understand the thermodynamic aspects of real systems through the usage of ideal systems. A classic example is Planck's analysis of radiation in thermodynamic equilibrium with matter (blackbody radiation) in which Planck considered idealized simple harmonic oscillators interacting with radiation. Planck considered simple harmonic oscillators not merely because they are good approximations of molecules but because the properties of radiation in thermal equilibrium with matter are universal, regardless of the particular nature of the matter with which the radiation interacts. The conclusions one arrives at using idealized oscillators and the laws of thermodynamics must also be valid for all other forms of matter, however complex.

In the modern context, the formulation summarized in Fig. 3.7 is fundamental for understanding thermodynamic aspects of self-organization, evolution of order and life that we see in Nature. When a system is isolated $d_eS = 0$. In this case the entropy of the system will continue to increase due to irreversible processes and reach the maximum possible value, which is the state of thermodynamic equilibrium. In the state of equilibrium, all irreversible processes cease. When a system begins to exchange entropy with the exterior then, in general, it is driven away from equilibrium and the entropy producing irreversible processes begins to operate. The exchange of entropy is due to exchange of heat and matter. The entropy flowing out of the system is always larger than the entropy flowing into the system, the difference arising due to entropy produced by irreversible processes within the system. As we shall see in the following chapters, systems that exchange entropy with their exterior do not simply increase the entropy of the exterior, but may undergo dramatic spontaneous transformations to "self-organization." *The irreversible processes that produce entropy create these organized states.* Such self-organized states range from convection patterns in fluids to life. Irreversible processes are the driving force that create this order.

References

1. Mendoza, E. (ed.) *Reflections on the Motive Force of Fire by Sadi Carnot and other Papers on the Second Law of Thermodynamics by E. Clapeyron and R. Clausius.* 1977, Glouster, MA: Peter Smith.
2. Kastler, A., L'Oeuvre posthume de Sadi Carnot, *in Sadi Carnot et l'Essor de la Thermodynamique*, A.N. Editor (ed.) 1974, Paris: CNRS.
3. Segré, E., *From Falling Bodies to Radio Waves.* 1984, New York: W.H. Freeman.
4. Clausius, R., *Mechanical Theory of Heat.* 1867, London: John van Voorst.
5. Planck, M., *Treatise on Thermodynamics*, 3rd ed. 1945, New York: Dover.
6. Bridgman, P. W., *The Nature of Thermodynamics.* 1943, Cambridge MA: Harvard University Press.
7. Bertrand, J. L. F., *Thermodynamique.* 1887, Paris: Gauthiers-Villars.
8. Duhem, P., *Energétique.* 1911, Paris: Gauthiers-Villars.
9. Brouzeng, P., Duhem's contribution to the development of modern thermodynamics, in *Thermodynamics: History and Philosophy*, K. Martinás, L. Ropolyi, and P. Szegedi, (eds) 1991, London: World Scientific, p. 72–80.
10. Natanson, L., *Z. Phys. Chem.*, **21** (1896) 193.

11. Lohr, E., *Math. Naturw. Klasse*, **339** (1916) 93.

12. Jaumann, G., *Math. Naturw. Klasse*, **120** (1911) 385.

13. Jaumann, G., *Math. Naturw. Klasse*, **95** (1918) 461.

14. de Donder, T., *Lecons de Thermodynamique et de Chimie-Physique*. 1920, Paris: Gauthiers-Villars.

15. de Donder, T., *L'Affinité*. 1927, Paris: Gauthiers-Villars.

16. de Donder, T. and Van Rysselberghe P., *Affinity*. 1936, Menlo Park CA: Stanford University Press.

17. Prigogine, I., *Etude Thermodynamique des Processus Irreversible*, 4th ed. 1967, Liège: Desoer.

18. Prigogine, I., *Introduction to Thermodynamics of Irreversible Processes*. 1967, New York: John Wiley.

19. Prigogine, I. and Defay, R., *Chemical Thermodynamics*, 4th ed. 1967, London: Longman.

20. Alder, B. J. and Wainright, T., Molecular dynamics by electronic computers, in *Transport Processes in Statistical Mechanics*. 1969, New York: Interscience.

21. Nernst, W., *A New Heat Theorem*. 1969, New York: Dover.

22. Prigogine, I., *The End of Certainty*. 1997, New York: Free Press.

Data Sources

[A] NBS table of chemical and thermodynamic properties. *J. Phys. Chem. Reference Data, 11*, suppl. 2, (1982).

[B] G. W. C. Kaye and Laby T. H. (eds), *Tables of Physical and Chemical Constants*. 1986, London: Longman.

[C] I. Prigogine and Defay R., *Chemical Thermodynamics*. 1967, London: Longman.

[D] J. Emsley, *The Elements*, 1989, Oxford: Oxford University Press.

[E] L. Pauling, *The Nature of the Chemical Bond*. 1960, Ithaca NY: Cornell University Press.

[F] D. R. Lide (ed.), *CRC Handbook of Chemistry and Physics*, 75th ed. 1994, Ann Arbor MI: CRC Press.

Examples

Example 3.1   Draw the *S–T* diagram for the Carnot cycle.

Solution: During the reversible adiabatic changes, the change in entropy is zero. Hence the *S–T* diagram is as shown.

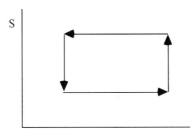

Example 3.2   A heat pump is used to maintain the inside temperature of a house at 20 °C when the outside temperature is 3.0 °C. What is the minimum amount of work necessary to transfer 100 J of heat to the inside of the house?

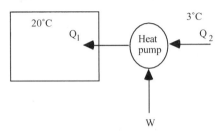

Solution: The ideal heat pump is the Carnot engine running in reverse, i.e. it uses work to pump heat from a lower temperature to a higher temperature. For an ideal pump $Q_1/T_1 = Q_2/T_2$. If $Q_1 = 100\,\text{J}$, $T_2 = 293\,\text{K}$ and $T_1 = 276\,\text{K}$, we obtain

$$Q_2 = 276\,\text{K}(100\,\text{J}/293\,\text{k}) = 94\,\text{J}$$

Thus the heat pump absorbs 94 J from the outside and delivers 100 J to the inside. From the first law it follows that the work $W = Q_1 - Q_2$ required is $100\,\text{J} - 94\,\text{J} = 6\,\text{J}$.

Example 3.3   The heat capacity of a solid is $C_p = 125.48\,\text{J}\,\text{K}^{-1}$. What is the change in its entropy if it is heated from 273.0 K to 373.0 K?

Solution: This is a simple case of heat transfer, $d_e S = dQ/T$, hence

$$S_{\text{final}} - S_{\text{initial}} = \int_{T_i}^{T_f} \frac{dQ}{T} = \int_{T_i}^{T_f} \frac{C_p\,dT}{T} = C_p \ln\left(\frac{T_f}{T_i}\right)$$

$$= 125.48\,\text{J}\,\text{K}^{-1} \ln(373/273) = 39.2\,\text{J}\,\text{K}^{-1}$$

Exercises

3.1   Show the equivalence between a perpetual motion machine of the second kind and Carnot's theorem.

3.2   A refrigerator operating reversibly extracts 45 kJ of heat from a thermal reservoir and delivers 67 kJ as heat to a reservoir at 300 K. Calculate the temperature of the reservoir from which heat was removed.

3.3   What is the maximum work that can be obtained from 1000.0 J of heat supplied to a steam engine with a high temperature reservoir at 120.0 °C if the condenser is at 25.0 °C?

3.4 The temperature of solar energy (sunlight) reaching the earth is approximately 6000 °C. (The temperature of light depends on the intensity of the colors, (wavelengths), not how hot it feels!)
(a) Assume the temperature of a solar energy cell is 298.15 °K and calculate the maximum possible efficiency of a solar energy cell in converting this energy to "useful work"?
(b) If 102 J of solar energy is incident on the solar cells of a calculator, what is the maximum energy available to run the calculator?

3.5 The heat of combustion of gasoline is approximately 47.0 kJ mol^{-1}. If a gasoline engine operates between 1500 K and 750 K, what is the maximum height that 5.0 moles of gasoline can lift an aircraft that weighs 400 kg?

3.6 The specific heat C_p of a substance is given by

$$C_p = a + bT$$

where $a = 20.35 \, \text{J K}^{-1}$ and $b = 0.20 \, \text{J K}^{-2}$. Calculate the change in entropy when increasing the temperature of this substance from 298.15 K to 304.0 K.

3.7 When 0.5 J of heat passes between two large bodies in contact at temperatures 70 °C and 25 °C, what is the change of entropy? If this occurs in 0.23 s, what is the rate of change of entropy, $d_i S/dt$?

3.8 What is the entropy of 1.0 L of $N_2(g)$ at $T = 350 \, \text{K}$ and $p = 2.0 \, \text{bar}$ given that $S_{om} = 191.61 \, \text{J K}^{-1} \, \text{mol}^{-1}$ at $T = 298.15 \, \text{K}$ and $p = 1 \, \text{bar}$?

3.9 (i) Find out how much solar energy reaches the surface of the Earth per square meter per second. The temperature of this radiation is approximately 6000 K. Use the Carnot efficiency theorem to estimate the maximum power (in Watts) that can be obtained by a 1 m^2 solar cell through solar energy conversion. (The solar constant* is about 1.3 kW/m^2.)
(ii) The present cost of electricity in the United States is in the range $0.8–0.15 per kilowatt-hour (1 kWh = $10^3 \times 3600$ J). Assume the efficiency of commercial solar cells is only about 5% and assume they can last 30 years, producing power for 5 hours per day on the average. How much should a 1 m^2 array of solar cells cost so that the total energy it can produce amounts to about $0.15 per kilowatt-hour. Make reasonable assumptions for any other quantities that are not specified.

---

* The solar constant is the energy flux incident on the Earth. About one third of it is reflected back to space, i.e. albedo $\simeq 0.3$.

4 ENTROPY IN THE REALM OF CHEMICAL REACTIONS

4.1 Chemical Potential and Affinity: the Driving Force of Chemical Reactions

Nineteenth-century chemists did not pay much attention to the developments in thermodynamics, while experiments done by chemists—such as Gay-Lussac's experiments on the expansion of a gas into vacuum—were taken up and discussed by the physicists for their thermodynamic implications. The interconversion of heat into other forms of energy was a matter of great interest, but mostly to the physicists. Among the chemists, the concept of heat as an indestructible caloric, a view supported by Lavoisier, largely prevailed [1]. As we noted in Chapter 2, the work of the Russian chemist Germain Hess on heats of reaction was an exception.

Motion is explained by the Newtonian concept of force, but what is the "driving force" for chemical change? Why do chemical reactions occur, and why do they stop at certain points? Chemists called the "force" that caused chemical reactions *affinity*, but it lacked a clear definition. For the chemists who sought quantitative laws, a definition of affinity, as precise as Newton's definition of mechanical force, was a fundamental problem. In fact, this centuries-old concept had different interpretations at different times. "It was through the work of the thermochemists and the application of the principles of thermodynamics as developed by the physicists," notes the chemistry historian Henry M. Leicester, "that a quantitative evaluation of affinity forces was finally obtained" [1, p. 203]. The thermodynamic formulation of affinity as we know it today is due to Théophile De Donder (1872–1957), the founder of the Belgian school of thermodynamics.

De Donder's formulation of chemical affinity [2, 3] was founded on the concept of *chemical potential*, one of the most fundamental and far-reaching concepts in thermodynamics that was introduced by Josiah Willard Gibbs (1839–1903). Already in the nineteenth century, the French chemist Marcellin Berthelot (1827–1907) and the Danish Chemist Julius Thomsen (1826–1909) attempted to quantify affinity using heats of reaction. After determining the heats of reaction for a large number of compounds, in 1875 Berthelot proposed a "principle of maximum work," according to which "all chemical changes occurring without intervention of outside energy tend toward the production of bodies or of a system of bodies which liberate more heat" [1, p. 205]. But this suggestion met with criticism from Hermann von Helmholtz. The controversy

J. Willard Gibbs (1839–1903) (Courtesy the E. F. Smith Collection, Van Pelt-Dietrich Library, University of Pennsylvania)

continued until the concept of a chemical potential formulated by Gibbs became known in Europe. Later it became clear that it was not the heat of reaction that characterized the evolution to the state of equilibrium, but another thermodynamic quantity called "free energy." As we shall describe in detail, De Donder used the idea of chemical potential and not only gave a precise definition for affinity, but also used it to obtain a relation between the rate of entropy change and chemical reaction rates. All chemical reactions drive the system to a state of equilibrium in which the affinities of the reactions vanish.

CHEMICAL POTENTIAL

Josiah Willard Gibbs introduced the idea of chemical potential in his famous work *On the Equilibrium of Heterogeneous Substances*, published in 1875 and

Théophile De Donder (1872–1957) (5th from the left, third row) at the historic 1927 Solvay conference. His book, *L'Affinité* was published the same year. *First row, L to R*: I. Langmuir, M. Planck, Mme Curie, H.A. Lorentz, A. Einstein, P. Langevin, Ch.E. Guye, C.T.R. Wilson, O.W. Richardson. *Second row, L to R*: P. Debye, M. Knudsen, W.L. Bragg, H.A. Kramers, P.A.M. Dirac, A.H. Compton, L. de Broglie, M. Born, N. Bohr. *Third row, L to R*: A. Picard, E. Henriot, P. Ehrenfest, Ed. Herzen, Th. De Donder, E. Schrödinger, E. Verschaffelt, W. Pauli, W. Heisenberg, R.H. Fowler, L. Brillouin.

1878 [4–6]. Gibbs published his work in the *Transactions of the Connecticut Academy of Sciences*, a journal that was not widely read. This work of Gibbs remained in relative obscurity until it was translated into German by Wilhelm Ostwald (1853–1932) in 1892 and into French by Henry-Louis Le Chatelier (1850–1936) in 1899 [1]. Much of present-day equilibrium thermodynamics can be traced back to this important work of Gibbs.

Gibbs considered a heterogeneous system (Fig. 4.1) that consisted of several homogeneous parts, each part containing various substances $s_1, s_2, \ldots, s_n$ of masses $m_1, m_2, \ldots, m_n$. His initial consideration did not include chemical reactions between these substances, but was restricted to their exchange between the different homogeneous parts of the system. Arguing that the change in energy dU of a homogeneous part must be proportional to changes in the masses of the substances, $dm_1, dm_2, \ldots, dm_n$, Gibbs introduced the equation

$$dU = T\,dS - p\,dV + \mu_1\,dm_1 + \mu_2\,dm_2 + \ldots + \mu_n\,dm_n \qquad (4.1.1)$$

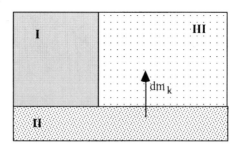

Figure 4.1 A heterogeneous system considered by Gibbs in which substances are exchanged between parts I, II and III. The change in energy dU of any part when matter is exchanged is given by (4.1.1).

for each homogeneous part. The coefficients μ_k are called the **chemical potentials**. The heterogeneous systems considered included different phases of a substance that exchanged matter. However, the considerations of Gibbs were restricted to transformations between states in equilibrium. This restriction is understandable from the viewpoint of the classical definition of entropy, which required the system to be in equilibrium and the transformations between equilibrium states to be reversible so that $dQ = T\,dS$. In the original formulation of Gibbs, the changes in the masses dm_k in equation (4.1.1) were due to exchange of the substances between the homogeneous parts, a situation encountered when various phases of a substance exchange matter and reach equilibrium.

It is more convenient to describe chemical reactions by the change in the mole numbers of the reactants rather than the change in their masses, because chemical reaction rates and the laws of diffusion are most easily formulated in terms of mole numbers. We shall therefore rewrite equation (4.1.1) in terms of the mole numbers N_k, of the constituent substances:

$$dU = T\,dS - p\,dV + \mu_1\,dN_1 + \mu_2\,dN_2 + \ldots + \mu_n\,dN_n$$

$$\boxed{dU = T\,dS - p\,dV + \sum_1^n \mu_k\,dN_k} \qquad (4.1.2)$$

Note how this equation implies that the energy is a function of S, V and N_k, and

$$\boxed{\left(\frac{\partial U}{\partial S}\right)_{V,N_k} = T} \qquad \boxed{\left(\frac{\partial U}{\partial V}\right)_{S,N_k} = -p} \qquad \boxed{\left(\frac{\partial U}{\partial N_k}\right)_{S,V,N_{j\neq k}} = \mu_k} \qquad (4.1.3)$$

CHEMICAL REACTIONS

Though Gibbs did not consider irreversible chemical reactions, equation (4.1.1) that he introduced included all that was needed for the consideration of irreversibility and entropy production in chemical processes. By making the important distinction between the entropy change $d_e S$ due to exchange of matter and energy with the exterior, and the irreversible increase of entropy $d_i S$ due to chemical reactions [2, 3], De Donder formulated the thermodynamics of irreversible chemical transformations. And we can now show he took the "uncompensated heat" of Clausius and gave it a clear expression for chemical reactions.

Let us look at equation (4.1.2) from the viewpoint of entropy flow $d_e S$ and entropy production $d_i S$, that was introduced in the previous chapter. To make a distinction between irreversible chemical reactions and reversible exchange with the exterior, we express the change in the mole numbers dN_k as a sum of two parts:

$$dN_k = d_i N_k + d_e N_k \tag{4.1.4}$$

in which $d_i N_k$ is the change due to irreversible chemical reactions and $d_e N_k$ is the change due to exchange of matter with the exterior. In equation (4.1.2) Gibbs considered *reversible* exchange of heat and matter. Because this corresponds to $d_e S$, we may use (3.4.10) and write

$$d_e S = \frac{dU + p\, dV}{T} - \frac{\sum_1^n \mu_k\, d_e N_k}{T} \tag{4.1.5}$$

De Donder recognized that, in a closed system, if the change of mole numbers dN_k were due to irreversible chemical reactions, then the resulting entropy production $d_i S$ could be written as

$$d_i S = - \frac{\sum_1^n \mu_k\, d_i N_k}{T} \tag{4.1.6}$$

This is the "uncompensated heat" of Clausius for chemical reactions. *The validity of this equation lies in the fact that chemical reactions occur in such a way that $d_i S$ is always positive in accordance with the Second Law.* For the total change in entropy dS we have

$$\boxed{dS = d_e S + d_i S} \tag{4.1.7}$$

in which

$$\boxed{d_e S = \frac{dU + p\, dV}{T} - \frac{1}{T}\sum_1^n \mu_k\, d_e N_k} \tag{4.1.8}$$

and

$$d_i S = -\frac{1}{T} \sum_1^n \mu_k \, d_i N_k > 0 \qquad (4.1.9)$$

For a closed system $d_e N_k = 0$. Since the rate of chemical reaction specifies dN_k/dt, the rate of entropy production can be written as

$$\frac{d_i S}{dt} = -\frac{1}{T} \sum_1^n \mu_k \frac{dN_k}{dt} > 0 \qquad (4.1.10)$$

If we sum (4.1.8) and (4.1.9) we recover (4.1.2):

$$dU = T \, dS - p \, dV + \sum_1^n \mu_k \, dN_k \qquad (4.1.11)$$

Further development of this theory relates the chemical potential to measurable state variables such as p, T and N_k. Thus, the pioneering work of De Donder established a clear connection between entropy production and irreversible chemical reactions. In a closed system, if initially the system is not in chemical equilibrium, chemical reactions will take place that irreversibly drive the system towards equilibrium. And, according to the Second Law of thermodynamics, this will happen in such a way that (4.1.10) is satisfied.

AFFINITY

De Donder also defined the affinity of a chemical reaction that enables us to write expression (4.1.10) in an elegant form, as the product of a thermodynamic force and a thermodynamic flow. The concept of affinity can be understood through the following simple example.

In a closed system, consider a chemical reaction of the form

$$X + Y \rightleftharpoons 2Z \qquad (4.1.12)$$

In this case the changes in the mole numbers dN_X, dN_Y and dN_Z of the components X, Y and Z are related by the stoichiometry. We can express this relation as

$$\frac{dN_X}{-1} = \frac{dN_Y}{-1} = \frac{dN_Z}{2} \equiv d\xi \qquad (4.1.13)$$

in which $d\xi$ is the change in the extent of reaction ξ, which was introduced in section 2.5. Using (4.1.11) the total entropy change and the entropy change due to irreversible chemical reactions can now be written as

$$dS = \frac{dU + p\,dV}{T} + \frac{1}{T}(\mu_X + \mu_Y - 2\mu_Z)d\xi \qquad (4.1.14)$$

$$d_iS = \frac{(\mu_X + \mu_Y - 2\mu_Z)}{T}d\xi > 0 \qquad (4.1.15)$$

For a chemical reaction

$$X + Y \rightleftharpoons 2Z$$

De Donder defined a new state variable called **affinity** [2, 3]:

$$A \equiv (\mu_X + \mu_Y - 2\mu_Z) \qquad (4.1.16)$$

This affinity is the driving force for chemical reactions (Fig. 4.2). A nonzero affinity implies that the system is not in thermodynamic equilibrium and that chemical reactions will continue to occur, driving the system towards equilibrium. In terms of affinity A, the rate of increase of entropy is written as

$$\boxed{\frac{d_iS}{dt} = \left(\frac{A}{T}\right)\frac{d\xi}{dt} > 0} \qquad (4.1.17)$$

As in the case of entropy production due to heat conduction, the entropy production due to a chemical reaction is a product of a thermodynamic force A/T, and a thermodynamic flow $d\xi/dt$. The flow in this case is the conversion of

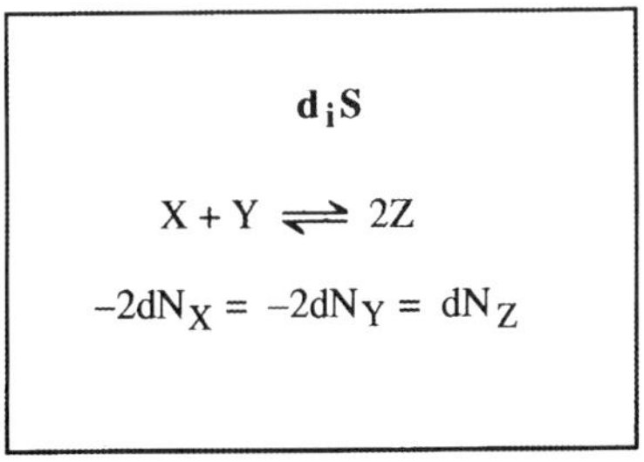

Figure 4.2 The changes in entropy d_iS due to irreversible chemical reactions are formulated using the concept of affinity. The illustrated reaction has affinity $A \equiv (\mu_X + \mu_Y - 2\mu_Z)$, in which μ are chemical potentials

reactants to products (or vice versa) which is caused by the force A/T. We shall refer to the thermodynamic flow $d\xi/dt$ as the **velocity of reaction**.

Note that, although a nonzero affinity means there is a driving force for chemical reactions, the velocity $d\xi/dt$ at which these chemical reactions will occur is not specified by the affinity's value. The rates of chemical reactions are usually known through empirical means and they depend on one or more mechanisms.

At equilibrium the thermodynamic flows and hence the entropy production must vanish. This implies that *in the state of equilibrium the affinity of a chemical reaction $A = 0$*. Thus we arrive at the conclusion that, at thermodynamic equilibrium, the chemical potentials of the compounds X, Y and Z will reach values such that

$$A \equiv (\mu_X + \mu_Y - 2\mu_Z) = 0 \tag{4.1.18}$$

In Chapter 9, devoted to the thermodynamics of chemical reactions, we will see how chemical potentials can be expressed in terms of experimentally measurable quantities such as concentrations and temperature. Equations such as (4.1.18) are specific predictions regarding the states of chemical equilibrium. These predictions have been amply verified by experiment, and today they are routinely used in chemistry.

For a general chemical reaction of the form

$$a_1 A_1 + a_2 A_2 + a_3 A_3 \ldots + a_n A_n \rightleftharpoons b_1 B_1 + b_2 B_2 + b_3 B_3 + \ldots + b_m B_m \tag{4.1.19}$$

the changes in the mole numbers of the reactants A_k and the products B_k are related in such a way that a change dX in one of the species (reactants or products) completely determines the corresponding changes in all the other species. Consequently, there is only one independent variable, which can be defined as

$$\frac{dN_{A1}}{-a_1} = \frac{dN_{A2}}{-a_2} = \ldots \frac{dN_{An}}{-a_n} = \frac{dN_{B1}}{b_1} = \frac{dN_{B2}}{b_2} \ldots \frac{dN_{Bm}}{b_m} = d\xi \tag{4.1.20}$$

The affinity A of the reaction (4.1.19) is defined as

$$A \equiv \sum_{k=1}^{n} \mu_{Ak} a_k - \sum_{k=1}^{m} \mu_{Bk} b_k \tag{4.1.21}$$

If several simultaneous reactions occur in a closed system, an affinity A_k and a degree of advancement ξ_k can be defined for each reaction, and the change of entropy is written as

$$dS = \frac{dU + p\,dV}{T} + \sum_k \frac{A_k}{T}\,d\xi_k \tag{4.1.22}$$

$$d_iS = \sum_k \frac{A_k}{T}\,d\xi_k \geq 0 \tag{4.1.23}$$

For the rate of entropy production we have the expression

$$\boxed{\frac{d_iS}{dt} = \sum_k \frac{A_k}{T}\frac{d\xi_k}{dt} \geq 0} \tag{4.1.24}$$

At thermodynamic equilibrium, the affinity A and the velocity $d\xi/dt$ of each reaction is zero. We will consider explicit examples of entropy production due to chemical reactions in Chapter 9.

In summary, when chemical reactions are included, the entropy is a function of the energy U, volume V, and the mole numbers N_k, $S = S(U, V, N_k)$. For a closed system, following equation (4.1.22) it can be written as a function of U, V, and the extent of reaction ξ_k, $S = S(U, V, \xi_k)$.

We conclude this section with a historical remark. In the next chapter we will introduce a quantity called the Gibbs free energy. The Gibbs free energy of one mole of X can also be interpreted as the chemical potential of X. The conversion of a compound X to a compound Z causes a decrease in the Gibbs free energy of X and an increase in the Gibbs free energy of Z. Thus, the affinity of a reaction, $X + Y \rightarrow 2Z$, defined as $A \equiv (\mu_X + \mu_Y - 2\mu_Z)$, can be interpreted as the negative of change in the Gibbs free energy when 1 mole of X and 1 mole of Y react to produce 2 moles of Z. This change in the Gibbs free energy, called the "Gibbs free energy of reaction," is related to affinity A by a simple negative sign, but there is a fundamental conceptual difference between the two: *affinity is a concept that relates irreversible chemical reactions to entropy, whereas Gibbs free energy is primarily used in connection with equilibrium states and reversible processes.* Nevertheless, in many texts the Gibbs free energy is used in place of affinity and no mention is made of the relation between entropy and reaction rates.[*] In his well-known book, *The Historical Background of Chemistry* [1, p. 206], Leicester traces the origin of this usage to the text of Gilbert Newton Lewis (1875–1946) and Merle Randall (1888–1950) [8]:

"The influential textbook of G.N. Lewis (1875–1946) and Merle Randall (1888–1950) which presents these ideas has led to the replacement of the term "affinity" by the term

---

[*] For some recent comments on this point see Gerhartl [7]

"free energy" in much of the English-speaking world. The older term has never been entirely replaced in thermodynamics literature, since after 1922 the Belgian school under Théophile De Donder (1872–1957) has made the concept of affinity still more precise."

De Donder's affinity has an entirely different conceptual basis; it relates entropy to irreversible chemical processes that occur in nature. It is clearly a more general view of entropy, one which does not restrict the idea of entropy to infinitely slow (quasistatic) reversible processes and equilibrium states.

4.2   General Properties of Affinity

Affinity of a reaction is a state function, completely defined by the chemical potentials. It can be expressed as a function of V, T and N_k. For a closed system, since all the changes in N_k can only be due to chemical reactions, it can be expressed in terms of V, T, ξ_k and the initial values of the mole numbers N_{k0}. There are some general properties of affinities that follow from the fact that chemical reactions can be interdependent when a reactant is involved in more than one reaction.

AFFINITY AND DIRECTION OF REACTION

The sign of affinity can be used to predict the direction of reaction. Consider the reaction $X + Y \rightleftharpoons 2Z$. The affinity is given by $A = (\mu_X + \mu_Y - 2\mu_Z)$. The velocity of reaction $(d\xi/dt)$ indicates the direction of reaction, i.e. whether the net conversion is from X,Y to Z or from Z to X,Y. From the definition of ξ it follows that if $(d\xi/dt) > 0$ then the reaction "proceeds to the right": $X + Y \rightarrow Z$; if $(d\xi/dt) < 0$ then the reaction "proceeds to the left": $2Z \rightarrow X + Y$. The Second Law requires that $A(d\xi/dt) \geq 0$. Thus we arrive at the following relation between the sign of A and the direction of the reaction:

- If $A > 0$ the reaction proceeds to the right.
- If $A < 0$ the reaction proceeds to the left.

ADDITIVITY OF AFFINITIES

A chemical reaction can be the net result of two or more successive chemical reactions. For instance

$$2C(s) + O_2(g) \rightleftharpoons 2CO(g) \quad A_1 \tag{4.2.1}$$

$$2CO(s) + O_2(g) \rightleftharpoons 2CO_2(g) \quad A_2 \tag{4.2.2}$$

$$\overline{2[C(s) + O_2(g) \rightleftharpoons CO_2(g)] \quad 2A_3} \tag{4.2.3}$$

which shows that reaction (4.2.3) is the net result or "sum" of the other two. From the definition of affinity, the affinities of the three reactions are

$$A_1 = 2\mu_C + \mu_{O_2} - 2\mu_{CO} \tag{4.2.4}$$

$$A_2 = 2\mu_{CO} + \mu_{O_2} - 2\mu_{CO_2} \tag{4.2.5}$$

$$A_3 = \mu_C + \mu_{O_2} - \mu_{CO_2} \tag{4.2.6}$$

From these definitions it is easy to see that

$$A_1 + A_2 = 2A_3 \tag{4.2.7}$$

Clearly this result can be generalized to many reactions. We thus have the general result: *the affinity of a net reaction is equal to the sum of the affinities of the individual reactions.*

COUPLING BETWEEN AFFINITIES

When several chemical reactions are taking place, an interesting possibility arises. Consider two reactions that are "coupled" by one or more reactants that take part in both reactions. For the total entropy production we have

$$\frac{d_i S}{dt} = \frac{A_1}{T}\frac{d\xi_1}{dt} + \frac{A_2}{T}\frac{d\xi_2}{dt} \geq 0 \tag{4.2.8}$$

For this inequality to be satisfied, it is not necessary that each term be positive. It may well be that

$$\frac{A_1}{T}\frac{d\xi_1}{dt} > 0, \quad \frac{A_2}{T}\frac{d\xi_2}{dt} < 0 \quad \text{but} \quad \frac{A_1}{T}\frac{d\xi_1}{dt} + \frac{A_2}{T}\frac{d\xi_2}{dt} > 0 \tag{4.2.9}$$

In this case the decrease in entropy due to one reaction is compensated by the increase in entropy due to the other. Such coupled reactions are common in biological systems. The affinity of a reaction is driven away from zero at the expense of another reaction whose affinity tends to zero.

4.3   Entropy Production due to Diffusion

The concepts of chemical potential and affinity not only describe chemical reactions but also the flow of matter from one region of space to another. With the concept of chemical potential, we are now in a position to obtain an expression for the entropy change due to diffusion, an example of an irreversible process we saw in the previous chapter (Fig. 3.8). The idea of chemical potential turns out to have wide applicability. Several other irreversible processes that can be described using a chemical potential will be discussed in Chapter 10.

When the chemical potentials of adjacent parts of a system are unequal, diffusion of matter takes place until the chemical potentials of the two parts become equal. The process is similar to flow of heat that occurs due to a difference in temperature. Diffusion is another irreversible process for which we can obtain the rate of increase in entropy in terms of the chemical potentials.

DISCRETE SYSTEMS

For simplicity, let us consider a system consisting of two parts of equal temperature T, one with chemical potential μ_1 and mole number N_1, and the other with chemical potential μ_2 and mole number N_2 as shown in Fig. 4.3. The flow of particles from one part to another can also be associated with an "extent of reaction," though no real chemical reaction is taking place here:

$$-dN_1 = dN_2 = d\xi \tag{4.3.1}$$

Following equations (4.1.14) the entropy change for this process can be written as

$$d_iS = \frac{dU + p\,dV}{T} - \left(\frac{\mu_2 - \mu_1}{T}\right)d\xi \tag{4.3.2}$$

$$= \frac{dU + p\,dV}{T} + \frac{A}{T}d\xi \tag{4.3.3}$$

If $dU = dV = 0$ then the transport of particles results in the change of entropy, given by

$$d_iS = -\left(\frac{\mu_2 - \mu_1}{T}\right)d\xi > 0 \tag{4.3.4}$$

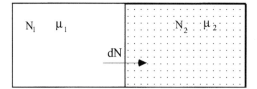

Figure 4.3 The irreversible process of diffusion can be described thermodynamically using chemical potential. The variation of chemical potential with location corresponds to an affinity that drives a flow of matter. The corresponding entropy production is given by (4.3.4)

The positivity of this quantity required by the Second Law implies that particle transport is from a region of high chemical potential to a region of low chemical potential. This is the process of diffusion of particles from a region of higher concentration to a region of lower concentration.

4.4   General Properties of Entropy

Entropy, as formulated here and in the previous chapter, encompasses all aspects of matter transformations: changes in energy, volume and composition. Thus, every system in Nature, be it a gas, an aqueous solution or a living cell, is associated with a certain entropy. We shall obtain explicit expressions for entropies of various systems in the following chapters and study how entropy production is related to irreversible processes. At this stage, however, we shall note some general properties of entropy as a function of state.

Entropy is a function of the total energy U, volume V and mole numbers N_k:

$$S = S(U, V, N_1, N_2, \ldots, N_s) \qquad (4.4.1)$$

For a function of many variables, we have the general relation

$$dS = \left(\frac{\partial S}{\partial U}\right)_{V, V_k} dU + \left(\frac{\partial S}{\partial V}\right)_{U, N_k} dV + \left(\frac{\partial S}{\partial N_k}\right)_{U, V, N_{j \neq k}} dN_k \qquad (4.4.2)$$

Furthermore, from the general relation $dU = T\, dS - p\, dV + \sum_1^n \mu_k\, dN_k$, it follows that

$$dS = \frac{1}{T} dU + \frac{p}{T} dV - \sum_k \frac{\mu_k}{T} dN_k \qquad (4.4.3)$$

(Here we have combined the change in N_k due to chemical reactions and the change due to exchange with the exterior.) Comparing (4.4.2) and (4.4.3) we immediately see that

$$\left(\frac{\partial S}{\partial U}\right)_{V, N_k} = \frac{1}{T} \qquad \left(\frac{\partial S}{\partial V}\right)_{U, N_k} = \frac{p}{T} \qquad \left(\frac{\partial S}{\partial N_k}\right)_{U, V, N_{j \neq k}} = -\frac{\mu_k}{T} \qquad (4.4.4)$$

If the change in mole number N_k is only due to a chemical reaction, the entropy can also be expressed as a function of U, V and ξ (Example 4.1). Then one can show that

$$\left(\frac{\partial S}{\partial \xi}\right)_{U,V} = \frac{A}{T} \tag{4.4.5}$$

In addition, for any function of many variables, the "cross derivatives" must be equal, i.e. we must have equalities of the type

$$\frac{\partial^2 S}{\partial V \, \partial U} = \frac{\partial^2 S}{\partial U \, \partial V} \tag{4.4.6}$$

The relations (4.4.4) then imply that

$$\left(\frac{\partial}{\partial V} \frac{1}{T}\right)_{U,N_k} = \left(\frac{\partial}{\partial U} \frac{p}{T}\right)_{V,N_k} \tag{4.4.7}$$

Many such relations can be similarly derived because entropy is a function of state.

For homogeneous systems, entropy is also assumed to be directly proportional to the size, i.e. *entropy is an extensive variable*. Mathematically, this means that entropy S is a homogeneous function of the variables U, V and N_k, i.e. it has the following property:

$$S(\lambda U, \lambda V, \lambda N_1, \lambda N_2, \ldots, \lambda N_s) = \lambda S(U,\ V,\ N_1,\ N_2, \ldots, N_s) \tag{4.4.8}$$

Differentiating (4.4.8) with respect to λ and setting $\lambda = 1$, we obtain the well-known **Euler theorem** for homogeneous functions:

$$S = \left(\frac{\partial S}{\partial U}\right)_{V,N_k} U + \left(\frac{\partial S}{\partial V}\right)_{U,N_k} V + \sum_k \left(\frac{\partial S}{\partial N_k}\right)_{U,V,N_{i \neq k}} N_k \tag{4.4.9}$$

Using relation (4.4.4) we can write this as

$$S = \frac{U}{T} + \frac{pV}{T} - \sum_k \frac{\mu_k N_k}{T} \tag{4.4.10}$$

In equations (4.4.9) and (4.4.10) we have expressed entropy as a function of U, V and N_k. Since U can be expressed as function of T, V and N_k, entropy can also be expressed as function of T, V and N_k : $S = S(T, V, N_k)$. (The temperature and volume dependence of the energy U and enthalpy H of each component is obtained by using the empirical values of the heat capacities, as described in Chapter 2.) Since T, V and N_k are directly measurable quantities, it is often more convenient to express thermodynamic variables such as entropy and energy as functions of these variables.

As a function of T, V and N_k, the derivatives of entropy can be obtained by expressing dU in (4.4.3) as a function of V, T and N_k:

$$T\,dS = dU + p\,dV - \sum_k \mu_k\,dN_k$$

$$= \left(\frac{\partial U}{\partial T}\right)_V dT + \left(\frac{\partial U}{\partial V}\right)_T dV + p\,dV - \sum_k \mu_k\,dN_k$$

$$+ \sum_k \left(\frac{\partial U}{\partial N_k}\right)_{V,T,N_{i\neq k}} dN_k \tag{4.4.11}$$

i.e.
$$dS = \frac{1}{T}\left[\left(\frac{\partial U}{\partial V}\right)_T + p\right] dV + \frac{1}{T}\left(\frac{\partial U}{\partial T}\right)_V dT - \sum_k \frac{\mu_k}{T}\,dN_k$$

$$+ \sum_k \left(\frac{\partial U}{\partial N_k}\right)_{V,T,N_{i\neq k}} dN_k$$

Equation (4.4.11) leads to

$$\left(\frac{\partial S}{\partial V}\right)_{T,N_k} = \frac{1}{T}\left(\frac{\partial U}{\partial V}\right)_T + \frac{p}{T} \tag{4.4.12}$$

$$\left(\frac{\partial S}{\partial T}\right)_{V,N_k} = \frac{1}{T}\left(\frac{\partial U}{\partial T}\right)_V = \frac{C_V}{T} \tag{4.4.13}$$

$$\left(\frac{\partial S}{\partial N_k}\right)_{V,T,N_{i\neq k}} = -\frac{\mu_k}{T} + \left(\frac{\partial U}{\partial N_k}\right)_{V,T,N_{i\neq k}} \frac{1}{T} \tag{4.4.14}$$

Similar relations can be derived for U as a function of T, V and N_k.

These relations are valid for homogeneous system with uniform temperature and pressure. They can be extended to inhomogeneous systems as long as one can associate a well-defined temperature to every location. The thermodynamics of an inhomogeneous system can be formulated in terms of entropy density $s(T(\mathbf{x}), n_k(\mathbf{x}))$, which is a function of the temperature and the mole number densities at the point $\mathbf{x}$. If $u(\mathbf{x})$ is the energy density, then following (4.4.4) we have the relations

$$\left(\frac{\partial s}{\partial u}\right)_{n_k} = \frac{1}{T(\mathbf{x})} \qquad \left(\frac{\partial s}{\partial n_k}\right)_u = -\frac{\mu(\mathbf{x})}{T(\mathbf{x})} \tag{4.4.15}$$

in which the positional dependence of the variables is explicitly shown.

An empirically more convenient way is to express both entropy and energy densities as functions of the local temperature $T(\mathbf{x})$ and mole number density

$n_k(\mathbf{x})$, both of which can be directly measured:

$$u = u(T(\mathbf{x}), n_k(\mathbf{x})) \qquad s = s(T(\mathbf{x}), n_k(\mathbf{x})) \tag{4.4.16}$$

The total entropy and energy of the system is obtained by integrating the entropy density over the volume of the system:

$$S = \int_V s(T(\mathbf{x}), n_k(\mathbf{x}))dV \qquad U = \int_V u(T(\mathbf{x}), n_k(\mathbf{x}))dV \tag{4.4.17}$$

Since the system as a whole is not in thermodynamic equilibrium, the total entropy S is generally not a function of the total energy U and the total volume V. Nevertheless, a thermodynamic description is still possible as long as the temperature is well defined at each location $\mathbf{x}$.

References

1. Leicester, H. M., *The Historical Background of Chemistry*. 1971, New York: Dover.
2. De Donder, T., *L'Affinité*. 1927, Paris: Gauthiers-Villars.
3. De Donder, T. and Van Rysselberghe P., *Affinity*. 1936, Menlo Park CA: Stanford University Press.
4. Gibbs, J. W., On the equilibrium of heterogeneous substances. *Trans. Conn. Acad. Sci.*, **III** (1878) 343–524.
5. Gibbs, J. W., On the equilibrium of heterogeneous substances. *Trans. Conn. Acad. Sci.*, **III** (1875) 108–248.
6. Gibbs, J. W., *The Scientific Papers of J. Willard Gibbs, Vol.1: Thermodynamics*, A. N. Editor (ed.). 1961, New York: Dover.
7. Gerhartl, F. J., *J. Chem. Ed.*, **71** (1994) 539–548.
8. Lewis, G. N. and Randall, M., *Thermodynamics and Free Energy of Chemical Substances*. 1923, New York: McGraw-Hill.

Example

Example 4.1 If the change in mole number is entirely due to one reaction, show that entropy is a function of V, U and ξ and that $(\partial S/\partial \xi)_{U,V} = A/T$.
Solution: Entropy is a function of U, V and N_k, $S(U, V, N_k)$. As shown in equation (4.4.3), for change in entropy dS we have

$$dS = \frac{1}{T}dU + \frac{p}{T}dV - \sum_k \frac{\mu_k}{T}dN_k$$

If ξ is the extent of the single reaction which causes N_k to change, then

$$dN_k = v_k \, d\xi \qquad (k = 1, 2, \ldots, s)$$

in which v_k are the stoichiometric coefficients of the s species that participate in the reaction; v_k is negative for the reactants and positive for the products. Any species which do not participate in the reaction have $v_k = 0$. The change in entropy dS can now be written as

$$dS = \frac{1}{T} dU + \frac{p}{T} dV - \sum_{k=1}^{s} \frac{\mu_k v_k}{T} d\xi$$

Now, the affinity of the reaction is $A = -\sum_{k=1}^{s} \mu_k v_k$ (note that v_k is negative for the reactants and positive for the products). Hence

$$dS = \frac{1}{T} dU + \frac{p}{T} dV + A \, d\xi$$

This shows that S is a function of U, V and ξ and that $(\partial S / \partial \xi)_{U,V} = A/T$.

If N_{10} is the mole number of N_1 at time $t = 0$ etc., and if we assume $\xi = 0$ at $t = 0$, then the mole numbers at any time t are: $N_{10} + v_1 \xi(t)$, $N_{20} + v_2 \xi(t), \ldots, N_{s0} + v_s \xi(t)$; all the others mole numbers are constant. Thus $S = S(U, V, N_{10} + v_1 \xi(t), N_{20} + v_2 \xi(t), \ldots, N_{s0} + \xi(t))$. Thus, for a given initial mole numbers N_{k0}, in a closed system with one reaction, entropy is a function of U, V and ξ.

Exercises

4.1 In a living cell, which is an open system that exchanges energy and matter with the exterior, the entropy can decrease, i.e. $dS < 0$. Explain how this is possible in terms of $d_e S$ and $d_i S$. How is the Second Law valid in this case?

4.2 In SI units, what are the units of entropy, chemical potential and affinity?

4.3 Explain which of the following are *not* extensive functions:

$$S_1 = (N/V)[s_0 + C_V \ln T + R \ln V]$$
$$S_2 = N[s_0 + C_V \ln T + R \ln(V/N)]$$
$$S_3 = N^2[s_0 + C_V \ln T + R \ln(V/N)]$$

4.4 Consider a reaction $A \rightarrow 2B$ in the gas phase (i.e. A and B are gases) occurring in a fixed volume V at a fixed temperature T. Use the ideal gas approximation. Assume $N_A(t)$ and $N_B(t)$ are the mole numbers at any time t.

(i) Write an expression for the total entropy.

(ii) Assume that at time $t = 0$, $N_A(0) = N_{A0}$, $N_B(0) = 0$ and the extent of reaction $\xi(0) = 0$. At any time t, express the mole numbers $N_A(t)$ and $N_B(t)$ in terms of $\xi(t)$.

(iii) At any time t, write the total entropy as a function of T, V and $\xi(t)$ (and N_{A0} which is a constant).

4.5 (a) Using the fact that S is a function of U, V and N_k, derive the relation

$$\left(\frac{\partial}{\partial V} \frac{\mu_k}{T} \right)_{U,N_k} + \left(\frac{\partial}{\partial N_k} \frac{p}{T} \right)_{U,V} = 0$$

(b) For an ideal gas, show that

$$\left(\frac{\partial}{\partial V} \frac{\mu_k}{T} \right)_{U,N_k} = -\frac{R}{V}$$

(c) For an ideal gas show that $(\partial S / \partial V)_{T,N_k} = nR$, where $n =$ moles per unit volulme.

PART II

EQUILIBRIUM THERMODYNAMICS

5 EXTREMUM PRINCIPLES AND GENERAL THERMODYNAMIC RELATIONS

Extremum Principles in Nature

For centuries we have been motivated by the belief that the laws of Nature are simple, and have been rewarded amply in our search. The laws of mechanics, gravitation, electromagnetism and thermodynamics can all be stated simply and expressed precisely with a few equations. The current search for a theory that unifies all the known fundamental forces between elementary particles is very much motivated by this belief. In addition to simplicity, Nature also seems to "optimize" or "economize": natural phenomena often occur in such a way that some physical quantity is minimized or maximized—or to use one word for both, "extremized". The French mathematician Pierre de Fermat (1601–1665) noticed that the bending of rays of light as they propagate through different media can all be precisely described using one simple principle: *light travels from one point to another along a path that minimizes the time of travel*. In fact, all the equations of motion in mechanics can be obtained by invoking the *principle of least action*, which states that, if a body is at a point x_1 at a time t_1, and at a point x_2 at time t_2, the motion occurs so as to minimize a quantity called the *action*. For an engaging exposition of these topics see *The Feynman Lectures on Physics* [1], Vol. I, Ch. 26 and Vol. II, Ch. 19.

Equilibrium thermodynamics too has its extremum principles. In this chapter we will see that the approach to equilibrium under different conditions is such that a **thermodynamic potential** is extremized. Following this, in preparation for the applications of thermodynamics in the subsequent chapters, we will derive general thermodynamic relations.

5.1 Extremum Principles and the Second Law

We have already seen that all isolated systems evolve to the state of equilibrium in which the entropy reaches its maximum value. This is the basic extremum principle of thermodynamics. But we don't always deal with isolated systems. In many practical situations, the physical or chemical system under consideration is subject to constant pressure or temperature or both. In these situations the positivity of entropy change due to irreversible processes, $d_iS > 0$, can also be expressed as the evolution of certain thermodynamic functions to their extremum values. Under each *constraint*, such as constant pressure, constant temperature or both, the evolution of the system to the state of equilibrium corresponds to the extremization of a thermodynamic quantity. These quantities are the *Gibbs free energy* the *Helmholtz free energy*, and *enthalpy* (that was introduced in Chapter 2). These functions which are associated with extremum principles are also called **thermodynamic potentials**, in analogy with the potentials associated with forces in mechanics whose minima are also points of stable equilibrium. *The systems we consider are either isolated or closed.*

MAXIMUM ENTROPY

As we have seen in the previous chapter, due to irreversible processes, the entropy of an *isolated system* continues to increase ($d_i S > 0$) until it reaches the maximum possible value. The state thus reached is the state of equilibrium. *Therefore, when U and V are constant, every system evolves to a state of maximum entropy.*

MINIMUM ENERGY

The Second Law also implies that *at constant S and V, every system evolves to a state of minimum energy.* This can be seen as follows. We have seen that for closed systems, $dU = dQ - pdV = Td_e S - pdV$. Because the total entropy change $dS = d_e S + d_i S$, we may write $dU = T\,dS - p\,dV - Td_i S$. Since S and V are constant, $dS = dV = 0$. We therefore have:

$$dU = -Td_i S \leq 0 \tag{5.1.1}$$

Thus, in systems whose entropy is maintained at a fixed value, driven by irreversible processes, the energy evolves to the minimum possible value.

To keep the entropy constant, the entropy, $d_i S$, produced by irreversible processes, has to be removed from the system. If a system is maintained at constant T, V and N_k, the entropy remains constant. The decrease in energy, $dU = -Td_i S$, is generally due to irreversible conversion of mechanical energy to heat, which is removed from the system to keep the entropy (temperature) constant. A simple example is the falling of an object to the bottom of a fluid (Fig. 5.1). Here $dU = -Td_i S$ is the heat produced as a result of fluid friction or viscosity. If this heat is removed rapidly so as to keep the temperature constant, the system will evolve to a state of minimum energy. Note that during the approach to equilibrium, $dU = -T\,d_i S < 0$ for every time interval dt. This represents a continuous conversion of mechanical energy (kinetic energy + potential energy) into heat; at no time does the conversion occur in the opposite direction.

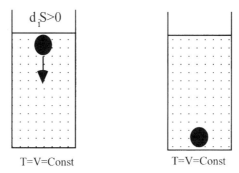

Figure 5.1 A simple illustration of the principle of minimum energy. In this example the entropy and the volume of the system remain essentially unchanged. The system evolves to a state of minimum energy

MINIMUM HELMHOLTZ FREE ENERGY

For systems maintained at constant T and V, a thermodynamic quantity called the **Helmholtz free energy**, F evolves to its minimum value. F is defined as:

$$\boxed{F = U - TS} \tag{5.1.2}$$

At constant T we have

$$\begin{aligned} dF &= dU - T\,dS = dU - T\,d_e S - Td_i S \\ &= dQ - p\,dV - Td_e S - T\,d_i S \end{aligned}$$

If V is also kept constant then $dV = 0$ and, for closed systems, $T\,d_e S = dQ$. Thus, at *constant T and V* we obtain the inequality:

$$dF = -Td_i S \leq 0 \tag{5.1.3}$$

as a direct consequence of the Second Law. This tells us that a closed system whose temperature and volume are maintained constant evolves such that the Helmholtz free energy is minimized.

An example of the minimization of F is a reaction, such as $2H_2(g) + O_2(g) \rightleftharpoons 2H_2O(g)$, that takes place at a fixed value of T and V (Fig. 5.2(a)). To keep T constant, the heat generated by the reaction has to be removed. In this case, following De Donder's identification of the entropy production in an irreversible chemical reaction (4.1.6), we have $T\,d_i S = -\sum_k \mu_k\,d_i N_k = -dF$. Another example is the natural evolution in the shape of a liquid drop (Fig. 5.2(b)). In the absence of gravity (or if the liquid drop is small enough that the change

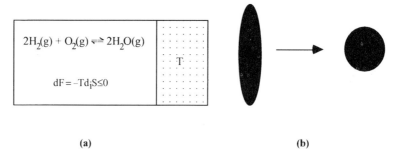

(a)   (b)

Figure 5.2   Minimization of Helmholtz free energy F. (a) If V and T are kept at a fixed value, a chemical reaction will progress to state of minimum F. In this case the irreversible production of entropy $T\,d_iS = -\sum_k \mu_k\,dN_k = -dF \geq 0$. (b) Similarly, for a liquid drop, if we assume that V and T are essentially constant, we see that the minimization of surface energy drives the system to the shape of a sphere — the shape that has the least surface area for a given volume. In this case we can identify $Td_iS = -dF \geq 0$

in gravitational energy due to change in shape is insignificant compared to other energies of the system), regardless of its initial shape, a liquid drop finally assumes the shape of a sphere. During this evolution from an arbitrary initial shape to a sphere, the volume and temperature remain fixed—though the entropy may change with the surface area. This happens because the molecules at the surface of the liquid have higher Helmholtz free energy than the molecules in the bulk of the liquid. The excess energy per unit area, γ, is called "surface tension." This excess surface energy is usually small, of the order of 10^{-2} J m^{-2}. For water, $\gamma = 7.7 \times 10^{-2}$ J m^{-2}. As the area of the drop irreversibly decreases, the surface energy is transformed into heat, which escapes to the surroundings (thus T is maintained constant). The entropy production in this irreversible processes is given by $Td_iS = -dF$. More about surface tension can be found as a special topic at the end of this chapter.

The minimization of Helmholtz free energy is a very useful principle. Many interesting features such as phase transitions and formation of complex patterns in equilibrium systems [2] can be understood using this principle. One can also show that the Helmholtz free energy F is the energy that is "free," available to do work in a reversible process (Example 5.1) — hence the name "free energy."

The Helmholtz free energy is a state function. We can show that F is a function of T, V and N_k and obtain its derivatives with respect to these quantities. From (5.1.2) it follows that $dF = dU - T\,dS - S\,dT$. For the change of entropy due to exchange of energy and matter, we have $T\,d_eS = dU + p\,dV - \sum_k \mu_k\,d_e N_k$. For the change of entropy due to irreversible chemical reaction, we have $T\,d_iS = -\sum_k \mu_k d_i N_k$. For the total change in entropy we have $T\,dS = T\,d_eS + Td_iS$. Substituting these expressions for dU and dS into the expression for dF, we obtain

$$dF = dU - T\left[\frac{dU + p\,dV}{T} - \frac{1}{T}\sum_k \mu_k d_e N_k\right] - T\frac{\sum_k \mu_k d_i N_k}{T} - S\,dT$$

$$= -p\,dV - S\,dT + \sum_k \mu_k(d_e N_k + d_i N_k) \tag{5.1.4}$$

Since $dN_k = d_e N_k + d_i N_k$ we may write equation (5.1.4) as

$$dF = -p\,dV - S\,dT + \sum_k \mu_k\,dN_k \tag{5.1.5}$$

This shows that F is a function of V, T and N_k. It also leads to the following identification of the derivatives of $F(V, T, N_k)$ with respect to V, T and N_k*:

$$\left(\frac{\partial F}{\partial V}\right)_{T,N_k} = -p \qquad \left(\frac{\partial F}{\partial T}\right)_{V,N_k} = -S \qquad \left(\frac{\partial F}{\partial N_k}\right)_{T,V} = \mu_k \tag{5.1.6}$$

It is straightforward to include surface or other contributions to the energy (equations 2.2.10 and 2.2.11) within the expression for F and obtain similar derivatives.

If the changes in N_k are only due to a chemical reaction, then F is a function of T, V and the extent of reaction ξ. Then it can easily be shown (Example 5.2) that

$$\left(\frac{\partial F}{\partial \xi}\right)_{T,V} = -A \tag{5.1.7}$$

MINIMUM GIBBS FREE ENERGY

If both the pressure and temperature of a closed system are maintained constant, the quantity that is minimized at equilibrium is the Gibbs free energy. The **Gibbs free energy** G is similar to the Helmholtz free energy and is defined by:

$$G = U + pV - TS = H - TS \tag{5.1.8}$$

where we have used the definition of enthalpy $H = U + pV$. Just as F evolves to a minimum when T and V are maintained constant, G evolves to a minimum when

---

*In this and the following chapters, for derivatives with respect to N_k, we assume the subscript $N_{i\neq k}$ is understood and drop its explicit use.

the pressure p and temperature T are maintained constant. For this case, in which p and T are kept constant, we can relate dG to d_iS as follows:

$$
\begin{aligned}
dG &= dU + p\,dV + V\,dp - T\,dS - S\,dT \\
&= dQ - p\,dV + p\,dV + V\,dp - T\,d_eS - T\,d_iS - S\,dT \\
&= -T\,d_iS \leq 0
\end{aligned}
\tag{5.1.9}
$$

where we have used the fact that p and T are constant and $Td_eS = dQ$ for closed systems.

The Gibbs free energy is mostly used to describe chemical processes because the usual laboratory situation corresponds to constant p and T. Using (4.1.23) the irreversible evolution of G to its minimum value can be related to the affinities A_k of the reactions and the reaction velocities $d\xi_k/dt$ (in which the index k identifies different reactions)

$$
\frac{dG}{dt} = -T\frac{d_iS}{dt} = -\sum_k A_k \frac{d\xi_k}{dt} \leq 0
\tag{5.1.10}
$$

or

$$
dG = -\sum_k A_k\,d\xi_k \leq 0
\tag{5.1.11}
$$

in which the equality holds at equilibrium. Equation (5.1.11) shows that, at constant p and T, G is a function of the state variable ξ_k, the extent of reaction for reaction k. It also follows that

$$
\boxed{-A_k = \left(\frac{\partial G}{\partial \xi_k}\right)_{p,T}}
\tag{5.1.12}
$$

In view of this relation, it is inappropriate to call the affinity the "Gibbs free energy of reaction," as is commonly done in many texts. As shown in Fig. 5.3(b), at constant p and T, the extents of reactions ξ_k will evolve to a value that minimizes $G(\xi_k, p, T)$.

Note that G evolves to its minimum value monotonically in accordance with the Second Law. Thus ξ cannot reach its equilibrium value, in an oscillatory manner, as a pendulum does. For this reason, an oscillatory approach to equilibrium in a chemical reaction is impossible. This does not mean that concentration oscillations in chemical systems are not possible (as it was once widely thought). Chapter 19 describes how concentration oscillations *can* occur in systems that are far from equilibrium. These oscillations occur about a nonequilibrium value of ξ.

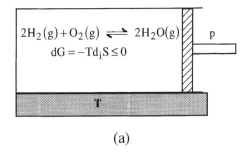

(a)

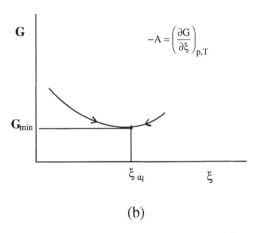

(b)

Figure 5.3   Minimization of the Gibbs free energy G. (a) Under conditions of constant p and temperature T, irreversible chemical reactions will drive the system to a state of minimum G. (b) The extent of reaction ξ evolves to ξ_{eq}, which minimizes G

We showed above that F is function of V, T and N_k. In a similar manner (exc. 5.3), it is straightforward to show that

$$dG = V\,dp - S\,dT + \sum_k \mu_k\,dN_k \qquad (5.1.13)$$

This expression shows that G is a function of p, T and N_k and that

$$\left(\frac{\partial G}{\partial p}\right)_{T,N_k} = V \qquad \left(\frac{\partial G}{\partial T}\right)_{p,N_k} = -S \qquad \left(\frac{\partial G}{\partial N_k}\right)_{T,p} = \mu_k \qquad (5.1.14)$$

One very useful property of the Gibbs free energy is its relation to the chemical potential. From a homogenous system we have shown (equation 4.4.10) that

$U = TS - pV + \sum_k \mu_k N_k$. Substituting this into the definition of G, (equation 5.1.8), we obtain:

$$\boxed{G = \sum_k \mu_k N_k}$$

(5.1.15)

For a pure compound (5.1.15) becomes $G = \mu N$. *Therefore, one might think of the chemical potential μ as the Gibbs free energy per mole of a pure compound.* For a multicomponent system, dividing (5.1.15) by N the total number of moles, we see that the molar Gibs free energy is

$$G_{\mathrm{m}} \equiv \frac{G}{N} = \sum_k \mu_k x_k$$

(5.1.16)

in which the x_k are the mole fractions. For a given mole fraction x_k, since G must be an extensive function of N, we see that G_m is a function of p, T and the mole fractions x_k. As shown in Example 5.3, at constant p and T, we have the relation[*]

$$(dGm)_{p,T} = \sum_k \mu_k \, dx_k$$

(5.1.17)

In a multicomponent system the chemical potential is a function of p, T and the mole fractions x_k : $\mu_k = \mu_k(p, T, x_k)$.

MINIMUM ENTHALPY

Chapter 2 introduced enthalpy

$$\boxed{H = U + pV}$$

(5.1.18)

Just as for the Helmholtz free energy F and the Gibbs free energy G, the enthalpy is also associated with an extremum principle: *at fixed entropy S and pressure p, the enthalpy H evolves to its minimum value.* This can be seen as before by relating the enthalpy change dH to d_iS. Since we assume that p is a constant, we have:

$$dH = dU + p \, dV = dQ$$

(5.1.19)

For a closed system, $dQ = Td_eS = T(dS - d_iS)$. Hence $dH = TdS - Td_iS$. But because the total entropy S is fixed, $dS = 0$. We therefore have the relation

$$dH = -Td_iS \leq 0$$

(5.1.20)

---

[*] Note that dx_k are not all independent because $\sum x_k = 1$.

in accordance with the Second Law. When irreversible chemical reactions take place, we normally don't encounter situations in which the total entropy remains constant, but here is an example.

Consider the reaction

$$H_2(g) + Cl_2(g) \rightleftharpoons 2HCl(g)$$

The total number of molecules does not change. As we have seen in section 3.7, the entropy of an ideal gas $S(V, T, N) = N[s_0 + R \ln(V/N) + C_V \ln T]$. Although there are considerable difference in the heat capacities of molecules with different numbers of atoms, the difference in the heat capacities of two diatomic molecules is relatively small. The difference in the term s_0 is also small for two diatomic molecules. If we ignore these small differences in the entropy between the three species of diatomic molecule, then the entropy which depends on N_k, V and T will essentially remain constant if T and V are maintained constant. At the same time, since the number of molecules does not change, the pressure p remains constant (assuming ideal gas behavior). Since this reaction is exothermic, the removal of heat produced by the reaction is necessary to keep T constant. Under these conditions, both p and S remain constant as the reaction proceeds, and the enthalpy reaches its minimum possible value when the system reaches the state of equilibrium. For an arbitrary chemical reaction, V and T have to be simultaneously adjusted so as to keep p and S constant, which is not a simple task.

Just as we derived $dF = -p\,dV - S\,dT + \sum_k \mu_k dN_k$ (5.1.5), it can easily be shown (exc 5.4) that

$$dH = T\,dS + V\,dp + \sum_k \mu_k dN_k \tag{5.1.21}$$

This equation shows that H can be expressed as a function of S, p and N_k. The derivatives of H with respect to these variables are:

$$\left(\frac{\partial H}{\partial p}\right)_{S,N_k} = V \qquad \left(\frac{\partial H}{\partial S}\right)_{p,N_k} = T \qquad \left(\frac{\partial H}{\partial N_K}\right)_{S,p} = \mu_k \tag{5.1.22}$$

Once again, if the change in N_k is only due to a chemical reaction, H is a function of p, S and ξ, and we have

$$\boxed{\left(\frac{\partial H}{\partial \xi}\right)_{p,S} = -A} \tag{5.1.23}$$

EXTREMUM PRINCIPLES AND STABILITY OF EQUILIBRIUM STATE

In thermodynamics the existence of extremum principles has an important consequence for the behavior of microscopic fluctuations. Since all macroscopic systems are made of a very large number of molecules, which are in constant random motion, thermodynamic quantities such as temperature, pressure and number density undergo small fluctuations. Why don't these fluctuations slowly drive the thermodynamic variables from one value to another, just as small random fluctuations in the positions of an object slowly move the object from one location to another (a phenomenon called Brownian motion)? The temperature or concentration of a system in thermodynamic equilibrium fluctuates about a fixed value but does not drift randomly. This is because the state of equilibrium is stable. As we have seen, irreversible processes drive the system to the equilibrium state in which one of the potentials is extremized. Thus, whenever a fluctuation drives a system away from equilibrium, irreversible processes bring it back to equilibrium. The tendency of the system to reach and remain at an extremum of thermodynamic potential keeps the system stable. In this way the stability of the equilibrium state is related to the existence of thermodynamic potentials.

Thermodynamic systems are not always stable. There are situations in which fluctuations can drive a system from one state to another. Then the initial state is said to be thermodynamically unstable. Some homogeneous mixtures become unstable when the temperature is decreased; driven by fluctuations, they then evolve to a state in which the components separate into two distinct phases, a phenomenon called "phase separation." In Chapters 12, 13 and 14 we shall discuss thermodynamic stability at greater length.

When a system is far from thermodynamic equilibrium, the state may not be characterized by an extremum principle and the irreversible processes do not always keep the system stable. The consequent instability within a far-from-equilibrium system drives it to states with a high level of organization, such as concentration oscillations and spontaneous formation of spatial patterns. We shall discuss far-from-equilibrium in-stability and the consequent "self-organization" in Chapters 18 and 19.

LEGENDRE TRANSFORMATIONS

The relations between the thermodynamic functions $F(T, V, N_k), G(T, p, N_k)$ and $H(S, p, N_k)$ and the total energy $U(S, V, N_k)$, expressed as a function of S, V and N_k, are a particular instance of a general class of relations called Legendre transformations. In a Legendre transformation, a function $U(S, V, N_k)$ is transformed to a function in which one or more of the independent variables S, V, and N_k are replaced by the corresponding partial derivatives of U. Thus $F(T, V, N_k)$ is a Legendre transform of U in which S is replaced by the corresponding derivative $(\partial U/\partial S)_{V,N_k} = T$. Similarly, $G(T, p, N_k)$ is a Legendre transform of U in which S and V are replaced by their corresponding derivatives $(\partial U/\partial S)_{V,N_k} = T$ and $(\partial U/\partial V)_{S,N_k} = -p$. We thus have the Legendre transforms shown in Table 5.1.

Table 5.1 Legendre transforms in thermodynamics

$U(S,V,N_k) \longrightarrow F(T,V,N_k) = U - TS$	S replaced by $\left(\dfrac{\partial U}{\partial S}\right)_{V,N_k} = T$
$U(S,V,N_k) \longrightarrow H(S,p,N_k) = U + pV$	V replaced by $\left(\dfrac{\partial U}{\partial V}\right)_{S,N_k} = -p$
$U(S,V,N_k) \longrightarrow G(S,p,N_k) = U + pV - TS$	S replaced by $\left(\dfrac{\partial U}{\partial S}\right)_{V,N_k} = T$ and
	V replaced by $\left(\dfrac{\partial U}{\partial V}\right)_{S,N_k} = -p$

Legendre transforms show us the general mathematical structure of thermodynamics. Clearly there are more Legendre transforms that can be defined, not only of $U(S,V,N_k)$ but also of $S(U,V,N_k)$ of $S(U,V,N_k)$. A detailed presentation of the Legendre transforms in thermodynamics can be found in the text written by Herbert Callen [3]. (Legendre transforms also appear in classical mechanics; the Hamiltonian is a Legendre transform of the Lagrangian.)

5.2   General Thermodynamic Relations

As Einstein noted (see the introduction to Chapter 1), it is remarkable that the two laws of thermodynamics are simple to state but they relate so many different quantities and have a wide range of applicability. Thermodynamics gives us many general relations between state variables which are valid for *any system in equilibrium*. In this section we shall present a few important general relations. We will apply them to particular systems in later Chapters. As we shall see in Chapters 15–17, some of these relations can also be extended to nonequilibrium systems that are locally in equilibrium.

THE GIBBS–DUHEM EQUATION:

One of the important general relations is the Gibbs–Duhem equation. This relation shows that the intensive variables T, p and μ_k are not all independent. It is obtained from the fundamental relation (4.1.2) through which Gibbs introduced the chemical potential

$$dU = T\,dS - p\,dV + \sum_k \mu_k dN_k \qquad (5.2.1)$$

and the relation (4.4.10), which can be rewritten as:

$$U = TS - pV + \sum_k \mu_k N_k \tag{5.2.2}$$

This relation follows from the assumption that entropy is an extensive function of U, V and N_k and from the Euler theorem. The differential of (5.2.2) is

$$dU = T\,dS + S\,dT - V\,dp - p\,dV + \sum_k (\mu_k\,dN_k + N_k\,d\mu_k) \tag{5.2.3}$$

This relation can be consistent with (5.2.1) only if

$$\boxed{S\,dT - V\,dp + \sum_k N_k\,d\mu_k = 0} \tag{5.2.4}$$

Equation (5.2.4) is called the **Gibbs–Duhem equation**. It shows that changes in the intensive variables T, p and μ_k cannot all be independent. We shall see in Chapter 7 that the Gibbs–Duhem equation can be used to understand the equilibrium between phases, and the variation of boiling point with pressure as described by the Clausius–Clapeyron equation.

At constant temperature and pressure, from (5.2.4) it follows that $\sum_k N_k (d\mu_k)_{p,T} = 0$. Since the change in the chemical potential $(d\mu_k)_{p,T} = \sum_i (\partial\mu_k/\partial N_i)dN_i$, we can write this expression as:

$$\sum_k \sum_i N_k \left(\frac{\partial\mu_k}{\partial N_i}\right)_{p,T} dN_i = \sum_i \left(\sum_k \left(\frac{\partial\mu_k}{\partial N_i}\right)_{p,T} N_k\right) dN_i = 0 \tag{5.2.5}$$

Since the dN_i are independent and arbitrary variations, equation (5.2.5) can be valid only if the coefficient of every dN_i is equal to zero. Thus, we have $\sum_k (\partial\mu_k/\partial N_i)_{p,T} N_k = 0$. Furthermore, since,

$$\left(\frac{\partial\mu_k}{\partial N_i}\right)_{p,T} = \left(\frac{\partial^2 G}{\partial N_i\,\partial N_k}\right)_{p,T} = \left(\frac{\partial^2 G}{\partial N_k \partial N_i}\right)_{p,T} = \left(\frac{\partial\mu_i}{\partial N_k}\right)_{p,T}$$

we can write

$$\boxed{\sum_k \left(\frac{\partial\mu_i}{\partial N_k}\right) N_k = 0} \tag{5.2.6}$$

Equation (5.2.6) is an important result that we will use in later chapters.

THE HELMHOLTZ EQUATION

We have seen that the entropy S is a state variable and that it can be expressed as a function of T, V and N_k. The Helmholtz equation follows from the fact that, for function of many variables, the second "cross derivatives" must be equal, i.e.

$$\left(\frac{\partial^2 S}{\partial T \, \partial V}\right) = \left(\frac{\partial^2 S}{\partial V \, \partial T}\right) \tag{5.2.7}$$

For closed systems in which no chemical reactions take place, the changes in entropy can be written as

$$dS = \frac{1}{T}dU + \frac{p}{T}dV \tag{5.2.8}$$

Since U can be expressed as a function of V and T, we have $dU = (\partial U/\partial V)_T \, dV + (\partial U/\partial T)_V \, dT$. Using this expression in (5.2.8) we obtain

$$\begin{aligned}
dS &= \frac{1}{T}\left(\frac{\partial U}{\partial V}\right)_T dV + \frac{1}{T}\left(\frac{\partial U}{\partial T}\right)_V dT + \frac{p}{T}dV \\
&= \left[\frac{1}{T}\left(\frac{\partial U}{\partial V}\right)_T + \frac{p}{T}\right]dV + \frac{1}{T}\left(\frac{\partial U}{\partial T}\right)_V dT
\end{aligned} \tag{5.2.9}$$

The coefficients of dV and dT can now be identified as the derivatives $(\partial S/\partial V)_T$ and $(\partial S/\partial T)_V$ respectively. As expressed in (5.2.7), since the second "cross derivatives" must be equal, we have

$$\left(\frac{\partial}{\partial T}\left[\frac{1}{T}\left(\frac{\partial U}{\partial V}\right)_T + \frac{p}{T}\right]\right)_V = \left(\frac{\partial}{\partial V}\left[\frac{1}{T}\left(\frac{\partial U}{\partial T}\right)\right]\right)_T \tag{5.2.10}$$

It is a matter of simple calculation (exc 5.5) to show that (5.2.10) leads to the **Helmholtz equation**:

$$\boxed{\left(\frac{\partial U}{\partial V}\right)_T = T^2\left(\frac{\partial}{\partial T}\frac{p}{T}\right)_V} \tag{5.2.11}$$

This equation enables us to determine the variation of the energy with volume if the equation of state is known. In particular, it can be used to conclude that, for an ideal gas, the equation $pV = NRT$ implies that, at constant T, the energy U is independent of the volume.

THE GIBBS–HELMHOLTZ EQUATION

The Gibbs–Helmholtz equation relates the temperature variation of the Gibbs free energy G to the enthalpy H. It provides us a way to determine the heats of chemical reaction if the Gibbs free energy is known as a function of temperature. The Gibbs–Helmholtz equation is obtained as follows. The Gibbs free energy $G = H - TS$ (5.1.8). First we note that $S = -(\partial G/\partial T)_{p,N_k}$ and we write

$$G = H + \left(\frac{\partial G}{\partial T}\right)_{p,N_k} T \qquad (5.2.12)$$

It is now simple to show (exc 5.7) that this equation can be rewritten as

$$\frac{\partial}{\partial T}\left(\frac{G}{T}\right) = -\frac{H}{T^2} \qquad (5.2.13)$$

For a chemical reaction this equation can be written in terms of the *changes* in G and H when the reactants are converted to the products. If the total Gibbs free energy and the enthalpy of the reactants are G_r and H_r and if the corresponding quantities for the products are G_p and H_p, then the changes due to the reactions will be $\Delta G = G_p - G_r$ and $\Delta H = H_p - H_r$. Applying equation (5.2.13) to reactants and products then subtracting one equation from the other, we obtain

$$\frac{\partial}{\partial T}\left(\frac{\Delta G}{T}\right) = -\frac{\Delta H}{T^2} \qquad (5.2.14)$$

In Chapter 9 we will see that a quantity called the "standard ΔG" of a reaction can be obtained by measuring the equilibrium concentrations of the reactants and products. If the equilibrium concentrations (hence ΔG) are measured at various temperatures, the data on the variation of ΔG with T can be used to obtain ΔH, which is the heat of reaction. Equations (5.2.13) and (5.2.14) are versions of the **Gibbs–Helmholtz equation**.

5.3   Gibbs Free Energy of Formation and Chemical Potential

Other than heat conduction, every irreversible process—chemical reactions, diffusion, the influence of electric, magnetic and gravitational fields, ionic conduction, dielectric relaxation, etc.—can be described in terms of suitable chemical potentials. Chapter 10 is devoted to the wide variety of processes described using the concept of a chemical potential. All these processes drive the system to the equilibrium state in which the corresponding affinity vanishes.

Because of its central role in the description of irreversible processes, we will derive a general expression for the chemical potential in this section.

For a *pure compound* the chemical potential is also equal to the Gibbs free energy of one mole of that compound, i.e., $\mu = $ *the molar Gibbs free energy of a pure compound*. In general, the Gibbs free energy and the chemical potential are related by

$$\left(\frac{\partial G}{\partial N_k} \right)_{p,T} = \mu_k \tag{5.3.1}$$

To obtain a general relation between μ_k and the enthalpy H (which bears a more direct relation to experimentally measured molar heat capacity), we differentiate the Gibbs–Helmholtz equation (5.2.13) with respect to N_k and use (5.3.1) to obtain

$$\frac{\partial}{\partial T} \left(\frac{\mu_k}{T} \right) = -\frac{H_{mk}}{T^2} \quad \text{where} \quad H_{mk} = \left(\frac{\partial H}{\partial N_k} \right)_{p,T,N_{i \neq k}} \tag{5.3.2}$$

H_{mk} is called the **partial molar enthalpy** of the compound k.

If the value of the chemical potential $\mu(p_0, T_0)$ at a temperature T_0 and pressure p_0 is known, then by integrating equation (5.3.2) we can obtain the chemical potential at any other temperature T, if the partial molar enthalpy $H_{mk}(p_0, T)$ is known as a function of T:

$$\frac{\mu(p_0, T)}{T} = \frac{\mu(p_0, T_0)}{T_0} + \int_{T_0}^{T} \frac{-H_{mk}(p_0, T')}{T'^2} \, dT' \tag{5.3.3}$$

As was shown in Chapter 2 (equations (2.4.10) and (2.4.11)), the molar enthalpy of a pure compound $H_m(T)$ can be obtained from the tabulated values of $C_p(T)$, the heat capacity at constant pressure. For ideal mixtures the H_{mk} are the same as that of a pure compound. For nonideal mixtures a detailed knowledge of the molar heat capacities of the mixture is needed to obtain the H_{mk}.

For a pure compound, knowing $\mu(p_0, T)$ at pressure p_0 and temperature T, the value of $\mu(p, T)$ at any other pressure p can be obtained using the expression $d\mu = -S_m dT + V_m dp$ which follows from the Gibbs–Duhem equation (5.2.4), where the molar quantities $S_m = S/N$ and $V_m = V/N$. Since T is fixed, $dT = 0$ and we may integrate this expression with respect to p to obtain

$$\mu(p, T) = \mu(p_0, T) + \int_{p_0}^{p} V_m(p', T) dp' \tag{5.3.4}$$

Thus, if the value of the chemical potential $\mu(p_0, T_0)$ is known at a standard pressure p_0 and T_0, equations (5.3.3) and (5.3.4) tell us that a knowledge of the molar volume $V_m(p, T)$ (or density) and the molar enthalpy $H_m(p, T)$ of a pure compound will enable us to calculate the chemical potential at any other pressure p and temperature T. An alternative and useful way of writing the chemical potential is due to G.N. Lewis (1875–1946), who introduced the concept of **activity** a_k of a compound k. The activity is defined by the expression

$$\mu_k(p, T) = \mu_k(p_0, T) + RT \ln a_k \qquad (5.3.5)$$

The concept of activity is most useful in relating chemical potentials to experimentally measurable quantities such as concentration and pressure. As an illustration, let us apply (5.3.4) to the case of an ideal gas. Since $V_m = RT/p$, we have

$$
\begin{aligned}
\mu(p, T) &= \mu(p_0, T) + \int_{p_0}^{p} \frac{RT}{p'} dp' \\
&= \mu(p_0, T) + RT \ln(p/p_0) \\
&= \mu_0 + RT \ln p
\end{aligned}
\qquad (5.3.6)
$$

which shows that the activity $a = (p/p_0)$ in the ideal gas approximation; μ_0 is the chemical potential at unit pressure. In Chapter 6 we will obtain the expression for the activity of gases when the molecular size and molecular forces are taken into account, as in the van der Waals equation.

TABULATION OF THE GIBBS FREE ENERGIES OF COMPOUNDS

The convention in Box 5.1 is used when tabulating the Gibbs free energies of compounds. It defines the **molar Gibbs free energy of formation** for a compound k, denoted by $\Delta G_f^0[k]$. Since chemical thermodynamics assumes there is no interconversion between the elements, the Gibbs free energy of elements may be used to define the "zero" with respect to which the Gibbs free energies of all other compounds are measured. The Gibbs free energy of formation of H_2O, written as $\Delta G_f^0[H_2O]$, is the Gibbs free energy *change*, ΔG, of the reaction

$$H_2(g) + \tfrac{1}{2}O_2(g) \rightarrow H_2O(l)$$

The Gibbs free energies of formation, $\Delta G_f^0 = \mu(p_0, T_0)$, of compounds are tabulated. We shall consider the use of ΔG_f^0 in more detail in Chapter 9 devoted to the thermodynamics of chemical reactions. From these values the chemical potentials of compounds can be calculated as explained above. We conclude this section by noting that substitution of (5.3.3) into (5.3.4) gives us a general expression for the computation of the chemical potential:

$$\mu(p, T) = \left(\frac{T}{T_0}\right)\mu(p_0, T_0) + \int_{p_0}^{p} V_m(p', T)dp' + T\int_{T_0}^{T} \frac{-H_m(p, T')}{T'^2} dT'$$

$$(5.3.7)$$

Box 5.1 Tabulation of Gibbs Free Energies of Compounds

For practical purposes, the **molar Gibbs free energy** $\mu(p_0, T_0)$ of a compound in its **standard state** (its state at pressure $p_0 = 1$ bar, $T_0 = 298.15$ K) may be *defined* as follows:

$\Delta G_f[k, T] = 0$ for all elements k at all temperatures T

$\mu_k(p_0, T_0) = \Delta G_f^0[k] =$ Standard molar Gibbs free energy of formation of

           compound k

       $=$ Gibbs free energy of formation of one mole of the

           compound from its constituent elements, all in their

           standard states

Since chemical thermodynamics assumes there is no interconversion between the elements, the Gibbs free energy of elements may be used to define the "zero" with respect to which the Gibbs free energies of all other compounds are measured.

     The molar Gibbs free energy at any other p and T can be obtained using (5.3.3) and (5.3.4) as shown in the figure below.

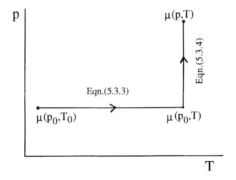

5.4   Maxwell Relations

The two laws of thermodynamics establish energy and entropy as functions of state, making them functions of many variables. As we have seen,

$U = U(S, V, N_k)$ and $S = S(U, V, N_k)$ are functions of the indicated variables. James Clerk Maxwell (1831–1879) used the rich theory of functions of many variables to obtain a large number of relations between thermodynamic variables. The methods he employed are general and the relations he obtained are called the Maxwell relations.

In Appendix 1.1 we have introduced the following result. If three variables x, y and z are such that each may be expressed as a function of the other two, $x = x(y, z)$, $y = y(x, z)$ and $z = z(x, y)$, then the theory of functions of many variables gives us the following fundamental relations:

$$\frac{\partial^2 x}{\partial y\, \partial z} = \frac{\partial^2 x}{\partial z\, \partial y} \tag{5.4.1}$$

$$\left(\frac{\partial x}{\partial y}\right)_z = \frac{1}{\left(\dfrac{\partial y}{\partial x}\right)_z} \tag{5.4.2}$$

$$\left(\frac{\partial x}{\partial y}\right)_z \left(\frac{\partial y}{\partial z}\right)_x \left(\frac{\partial z}{\partial x}\right)_y = -1 \tag{5.4.3}$$

Also, if we consider $z = z(x, y)$ and $w = w(x, y)$, two functions of x and y, the partial derivative $(\partial z / \partial x)_w$, in which the derivative is evaluated at constant w, is given by:

$$\left(\frac{\partial z}{\partial x}\right)_w = \left(\frac{\partial z}{\partial x}\right)_y + \left(\frac{\partial z}{\partial y}\right)_x \left(\frac{\partial y}{\partial x}\right)_w \tag{5.4.4}$$

We have already seen how (5.4.1) can be used to derive the Helmholtz equation (5.2.11) in which entropy S is considered as a function of T and V. In most cases, (5.4.1) to (5.4.4) are used to write thermodynamic derivatives in a form that can easily be related to experimentally measurable quantities. For example, using the fact that the Helmholtz energy $F(V, T)$ is a function of V and T, equation (5.4.1) can be used to derive the relation $(\partial S / \partial V)_T = (\partial p / \partial T)_V$ in which the derivative on the right-hand-side is clearly more easily related to the experiment.

Often it is also convenient to express thermodynamic derivatives in terms of quantities such as the **isothermal compressibility**

$$\kappa_T \equiv -\frac{1}{V}\left(\frac{\partial V}{\partial p}\right)_T \tag{5.4.5}$$

and **coefficient of volume expansion**

$$\alpha \equiv \frac{1}{V}\left(\frac{\partial V}{\partial T}\right)_p \tag{5.4.6}$$

For example, the **pressure coefficient** $(\partial p / \partial T)_V$ can be expressed in terms of κ_T and α. From (5.4.3) it follows that

$$\left(\frac{\partial p}{\partial T}\right)_V = \frac{-1}{\left(\dfrac{\partial V}{\partial p}\right)_T \left(\dfrac{\partial T}{\partial V}\right)_p}$$

Now using (5.4.2) and dividing the numerator and the denominator by V, we obtain

$$\left(\frac{\partial p}{\partial T}\right)_V = \frac{-\dfrac{1}{V}\left(\dfrac{\partial V}{\partial T}\right)_p}{\dfrac{1}{V}\left(\dfrac{\partial V}{\partial p}\right)_T} = \frac{\alpha}{\kappa_T} \tag{5.4.7}$$

GENERAL RELATION BETWEEN C_p AND C_V

As another example of the application of Maxwell's relations, we will derive a general relation between C_p and C_V in terms of α, κ_T, the molar volume V_m and T — all of which can be experimentally measured. We start with equation (2.3.5) derived in Chapter 2:

$$C_p - C_V = \left[p + \left(\frac{\partial U}{\partial V}\right)_T \right]\left(\frac{\partial V_m}{\partial T}\right)_p \tag{5.4.8}$$

where we have indicated explicitly molar volume with the subscript m because the heat capacities are molar heat capacities. The first step is to write the derivative $(\partial U / \partial V)_T$ in terms of the derivatives involving p, V and T, so that we can relate it to α and κ_T. From the Helmholtz equation (5.2.11) it is easy to see that $(\partial U / \partial V)_T + p = T(\partial p / \partial T)_V$. We can therefore write (5.4.8) as

$$C_p - C_V = T\left(\frac{\partial p}{\partial T}\right)_V \alpha V_m \tag{5.4.9}$$

in which we have used the definition (5.4.6) for α. We already obtained the Maxwell relation $(\partial p / \partial T)_V = \alpha / \kappa_T$ (equation 5.4.7); substituting in (5.4.9) we obtain the general relation

$$\boxed{C_p - C_V = \frac{T\alpha^2 V_m}{\kappa_T}} \tag{5.4.10}$$

5.5 Extensivity and Partial Molar Quantities

In multicomponent systems, thermodynamic functions such as volume V, Gibbs free energy G, and many other thermodynamic functions that can be expressed as functions of p, T and N_k are extensive functions of N_k. This extensivity gives us general thermodynamic relations, some of which we will discuss in this section. Consider the volume of a system as a function of p, T and $N_k : V = V(p, T, N_k)$. At constant p and T, if all the mole numbers were increased by a factor λ, the volume V would also increase by the same factor. This is the property of extensivity we have already discussed several times. In mathematical terms, we have

$$V(p, T, \lambda N_k) = \lambda V(p, T, N_k) \tag{5.5.1}$$

At constant p and T using Euler's theorem, as in Section 4.4, we can arrive at the relation

$$V = \sum_k \left(\frac{\partial V}{\partial N_k} \right)_{p,T} N_k \tag{5.5.2}$$

It is convenient to define **partial molar volumes** as the derivatives

$$V_{mk} \equiv \left(\frac{\partial V}{\partial N_k} \right)_{p,T} \tag{5.5.3}$$

Using this definition, equation (5.5.2) can be written as

$$\boxed{ V = \sum_k V_{mk} N_k } \tag{5.5.4}$$

Partial molar volumes are intensive quantities. As in the case of the Gibbs–Duhem relation, we can derive a relation between V_{mk} by noting that, at constant p and T,

$$dV = \sum_k \left(\frac{\partial V}{\partial N_k} \right)_{p,T} dN_k = \sum_k V_{mk} dN_k \tag{5.5.5}$$

Comparing dV obtained from (5.5.4) and (5.5.5), we see that $\sum_k N_k (dV_{mk})_{p,T} = 0$, in which we have explicitly noted that the change dV_{mk} is at constant p and T. From this relation it follows that[*]

$$\boxed{ \sum_k N_k \left(\frac{\partial V_{mk}}{\partial N_i} \right)_{p,T} = 0 } \quad \text{or} \quad \boxed{ \sum_k N_k \left(\frac{\partial V_{mi}}{\partial N_k} \right)_{p,T} = 0 } \tag{5.5.6}$$

---

[*] This is because $dV_{mk} = \sum (\partial V_{mk}/\partial Ni)dNi$ in which dNi are arbitrary.

where we have used the property $(\partial V_{mk}/\partial N_i) = (\partial^2 V/\partial N_i \partial N_k) = (\partial V_{mi}/\partial N_k)$. Relations similar to (5.5.4) and (5.5.6) can be obtained for all other functions that are extensive in N_k. For G, the equation corresponding to (5.5.4) is

$$G = \sum_k \left(\frac{\partial G}{\partial N_k}\right)_{p,T} N_k = \sum_k G_{mk} N_k = \sum_k \mu_k N_k \qquad (5.5.7)$$

in which we recognize the **partial molar Gibbs free energy** G_{mk} as the chemical potentials μ_k (5.1.14). The equation corresponding to (5.5.6) follows from the Gibbs–Duhem relation (5.2.4) when p and T are constant

$$\boxed{\sum_k N_k \left(\frac{\partial \mu_i}{\partial N_k}\right)_{p,T} = 0} \qquad (5.5.8)$$

For the Helmholtz free energy F and the enthalpy H we can obtain the following relations:

$$\boxed{F = \sum_k F_{mk} N_k} \qquad \boxed{\sum_k N_k \left(\frac{\partial F_{mi}}{\partial N_k}\right)_{p,T} = 0} \qquad (5.5.9)$$

$$\boxed{H = \sum_k H_{mk} N_k} \qquad \boxed{\sum_k N_k \left(\frac{\partial H_{mi}}{\partial N_k}\right)_{p,T} = 0} \qquad (5.5.10)$$

in which the **partial molar Helmholtz free energy** $F_{mk} = (\partial F/\partial N_k)_{p,T}$ and the **partial molar enthalpy** $H_{mk} = (\partial H/\partial N_k)_{p,T}$. Similar relations can be obtained for entropy S and the total internal energy U.

5.6   Surface Tension

In this section we shall consider some elementary thermodynamics relations involving interfaces [4]. Since molecules at an interface are in a different environment from molecules in the bulk, their energies and entropies are different. Molecules at a liquid–air interface, for example, have larger Helmholtz free energy than those in the bulk. At constant V and T, since every system minimizes its Helmholtz free energy, the interfacial area shrinks to its minimum possible value, thus increasing the pressure in the liquid (Fig. 5.4).

The thermodynamics of such a system can be formulated as follows. Consider a system with two parts, separated by an interface of area A (Fig. 5.4). For this system we have,

$$dU = TdS - p'' dV'' - p' dV' + \gamma dA \qquad (5.6.1)$$

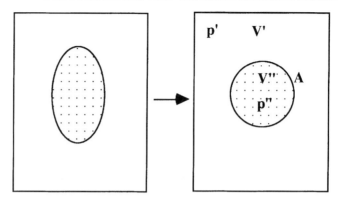

Figure 5.4 To minimize the interfacial Helmholtz free energy a liquid drop shrinks its surface area to the least possible value. As a result, the pressure p'' inside the drop is larger than the external pressure p'. The excess pressure $(p'' - p') = 2\gamma/r$.

in which p' and V' are the pressure and the volume of one phase and p'' and V'' are the pressure and the volume of the other; A is the interfacial area and the coefficient γ is called the surface tension. Since $dF = dU - TdS - SdT$,

$$dF = -SdT - p''dV'' - p'dV' + \gamma dA \qquad (5.6.2)$$

From this it follows that:

$$\left(\frac{\partial F}{\partial A}\right)_{T,V',V''} = \gamma \qquad (5.6.3)$$

Thus surface tension γ is the change of F per unit extension of the interfacial area at constant T, V' and V''. This energy is small, usually of the order of $10^{-2}\,\mathrm{J\,m^{-2}}$. Since enlarging an interfacial area increases its free energy, work needs to be done. As shown in Fig. 5.5, this means a force f is needed to stretch the surface by an amount dx, i.e., the liquid surface behaves like an elastic sheet.

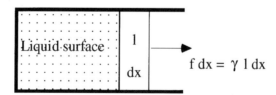

Figure 5.5 Energy is required to enlarge a surface of a liquid. The force per unit length is γ

The work done, $(f\,dx)$, equals the increase in the surface energy, $\gamma\,dA = (\gamma l\,dx)$, in which l is the width of the surface (Fig. 5.5). We see then that the force per unit length $(f/l) = \gamma$. For this reason, γ is called the "surface tension".

EXCESS PRESSURE IN A LIQUID DROP

In the case of a liquid drop in air shown in Fig. 5.4, the difference in the pressures $(p'' - p') = \Delta p$ is the excess pressure inside the liquid drop. An expression for the excess pressure Δp in a spherical liquid drop can be obtained as follows. As shown in Section 5.1, if the total volume of a system and its temperature are constant, then the irreversible approach to equilibrium is described by $-T\,d_i S = dF \leq 0$. Now consider an irreversible contraction of the volume V'' of the liquid drop to its equilibrium value when the total volume $V = V' + V''$ and T are constant. Then, since $dV' = -dV''$, we have

$$-T\frac{d_i S}{dt} = \frac{dF}{dt} = -(p'' - p')\frac{dV''}{dt} + \gamma\frac{dA}{dt} \tag{5.6.4}$$

For a spherical drop of radius r, $dV'' = (4\pi/3)3r^2\,dr$ and $dA = 4\pi\,2r\,dr$; hence, the above equation can be written as

$$-T\frac{d_i S}{dt} = \frac{dF}{dt} = \left(-(p'' - p')\,4\pi r^2 + \gamma\,8\pi r\right)\frac{dr}{dt} \tag{5.6.5}$$

We see that this expression is a product of a "thermodynamic force" $(-(p'' - p')4\pi r^2 + \gamma 8\pi r)/T$ that causes the "flow" dr/dt. At equilibrium both must vanish. Hence, $(-(p'' - p')4\pi r^2 + \gamma 8\pi r) = 0$. This gives us the well known **Laplace equation** for the excess pressure inside a liquid drop of radius r:

$$\boxed{\Delta p \equiv (p'' - p') = \frac{2\gamma}{r}} \tag{5.6.6}$$

CAPILLARY RISE

Another consequence of surface tension is the phenomenon of "capillary rise": in narrow tubes or capillaries, most liquids rise to a height h (Fig. 5.5) that depends on the radius of the capillary. The smaller the radius, the higher the rise, The liquid rises because increase in the area of liquid–glass interface lowers the free energy. The relation between the height h, the radius r and the surface tension can be derived as follows. As shown in Fig. 5.6(c), the force of surface tension of the liquid–air interface pulls the surface down while the force at the liquid–glass interface pulls the liquid up. Let the "contact angle", i.e., the angle at which the liquid is in contact with the wall of the capillary, be θ. When these

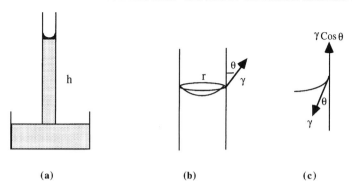

(a)   (b)   (c)

Figure 5.6 Capillary rise due to surface tension. (a) The height h to which the liquid rises depends on the contact angle θ, the surface tension γ and the radius r. (b) The contact angle θ specifies the direction in which the force due to the liquid–air interface acts. (c) The vertical component of the force due to the liquid–air interface balances the net force due to the liquid–glass and glass–air interfaces.

two forces balance each other along the vertical direction the force per unit length generated by the liquid–glass interface must be $\gamma \cos \theta$. As the liquid moves up, the liquid–glass interface is increasing while the glass–air interface is decreasing; $\gamma \cos \theta$ is the net force per unit length due to these two factors. Since the force per unit length is equal to the interfacial energy per unit area, as the liquid moves up, the decrease in the interfacial energy is $\gamma \cos \theta$ per unit area. Hence, the decrease in the free energy due to the glass–liquid interface is $(\gamma \cos \theta)$ per unit area. On the other hand, as the liquid rises in the capillary, there is an increase in the potential energy of the liquid due to gravity. A liquid layer of thickness dh, and density ρ, has the mass $(\pi r^2 dh\rho)$ and potential energy at a height h equal to $(\pi r^2 dh\rho)gh$. For the entire liquid column, this expression has to be integrated from 0 to h. The change in the free energy, ΔF, as the liquid rises is the sum of the potential energy and glass–liquid interfacial energy.

$$
\begin{aligned}
\Delta F(h) &= \int_0^h gh\rho\pi r^2 dh - 2\pi rh(\gamma \cos \theta) \\
&= \frac{\pi\rho gr^2h^2}{2} - 2\pi rh(\gamma \cos \theta)
\end{aligned}
\tag{5.6.7}
$$

The value of h that minimizes F is obtained by setting $\partial \Delta F(h)/\partial h = 0$ and solving for h. This leads to the expression:

$$
\boxed{h = \frac{2\gamma \cos \theta}{\rho gr}}
\tag{5.6.8}
$$

Table 5.2 Examples of surface tension and contact angles

	$\gamma/\mathrm{J\,m}^{-2}$ or N m^{-1}	Interface	Contact angle
Methanol	2.26×10^{-2}	Glass–water	$0°$
Benzene	2.89×10^{-2}	Glass–many organic liquids*	$0°$
Water	7.73×10^{-2}	Glass–kerosene	$26°$
Mercury	47.2×10^{-2}	Glass–mercury	$140°$
Soap solution	2.3×10^{-2} (approximately)	Paraffin–water	$107°$

More extensive data may be found in data source [F].
* Not all organic liquids have a contact angle value of $0°$, as is clear in the case of kerosene.

The same result can also be derived by balancing the forces of surface tension and the weight of the liquid column. As shown in Fig. 5.6(b), the liquid column of height h is held at the surface by the surface tension. The total force due to the surface tension of the liquid along the circumference is $2\pi r\gamma \cos\theta$. Since this force holds the weight of the liquid column we have:

$$2\pi r\gamma \cos\theta = \rho\, g h \pi r^2 \tag{5.6.9}$$

from which (5.6.8) follows.

The contact angle θ depends on the interface. For the glass–water interface the contact angle is nearly zero as it is for many, though not all, organic liquids. For the glass–kerosene interface, θ is $26°$. The contact angle can be greater than $90°$, as in the case of the mercury–glass interface for which θ is about $140°$ and for the paraffin–water interface for which it is about $107°$. When θ is greater than $90°$, the liquid surface in the capillary is lowered.

References

1. Feynman, R. P., Leighton R. B., and Sands M., *The Feynman Lectures on Physics (Vol. I, II and III)*. 1964, Reading, MA: Addison-Wesley.
2. Seul, M. and Andelman D., Domain shapes and patterns: the phenomenology of modulated phases. Science, **267** (1995). 476–483.
3. Callen, H. B., *Thermodynamics*, 2d ed. 1985, New York: John Wiley.
4. Defay, R., Prigogine, I., and Bellemans, A., *Surface Tension and Adsorption*. 1966, New York: John Wiley.

Data Sources

[A] NBS table of chemical and thermodynamic properties. *J. Phys. Chem. Reference Data*, **11**, Supl. 2 (1982).

[B] G. W. C. Kaye and T. H. Laby (eds.), *Tables of Physical and Chemical Constants*. 1986, London: Longman.

[C] I. Prigogine and R. Defay, *Chemical thermodynamics*. 1967, London: Longman.

[D] J. Emsley, *The Elements*, 1989, Oxford: Oxford University Press.

[E] L. Pauling, *The Nature of the Chemical Bond*. 1960, Ithaca NY: Cornell University Press.

[F] D. R. Lide (ed.) *CRC Handbook of Chemistry and Physics*, 75th ed. 1994, Ann Arbor MI: CRC Press.

Examples

Example 5.1   Show that the change in the value of the Helmholtz free energy F corresponds to the work done when T and N_k are constant, thus justifying the name "free energy" (available for doing work).

Solution: As a specific example, consider an ideal gas in contact with a thermal reservoir at temperature T. Work can be done by expanding this gas. Starting with $F = U - TS$, we can show that the change in F corresponds to the work done at constant T and N_k. It follows from (5.1.5) that:

$$dF = -p\,dV - S\,dT + \sum_k \mu_k\,dN_k$$

At constant T and N_k, $dF = -p\,dV$. Integrating both sides, we see that

$$\int_{F_1}^{F_2} dF = F_2 - F_1 = \int_{V_1}^{V_2} -p\,dV$$

which shows that the change in F is equal to the work done by the gas. The same will be true for any other system.

Example 5.2   For a closed system with one chemical reaction, show that $(\partial F/\partial \xi)_{T,V} = -A$.

Solution: The change in F (5.1.5) is given by

$$dF = -p\,dV - S\,dT + \sum_k \mu_k\,dN_k$$

Since the system is closed, the changes in N_k are due to chemical reaction, so we have $dN_k = \nu_k\,d\xi$ in which no ν_k are stoichiometric coefficients (negative for the reactants and positive for the products). Thus

$$dF = -p\,dV - S\,dT + \sum_k \nu_k \mu_k \, d\xi$$

Since $\sum_k \nu_k \mu_k = -A$, we have

$$dF = -p\,dV - S\,dT - A\,d\xi$$

When F is considered as a function of V, T and ξ, then

$$dF = (\partial F/\partial V)_{T,\xi}\,dV + (\partial F/\partial T)_{V,\xi}\,dT + (\partial F/\partial \xi)_{T,V}\,d\xi$$

so we see that $(\partial F/\partial \xi)_{T,V} = -A$.

Example 5.3 Using the Gibbs–Duhem relation, show that at constant p and T, $(dG_m)_{p,T} = \sum \mu_k \, dx_k$.
Solution: The molar Gibbs free energy $G_m = \sum x_k \mu_k$, in which x_k is the mole fraction, hence:

$$dG_m = \sum_k dx_k \mu_k + \sum_k x_k \, d\mu_k$$

The Gibbs–Duhem relation is

$$S\,dT - V\,dp + \sum_k N_k \, d\mu_k = 0$$

Since p and T are constant, $dT = dp = 0$. Furthermore, $x_k = N_k/N$, in which N is the total number of moles. Dividing the Gibbs–Duhem equation by N and setting $dp = dT = 0$, we have $\sum_k x_k \, d\mu_k = 0$. Using this result in the expression for dG_m, for constant p and T, we see that

$$(dG_m)_{p,T} = \sum_k \mu_k \, dx_k.$$

Exercises

5.1 Use the expression $T\,d_i S = -\gamma dA$ and $T\,d_e S = dU + p\,dV$ in the general expressions for the First and Second Laws, hence obtain $dU = T\,dS - pdV + \gamma dA$ (assuming $dN_k = 0$).

5.2 In an isothermal expansion of a gas from volume V_i to V_f, what is the change in the Helmholtz free energy F?

5.3 Use the relations $dU = dQ - p\,dV$, $T\,d_eS = dQ$ and $T\,d_iS = -\sum_k \mu_k\,dN_k$ to derive

$$dG = V\,dp - S\,dT + \sum_k \mu_k\,dN_k$$

which is equation (5.1.13).

5.4 Use the relations $dU = dQ - pdV$, $Td_eS = dQ$ and $T\,d_iS = -\sum_k \mu_k\,dN_k$ to derive

$$dH = T\,dS + V\,dp + \sum_k \mu_k\,dN_k$$

which is equation (5.1.21).

5.5 Obtain the Helmholtz equation (5.2.11) from (5.2.10).

5.6 (a) Use the Helmholtz equation (5.2.11) to show that, at constant T, the energy of an ideal gas is independent of volume.
(b) Use the Helmholtz equation (5.2.11) to calculate $(\partial N/\partial V)_T$ for N moles of a gas using the van der Waals equation.

5.7 Obtain (5.2.13) from (5.2.12).

5.8 Assume that ΔH changes little with temperature, integrate the Gibbs–Helmholtz equation (5.2.14) and express ΔG_f at temperature T_f in terms of $\Delta H, \Delta G_i$ and the corresponding temperature T_i.

5.9 Obtain an explicit expression for the Helmholz free energy of an ideal gas as a function of T, V and N.

5.10 The variation of Gibbs free energy of a substance with temperature is given by $G = aT + b + c/T$. Determine how the entropy and enthalpy of this substance vary with temperature.

5.11 Show that (5.4.10) reduces to $C_p - C_V = R$ for an ideal gas.

5.12 (a) By minimizing the free energy $\Delta F(h)$ given by (5.6.1) as a function of h, obtain the expression

$$h = \frac{2\gamma \cos \theta}{\rho g r}$$

for the height of capillary rise due to surface tension.

(b) Assume that the contact angle θ between water and glass is nearly zero and calculate the height of water in a capillary of diameter 0.1 mm.

5.13 (a) Due to surface tension, the pressure inside a bubble is larger than the pressure outside. Let this excess pressure be Δp. By equating the work done $\Delta p dV$ due to an infinitesimal increase dr in the radius r to the increase in surface energy γdA show that $\Delta p = 2\gamma/r$.

(b) Calculate the excess pressure inside water bubbles of radius 1.0 mm and 1.0 μm.

.

6 BASIC THERMODYNAMICS OF GASES, LIQUIDS AND SOLIDS

Introduction

The formalism and general thermodynamic relations that we have seen in the previous chapters have a wide applicability. In this chapter we will see how thermodynamic quantities can be calculated for gases, liquids and solids. We will also study some basic features of equilibrium between different phases.

6.1 Thermodynamics of Ideal Gases

Many thermodynamic quantities, such as total internal energy, entropy, chemical potential, etc., for an ideal gas have been derived in the preceeding chapters as examples. In this section we will bring all these results together and list the thermodynamic properties of gases in the ideal gas approximation. In the following section we will see how these quantities can be calculated for "real gases," for which we take into account molecular size and molecular forces.

THE EQUATION OF STATE

Our starting point is the equation of state, *the ideal gas law*

$$\boxed{pV = NRT} \tag{6.1.1}$$

As we saw in Chapter 1, for most gases this approximation is valid for densities less than about 1 mol/L. At these densities and a temperature of about 300 K, the pressure of $N_2(g)$ obtained using the ideal gas equation is 24.76 atm whereas the value predicted using the more accurate van der Waals equation is 24.36 atm, a difference of only a few percent.

TOTAL INTERNAL ENERGY

Through thermodynamics, we can see that the *ideal gas law* (6.1.1) *implies the total internal energy U is independent of the volume at fixed T*, i.e. the energy of an ideal gas depends only on the temperature. One arrives at this conclusion using the Helmholtz equation (5.2.11):

$$\left(\frac{\partial U}{\partial V}\right)_T = T^2 \left(\frac{\partial}{\partial T}\left(\frac{p}{T}\right)\right)_V \qquad (6.1.2)$$

(We remind the reader that the Helmholtz equation is a consequence of the fact that entropy is a state function of V, T and N_k.) Since the ideal gas equation implies that the term $p/T = (NR/V)$ is independent of T, it immediately follows from (6.1.2) that $(\partial U/\partial V)_T = 0$. Thus, the total internal energy $U(T, V, N)$ of an ideal gas is independent of the volume. We can get a more explicit expression for U. Since the molar heat capacity $C_V = (\partial U_m/\partial T)_V$ is independent of T, we can write

$$U_{\text{ideal}} = NU_0 + N\int_0^T C_V\, dT = NU_0 + NC_V T$$
$$= N(U_0 + C_V T) \qquad (6.1.3)$$

(The constant U_0 is not defined in classical thermodynamics, but using the definition of energy that the theory of relativity gives us, we may set $NU_0 = MNc^2$, in which M is the molar mass, N is the number of moles and c is the velocity of light. In thermodynamic calculations of changes of energy, U_0 does not explicitly appear.)

HEAT CAPACITIES AND ADIABATIC PROCESSES

We have seen earlier that there are two molar heat capacities, C_V and C_p, at constant volume and constant pressure respectively. We have also seen in Chapter 2 that the First Law gives us the following relation between molar heat capacities:

$$\boxed{C_p - C_V = R} \qquad (6.1.4)$$

For an *adiabatic process*, the First Law also gives us the relation

$$\boxed{TV^{(\gamma-1)} = \text{constant}} \quad \text{or} \quad \boxed{pV^\gamma = \text{constant}} \qquad (6.1.5)$$

in which $\gamma = C_p/C_V$. In an adiabatic process, by definition, $d_eS = dQ/T = 0$. If the process occurs such that $d_iS \approx 0$, the entropy of the system remains constant because $dS = d_iS + d_eS$.

ENTROPY, ENTHALPY AND FREE ENERGIES

We have already seen that the entropy $S(V, T, N)$ of an ideal gas (3.7.4) is

$$\boxed{S = N[s_0 + C_V \ln T + R \ln (V/N)]} \qquad (6.1.6)$$

From the equation of state (6.1.1) and the expressions for U_{ideal} and S it is straightforward to obtain explicit expressions for the enthalpy $H = U + pV$, the Helmholtz free energy $F = U - TS$ and the Gibbs free energy $G = U - TS + pV$ of an ideal gas (exc 6.1).

CHEMICAL POTENTIAL

For the chemical potential of an ideal gas we obtained the following expression (5.3.6) Section 5.3

$$\begin{aligned} \mu(p, T) &= \mu(p_0, T) + RT \ln (p/p_0) \\ &= \mu_0 + RT \ln p \end{aligned}$$

(6.1.7)

in which μ_0 is the chemical potential at unit pressure. For a mixture of gases the total energy is the sum of the energies of each component. The same is true for the entropy. The chemical potential of a component k can be expressed in terms of the partial pressures p_k as

$$\begin{aligned} \mu_k(p, T) &= \mu(p_0, T) + RT \ln (p_k/p_0) \\ &= \mu_{0k}(T) + RT \ln p_k \end{aligned}$$

(6.1.8)

Alternatively, if x_k is the mole fraction of the component k, since $p_k = x_k p$, the chemical potential can be written as

$$\boxed{\mu_k(x_k, p, T) = \mu_k(p, T) + RT \ln x_k}$$

(6.1.9)

in which $\mu_k(p, T) = \mu_{0k}(T) + RT \ln p$ is the chemical potential of a pure gas. This form of the chemical potential is generally used for muti-component systems that are considered "ideal mixtures".

ENTROPY OF MIXING AND THE GIBBS PARADOX

Using the expression for the entropy of an ideal gas, we can calculate the increase in entropy when two gases mix. Consider two nonidentical gases in chambers of volume V separated by a wall (Fig. 6.1). Let us assume that the two chambers contain the same number of moles, N, of the two gases. The total initial entropy of the system is the sum of the entropies of the two gases:

$$S_{\text{int}} = N[s_{01} + C_{V1} \ln T + R \ln (V/N)] + N[s_{02} + C_{V2} \ln T + R \ln (V/N)]$$

(6.1.10)

Now if the wall separating the two chambers is removed, the two gases will

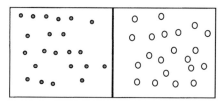

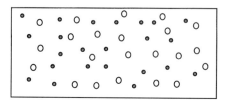

Figure 6.1 The entropy of mixing of two nonidentical gases, however small the difference between the gases, is given by (6.1.12). If the two gases are identical, there is no change in the entropy

irreversibly mix and the entropy will increase. When the two gases have completely mixed and the system has reached a new equilibrium, each gas will be occupying a volume of $2V$. Hence the total final entropy after the mixing is

$$S_{\text{fin}} = N[s_{01} + C_{V1} \ln T + R \ln (2V/N)] + N[s_{02} + C_{V2} \ln T + R \ln (2V/N)]$$

$$(6.1.11)$$

The difference between (6.1.10) and (6.1.11) is the entropy of mixing $\Delta S_{\text{mix}} = S_{\text{fin}} - S_{\text{int}}$. It is easy to see that

$$\Delta S_{\text{mix}} = 2NR \ln 2 \qquad (6.1.12)$$

The generalization of this result to unequal volumes and mole numbers is left as an exercise. It can be shown that, if initially the densities of the two gases are the same, i.e. $(N_1/V_1) = (N_2/V_2)$, then the entropy of mixing (exc 6.2) can be written as

$$\Delta S_{\text{mix}} = -RN[x_1 \ln x_1 + x_2 \ln x_2] \qquad (6.1.13)$$

where x_1 and x_2 are the mole fractions and $N = N_1 + N_2$.

Gibbs noted a curious aspect of this result. If the two gases are identical, then the state of the gas before and after the removal of the wall is indistinguishable.

Hence there is no change in entropy because the initial and final states are the same. But for two nonidentical gases, however small the difference, the change of entropy is given by (6.1.12). Generally, in most physical systems, a small change in one quantity results in a small change in another dependent quantity. Not so with the entropy of mixing; even the smallest difference between two gases leads to an entropy difference of $2NR \ln 2$. If the difference vanishes, S_{mix} abruptly drops to zero. This discontinuous behavior of the entropy of mixing is often called **the Gibbs paradox**.

The entropy of mixing (6.1.13) can also be obtained using the statistical formula $S = k_B \ln W$ introduced in Chapter 3 (Box 3.1). Consider a gas containing $(N_1 + N_2)$ moles or $(N_1 + N_2)N_A$ molecules. For this gas, interchange of molecules does not correspond to distinct microstates because the molecules are indistinguishable. However, if N_2 moles of the gas are replaced by another gas then an interchange of molecules of the two different gases corresponds to a distinct microstate. Thus the gas mixture with N_1 moles of one gas and N_2 of another has additional microstates in comparison with $(N_1 + N_2)$ moles of one gas. That these additional microstates when used in the formula $S = k_B \ln W$ give the entropy of mixing (6.1.13) can be seen as follows. The number of additional microstates in the mixture is:

$$W_{mix} = \frac{(N_A N_1 + N_A N_2)!}{(N_A N_1)!(N_A N_2)!} \tag{6.1.14}$$

in which we have introduced the Avogadro number N_A to convert mole numbers to number of molecules. Using the Stirling approximation $\ln N! \approx N \ln N - N$, it can easily be shown that (exc 6.2):

$$\Delta S_{mix} = k_B \ln W_{mix} = -k_B N_A (N_1 + N_2)[x_1 \ln x_1 + x_2 \ln x_2] \tag{6.1.15}$$

in which x_1 and x_2 are mole fractions. (6.1.15) is identical to (6.1.13) because $R = k_B N_A$ and $N = N_1 + N_2$. This derivation shows that the expression (6.1.13) for the entropy of mixing is entirely a consequence of distinguishability of the two components of the system.

6.2   Thermodynamics of Real Gases

Useful though it may be, the ideal gas approximation ignores the finite size of the molecules and the intermolecular forces. Consequently, as the gas becomes denser, the ideal gas equation does not predict the relation between the volume, pressure and temperature with good accuracy; one has to use other equations of state that provide a better description. If the molecular size and forces are included in the theory, one refers to it as a theory of a "real gas."

As a result of molecular forces, the total internal energy U, the relation between the molar heat capacities C_p and C_V, the equation for adiabatic processes and other themodynamic quantities will differ from those for the ideal gas. In this section we shall see how the thermodynamic quantities of a real gas can be obtained from an equation of state that takes molecular size and forces into account.

The van der Waals equation, which takes into account the intermolecular forces and molecular size, and the critical constants p_c, V_{mc} and T_c were introduced in Chapter 1:

$$\boxed{\left(p + \frac{a}{V_m^2}\right)(V_m - b) = RT}$$ (6.2.1a)

$$p_c = \frac{a}{27b^2} \qquad V_{mc} = 3b \qquad T_c = \frac{8a}{27bR}$$ (6.2.1b)

in which V_m is the molar volume. This is not the only equation that has been proposed for gases. Some of the other equations that have been proposed and the corresponding critical constants are:

The Berthelot equation:

$$\boxed{p = \frac{RT}{V_m - b} - \frac{a}{TV_m^2}}$$ (6.2.2a)

$$p_c = \frac{1}{12}\left(\frac{2aR}{3b^3}\right)^{1/2} \qquad V_{mc} = 3b \qquad T_c = \frac{2}{3}\left(\frac{2a}{3bR}\right)^{1/2}$$ (6.2.2b)

The Dieterici equation:

$$\boxed{p = \frac{RT\,e^{-a/RTV_m}}{V_m - b}}$$ (6.2.3a)

$$p_c = \frac{a}{4e^2b^2} \qquad V_{mc} = 2b \qquad T_c = \frac{a}{4Rb}$$ (6.2.3b)

in which a and b are constants similar to the van der Waals constants which can be related to the critical constants as shown. Another equation that is often used is the **virial expansion**, proposed by Kammerlingh Onnes. It expresses the pressure as a power series in the mole number density (N/V):

$$\boxed{p = RT\frac{N}{V}\left[1 + B(T)\left(\frac{N}{V}\right) + C(T)\left(\frac{N}{V}\right)^2 + \dots\right]}$$ (6.2.4)

in which $B(T)$ and $C(T)$ are functions of temperature called the **virial coefficients** which are experimentally measured and tabulated. As expected, equation (6.2.4) reduces to the ideal gas equation at low densities. The van der Waals constants, a and b, can be related (exc 6.4) to the virial coefficients $B(T)$ and $C(T)$. Thus the virial coefficients can be calculated from the van der Waals constants. An extensive listing of $B(T)$ and $C(T)$ can be found in data source [F]. Since the ideal gas equation is valid at low pressures, the virial equation may also be written as

$$p = RT\frac{N}{V}\left[1 + B'(T)p + C'(T)p^2 + \ldots\right] \tag{6.2.5}$$

Comparing (6.2.4) and (6.2.5) reveals that $B = B'RT$ to leading approximation.

TOTAL INTERNAL ENERGY

For real gases, due to the molecular interaction, the energy is no longer only a function of the temperature. Because the interaction energy of the molecules depends on the distance between the molecules, a change in volume (at a fixed temperature) causes a change in energy, i.e. the term $(\partial U/\partial V)_T$ does not vanish for a real gas. Molecular forces have a short range. At low densities, since the molecules are far apart, the force of interaction is small. As the density approaches zero, the energy of a real gas, U_{real}, approaches the energy of an ideal gas, U_{ideal}. We can obtain an explicit expression for U_{real} through the Helmholtz equation $(\partial U/\partial V)_T = T^2[\partial(p/T)/\partial T]_V$, which is valid for all systems (not only for gases). Upon integration, this equation yields

$$U_{\text{real}}(T, V, N) = U_{\text{real}}(T, V_0, N) + \int_{V_0}^{V} T^2\left(\frac{\partial}{\partial T}\frac{p}{T}\right)_V dV \tag{6.2.6}$$

To write this in a convenient form, first we note that, for a fixed N, as the volume $V_0 \to \infty$, the density approaches zero, and U_{real} approaches the energy of an ideal gas, U_{ideal}, given by (6.1.3). Hence equation (6.2.6) can be written as

$$U_{\text{real}}(T, V, N) = U_{\text{ideal}}(T, N) + \int_{\infty}^{V} T^2\left(\frac{\partial}{\partial T}\frac{p}{T}\right)_V dV \tag{6.2.7}$$

If $[\partial(p/T)/\partial T]_V$ can be calculated using the equation of state, explicit expressions for U_{real} may be derived. For example, using the van der Waals equation, we can obtain the energy of a real gas. From (6.2.1) it is easy to see that $p/T = NR/(V - Nb) - a(N/V)^2(1/T)$. Substituting this expression into

(6.2.7) we get

$$U_{\text{VW}}(T, V, N) = U_{\text{ideal}}(T, N) + \int_{\infty}^{V} a\left(\frac{N}{V}\right)^2 dV$$

where U_{VW} is the energy of a van der Waals gas. Evaluation of the integral gives

$$U_{\text{VW}}(V, T, N) = U_{\text{ideal}} - a\left(\frac{N}{V}\right)^2 V \tag{6.2.8}$$

Writing the energy in this form shows us that the energy due to molecular interactions is equal to $-a(N/V)^2$ per unit volume. As the volume increases U_{VW} approaches U_{ideal}.

HEAT CAPACITIES C_V AND C_p

If the molar internal energy U_m of a gas is known, then the heat capacity at constant volume $C_V = (\partial U_m/\partial T)_V$ can be calculated. For a real gas we can use (6.2.7) to obtain the following expression for C_V:

$$C_{V,\text{real}} = \left(\frac{\partial U_{m,\text{real}}}{\partial T}\right)_V = \left(\frac{\partial U_{m,\text{ideal}}}{\partial T}\right)_V + \frac{\partial}{\partial T} \int_{\infty}^{V} T^2 \left(\frac{\partial}{\partial T}\frac{p}{T}\right)_V dV$$

which upon explicit evaluation of the derivatives in the integral gives

$$C_{V,\text{real}} = C_{V,\text{ideal}} + \int_{\infty}^{V} T\left(\frac{\partial^2 p}{\partial T^2}\right)_V dV \tag{6.2.9}$$

Given an equation of state, such as the van der Waals equation, the above integral can be evaluated to obtain an explicit expression for C_V. Equation (6.2.9) shows that, *for any equation of state in which p is a linear function of T,* $C_{V,\text{real}} = C_{V,\text{ideal}}$. This is true for the van der Waals equation. The energy due to the molecular interactions depends on the intermolecular distance or density (N/V). Because this does not change at constant V, the value of C_V is unaffected by the molecular forces; C_V is the change in kinetic energy of the molecules per unit change in temperature.

Also, given the equation of state, the isothermal compressibility κ_T, and the coefficient of volume expansion α (defined by equations 5.4.5 and 5.4.6 respectively) can be calculated. Then, using the general relation

$$\boxed{C_p - C_V = \frac{TV_m\alpha^2}{\kappa_T}} \tag{6.2.10}$$

(see equation 5.4.10) it is also possible to obtain C_p. Thus using (6.2.9) and (6.2.10) the two molar heat capacities of a real gas can be calculated from the equation of state.

ADIABATIC PROCESSES

For an ideal gas, we have seen in Chapter 2 that in an adiabatic process $TV^{\gamma-1} = $ constant or $pV^{\gamma} = $ constant (equations 2.3.11 and 2.3.12) in which $\gamma = C_p/C_V$. One can obtain a similar equation for a real gas. An adiabatic process is defined by $dQ = 0 = dU + p\,dV$. By considering U as a function of V and T, this equation can be written as

$$\left(\frac{\partial U}{\partial V}\right)_T dV + \left(\frac{\partial U}{\partial T}\right)_V dT + p\,dV = 0 \tag{6.2.11}$$

Since $(\partial U/\partial T)_V = NC_V$, where N is the number of moles of the gas, this equation becomes

$$\left[\left(\frac{\partial U}{\partial V}\right)_T + p\right]dV = -NC_V\,dT \tag{6.2.12}$$

By evaluating the derivative on the right-hand side of the Helmholtz equation (5.2.11), it is easy to see that $[(\partial U/\partial V)_T + p] = T(\partial p/\partial T)_V$ (exc 6.5). Furthermore, we have also seen in Chapter 5 (5.4.7) that $(\partial p/\partial T)_V = \alpha/\kappa_T$. Using these two relations, equation (6.2.12) can be written as

$$\frac{T\alpha}{\kappa_T}dV = -NC_V\,dT \tag{6.2.13}$$

To write this expression in terms of the ratio $\gamma = C_p/C_V$, we use the following general relation also obtained in Chapter 5 (5.4.10):

$$C_p - C_V = \frac{T\alpha^2 V_{\mathrm{m}}}{\kappa_T} \tag{6.2.14}$$

in which V_m is the molar volume. From (6.2.13) and (6.2.14), we obtain

$$N\frac{C_p - C_V}{V\alpha}dV = -NC_V\,dT \tag{6.2.15}$$

where we have made the substitution $V_m = V/N$ for the molar volume. Dividing both sides of this expression by C_V and using the definition $\gamma = C_p/C_V$, we obtain the simple expression

$$\frac{\gamma - 1}{V} dV = -\alpha \, dT \tag{6.2.16}$$

Generally γ varies little with volume or pressure, so it may be treated as a constant and equation (6.2.16) can be integrated to obtain

$$(\gamma - 1)\ln V = -\int \alpha(T)dT + C \tag{6.2.17}$$

in which we have written α as an explicit function of T, C is the integration constant. An alternative way of writing this expression is

$$\boxed{V^{(\gamma-1)}e^{\int \alpha(T)dT} = \text{constant}} \tag{6.2.18}$$

This relation is valid for all gases. For an ideal gas $\alpha = (1/V)(\partial V/\partial T)_p = 1/T$. When this is substituted into (6.2.16), we obtain the familiar equation $TV^{\gamma-1} = $ constant. If p is a linear function of T, as is with the van der Waals equation, since $C_{V,\text{real}} = C_{V,\text{ideal}}$ (6.2.9), from (6.2.14) it follows that

$$\gamma - 1 = \frac{T\alpha^2 V_{\text{m}}}{C_{V,\text{ideal}}\kappa_T} \tag{6.2.19}$$

If the equation for state of a real gas is known, then α and γ can be evaluated (numerically, if not analytically) as a function of T, and the relation (6.2.18) between V and T can be made explicit for an adiabatic process.

HELMHOLTZ AND GIBBS FREE ENERGIES

The method used to obtain a relation (6.2.7) between U_{ideal} and U_{real} can also be used to relate the corresponding Helmholtz and Gibbs free energies. The main idea is that as $p \to 0$ or $V \to \infty$, the thermodynamic quantities for a real gas approach those of an ideal gas. Let us consider the Helmholtz free energy F. Since $(\partial F/\partial V)_T = -p$ (5.1.6) we have the general expression

$$F(T, V, N) = F(T, V_0, N) - \int_{V_0}^{V} p \, dV \tag{6.2.20}$$

The difference between real and ideal gas values of F at any T, V and N can be obtained as follows. Writing this equation for a real gas and an ideal gas then subtracting one from the other, it is easy to see that

$$F_{real}(T, V, N) - F_{ideal}(T, V, N) = F_{real}(T, V_0, N) - F_{ideal}(T, V_0, N)$$
$$- \int_{V_0}^{V} (p_{real} - p_{ideal}) dV \qquad (6.2.21)$$

Now since $\lim_{V_0 \to \infty} [F_{real}(V_0, T, N) - F_{ideal}(V_0, T, N)] = 0$, we can write the above expression as

$$\boxed{F_{real}(T, V, N_k) - F_{ideal}(T, V, N_k) = -\int_{\infty}^{V} (p_{real} - p_{ideal}) dV} \qquad (6.2.22)$$

where we have explicitly indicated the fact that this expression is valid for a multicomponent system by replacing N by N_k. Similarly we can also show that

$$\boxed{G_{real}(T, p, N_k) - G_{ideal}(T, p, N_k) = \int_{0}^{p} (V_{real} - V_{ideal}) dp} \qquad (6.2.23)$$

As an example, let us calculate F using the van der Waals equation. For the van der Waals equation, we have $p_{real} = p_{VW} = NRT/(V - bN) - aN^2/V^2$. Substituting this expression for p_{real} into (6.2.22) and performing the integration, one can obtain (exc. 6.10)*

$$F_{VW}(T, V, N) = F_{ideal}(T, V, N) - a\left(\frac{N}{V}\right)^2 V - NRT \ln\left(\frac{V - Nb}{V}\right) \qquad (6.2.24)$$

where

$$\begin{aligned} F_{ideal} &= U_{ideal} - TS_{ideal} \\ &= U_{ideal} - TN[s_0 + C_V \ln T + R \ln(V/N)] \end{aligned} \qquad (6.2.25)$$

Substituting (6.2.25) into (6.2.24) and simplifying (exc 6.10), we obtain

$$\begin{aligned} U_{VW} &= U_{ideal} - a(N/V)^2 V - TN\{s_0 + C_V \ln T + R \ln[(V - Nb)/N]\} \\ &= U_{VW} - TN\{s_0 + C_V \ln T + R \ln[(V - Nb)/N]\} \end{aligned} \qquad (6.2.26)$$

where we have used the expression $U_{VW}(V, T, N) = U_{ideal} - a(N/V)^2 V$ for the energy of a van der Waals gas (6.2.8). Similarly, the Gibbs free energy of a real gas can be calculated using the van der Waals equation.

---

* The limit $V \to \infty$ can be applied when the integral is written in this form.

ENTROPY

The entropy of a real gas can be obtained using expressions (6.2.7) and (6.2.21) for U_{real} and F_{real} because $F_{\text{real}} = U_{\text{real}} - TS_{\text{real}}$. Using the van der Waals equation, the entropy S_{VW} of a real gas can be identified in (6.2.26) as

$$S_{\text{VW}}(T, V, N) = N\{s_0 + C_V \ln T + R \ln[(V - Nb)/N]\} \qquad (6.2.27)$$

A comparison of (6.2.27) with the entropy of an ideal gas (6.1.6) shows that the only difference is that the volume term in the entropy of van der Waals gas is $(V - Nb)$ instead of V.

CHEMICAL POTENTIAL

The chemical potential for a real gas can be derived from the expression (6.2.23) for the Gibbs free energy. Since the chemical potential of the component k is $\mu_k = (\partial G/\partial N_k)_{p,T}$, by differentiating (6.2.23) with respect to N_k, we obtain

$$\mu_{k,\text{real}}(T, p) - \mu_{k,\text{ideal}}(T, p) = \int_0^p (V_{\text{m}k,\text{real}} - V_{\text{m}k,\text{ideal}})dp \qquad (6.2.28)$$

in which $V_{\text{m}k} \equiv (\partial V/\partial N_k)_{p,T}$ is the partial molar volume of the component k. For simplicity, let us consider a single gas. To compare the molar volume of the ideal gas $V_{\text{m,ideal}} = RT/p$ to that of a real gas $V_{\text{m,real}}$, a **compressibility factor** Z is defined:

$$\boxed{V_{\text{m,real}} = ZRT/p} \qquad (6.2.29)$$

For an ideal gas $Z = 1$; a deviation of the value of Z from 1 indicates nonideality. In terms of Z the chemical potential can be written as

$$\begin{aligned}
\mu_{\text{real}}(T, p) &= \mu_{\text{ideal}}(T, p) + RT \int_0^p \left(\frac{Z-1}{p}\right) dp \\
&= \mu_{\text{ideal}}(p_0, T) + RT \ln\left(\frac{p}{p_0}\right) + RT \int_0^p \left(\frac{Z-1}{p}\right) dp
\end{aligned} \qquad (6.2.30)$$

in which we have used the expression $\mu(p, T) = \mu_0 + RT \ln p$ for the chemical potential of an ideal gas (6.1.7). The chemical potential is also expressed in terms of **fugacity** f, which was introduced by G.N. Lewis. To keep the form of the chemical potential for a real gas similar to the potential for an ideal gas, G.N. Lewis introduced the fugacity through the definition

$$\mu_{\text{real}}(p, T) = \mu_{\text{ideal}}(p, T) + RT \ln\left(\frac{f}{p}\right) \qquad (6.2.31)$$

G. N. Lewis (Reproduced, by permission, from the Emilio Segré Visual Archives of the American Institute of Physics)

in which f is the fugacity. Indeed we must have $\lim_{p \to 0}(f/p) = 1$ to recover the expression for the ideal gas at a very low pressure. Comparing (6.2.30) and (6.2.31) we see that

$$\ln\left(\frac{f}{p}\right) = \int_0^p \left(\frac{Z-1}{p}\right) dp \qquad (6.2.32)$$

It is possible to obtain Z explicitly for various equations such as the van der Waals equation or the virial equation (6.2.5). For example, if we use the virial equation, it gives

$$Z = \frac{pV_m}{RT} = \left[1 + B'(T)p + C'(T)p^2 + ...\right] \qquad (6.2.33)$$

Substituting this expression into (6.2.32), to the second order in p, we find that

$$\ln\left(\frac{f}{p}\right) = B'(T)p + \frac{C'(T)p^2}{2} + ... \qquad (6.2.34)$$

Generally, terms of the order p^2 are small and may be ignored. Then (6.2.34) can be used to obtain the chemical potential of a real gas, μ_{real}, by using it in (6.2.31):

$$\mu_{\text{real}}(p, T) = \mu_{\text{ideal}}(p, T) + RT \ln\left(\frac{f}{p}\right)$$

$$= \mu_{\text{ideal}}(p, T) + RT B'(T)p + \dots \tag{6.2.35}$$

This expression can also be written in terms of the virial coefficient B of equation (6.2.4) by noting the relation $B = B'RT$ to leading approximation. Thus

$$\mu_{\text{real}} = \mu_{\text{ideal}}(p, T) + Bp + \dots \tag{6.2.36}$$

Equations (6.2.34) and (6.2.35) give us the chemical potential of a real gas in terms of its virial coefficients. A similar computation can be performed using the van der Waals equation.

We can also obtain explicit expressions for μ using $(\partial F/\partial N)_{T,V} = \mu(V, T)$. Using the van der Waals equation, we can write the chemical potential as a function of the number density $n = (N/V)$ and temperature T (exc 6.9):

$$\mu(n, T) = (U_0 - 2an) + \left(\frac{C_V}{R} + \frac{1}{1 - nb}\right)RT - T\left[s_0 + C_V \ln T - R\ln\left(\frac{n}{1 - bn}\right)\right] \tag{6.2.37}$$

CHEMICAL AFFINITIES

Finally, to understand the nature of chemical equilibrium of real gases, it is useful to obtain affinities for chemically reacting real gases. The affinity of a reaction* $A = -\sum v_k \mu_k$ for a real gas can be written using expression (6.2.28) for the chemical potential:

$$A_{\text{real}} = A_{\text{ideal}} - \sum_k v_k \int_0^p (V_{k,\text{m,real}} - V_{k,\text{m,ideal}})dp \tag{6.2.38}$$

This expression can be used to calculate the equilibrium constants for reacting real gases. The partial molar volume $V_{k,\text{m,ideal}}$ for all gases is RT/p. Hence the above expression becomes

$$A_{\text{real}} = A_{\text{ideal}} - \sum_k v_k \int_0^p \left(V_{k,\text{m,real}} - \frac{RT}{p}\right)dp \tag{6.2.39}$$

---

* Here v_k are stoichiometric coefficients, negative for reactants and positive for products.

With the above quantities, all the thermodynamics of real gases can be described, once the real gas parameters, such as the van der Waals constants or the virial coefficients, are known.

6.3 Thermodynamic Quantities for Pure Liquids and Solids

EQUATION OF STATE

For pure solids and liquids, jointly called *condensed phases*, the volume is determined by the molecular size and molecular forces; it does not change much with changes in p and T. Since the molecular size and forces are very specific to a compound, there is no generally valid equation of state. A relation between V, T and p is expressed in terms of the coefficient of thermal expansion α and the isothermal compressibility κ_T defined by (5.4.5) and (5.4.6). If we consider V as a function of p and T, $V(p, T)$, we can write

$$dV = \left(\frac{\partial V}{\partial T}\right)_p dT + \left(\frac{\partial V}{\partial p}\right)_T dp = \alpha V \, dT - \kappa_T V \, dp \qquad (6.3.1)$$

The values of α and κ_T are small for solids and liquids. For liquids the coefficient of thermal expansion α is in the range 10^{-3} to $10^{-4} \, \text{K}^{-1}$ and the isothermal compressibility κ_T is about $10^{-5} \, \text{atm}^{-1}$. Solids have α in the range 10^{-5} to $10^{-6} \, \text{K}^{-1}$ and κ_T in the range 10^{-6} to $10^{-7} \, \text{atm}^{-1}$. Table 6.1 lists the values of α and κ_T for some liquids and solids. Furthermore, the values of α and κ_T are almost constant for temperature variations of about 100 K and pressure variations of about 50 atm. Therefore (6.3.1) can be integrated to obtain the following equation of state:

$$\begin{aligned} V(p, T) &= V(p_0, T_0) \exp\left[\alpha(T - T_0) - \kappa_T(p - p_0)\right] \\ &\approx V(p_0, T_0)[1 + \alpha(T - T_0) - \kappa_T(p - p_0)] \end{aligned} \qquad (6.3.2)$$

Table 6.1 Coefficient of thermal expansion α and isothermal compressibility κ_T for some liquids and solids

Compound	$\alpha/10^{-4} \, (\text{K}^{-1})$	$\kappa_T/10^{-6} \, (\text{atm}^{-1})$
Water	2.1	49.6
Benzene	12.4	92.1
Mercury	1.8	38.7
Ethanol	11.2	76.8
Terachloromethane	12.4	90.5
Copper	0.501	0.735
Diamond	0.030	0.187
Iron	0.354	0.597
Lead	0.861	2.21

THERMODYNAMIC QUANTITIES

Thermodynamically, the characteristic feature of solids and liquids is that μ, S, and H change very little with pressure, hence they are essentially functions of T for a given N. If entropy is considered as a function of p and T, then

$$dS = \left(\frac{\partial S}{\partial T}\right)_p dT + \left(\frac{\partial S}{\partial p}\right)_T dp \tag{6.3.3}$$

The first term $(\partial S/\partial T)_p = NC_p/T$, which relates it to the experimentally measurable C_p. The second term can be related to α as follows:

$$\left(\frac{\partial S}{\partial p}\right)_T = -\left(\frac{\partial}{\partial p}\left(\frac{\partial G(p,T)}{\partial T}\right)_p\right)_T = -\left(\frac{\partial}{\partial T}\left(\frac{\partial G(p,T)}{\partial p}\right)_T\right)_p = -\left(\frac{\partial V}{\partial T}\right)_p = -V\alpha \tag{6.3.4}$$

With these observations, we can now rewrite (6.3.3) as

$$dS = \frac{NC_p}{T}dT - \alpha V\, dp \tag{6.3.5}$$

Upon integration, this equation yields

$$S(p, T) = S(0,0) + N\int_0^T \frac{C_p}{T}dT - N\int_0^p \alpha V_m\, dp \tag{6.3.6}$$

where we have used $V = NV_m$. (That $S(0,0)$ is well defined is guaranteed by the Nernst theorem.) Since V_m and α do not change much with p, the third term in (6.3.6) can be approximated to $N\alpha V_m p$. For $p = 1–10$ atm this term is small compared to the second term. In the case of water $V_m = 18.0 \times 10^{-6}\,\text{m}^3\,\text{mol}^{-1}$ and $\alpha = 2.1 \times 10^{-4}\,\text{K}^{-1}$. For $p = 10$ bar $= 10 \times 10^5$ Pa, the term $\alpha V_m p$ is about $3.6 \times 10^{-3}\,\text{J}\,\text{K}^{-1}\,\text{mol}^{-1}$. The value of C_p, on the other hand, is about $75\text{J}\,\text{K}^{-1}\,\text{mol}^{-1}$. Though C_p approaches zero, so S is finite as $T \to 0$, the molar entropy of water at $p = 1$ atm and $T = 298$ K is about $70\,\text{J}\,\text{K}^{-1}$. Thus, it is clear that the third term in (6.3.6) that contains p is insignificant compared to the second term. Since this is generally true for solids and liquids, we may write

$$\boxed{S(p, T) = S(0,0) + N\int_0^T \frac{C_p(T)}{T}dT} \tag{6.3.7}$$

where we have written C_p explicitly as a function of T. A knowledge of $C_p(T)$ will enable us to obtain the value of entropy for a pure solid or liquid. Note that the integral in (6.3.7) is $\int_0^T d_eS$ because $(NC_p\, dT/T) = dQ/T = d_eS$.

The chemical potential of condensed phases can be obtained from the Gibbs–Duhem equation $d\mu = -S_m\,dT + V_m\,dp$ (5.2.4). Integrating this expression we obtain:

$$\mu(p,\,T) = \mu(p_0, T_0) - \int_{T_0}^{T} S_m(T)dT + \int_{p_0}^{p} V_m\,dp \qquad (6.3.8)$$

$$= \mu(T) + V_m p \equiv \mu(T) + RT\ln a$$

where we assumed that V_m was essentially a constant. Once again it can be shown that the term containing p is small compared to the first term, which is a function of T. For water $V_m p = 1.8\,\mathrm{J\,mol^{-1}}$ when $p = 1$ atm, whereas the first term is of the order $280\,\mathrm{kJ\,mol^{-1}}$. Following the definition of activity a, if we write $V_m p = RT\ln a$ then we see that for *liquids and solids the activity is nearly equal to 1*.

In a similar manner, one can obtain other thermodynamic quantities such as enthalpy H and Helmholtz free energy F.

HEAT CAPACITIES

It is clear that one needs to know the heat capacities of a substance as a function of temperature and pressure in order to calculate the entropy and other thermodynamic quantities. A detailed understanding of the theory of heat capacities (which requires statistical mechanics) is beyond the scope of this book. Here we shall only give a brief outline of Peter Debye's theory for the heat capacities of solids, an approach that leads to an approximate general theory. The situation is more complex in liquids because there is neither complete molecular disorder, as in a gas, nor is there long-range order, as in a solid.

According to the Debye theory, the heat capacity C_V of a pure solid has the form

$$C_V = 3RD(T/\theta) \qquad (6.3.9)$$

in which D is a function of the ratio (T/θ). The parameter θ depends mainly on the chemical composition of the solid and, to a very small extent, varies with the pressure. As the ratio (T/θ) increases, the "Debye function" $D(T/\theta)$ tends to unity, and the heat capacities of all solids become $C_V = 3R$. The fact that the heat capacities of solids tend to have the same value had been observed long before Debye's theory; it was called the law of Dulong and Petit. At much lower temperatures, when $(T/\theta) < 0.1$,

$$D\left(\frac{T}{\theta}\right) \approx \frac{4\pi^4}{5}\left(\frac{T}{\theta}\right)^3 \qquad (6.3.10)$$

Thus Debye's theory predicts that the heat capacities at low temperatures will be proportional to the third power of the temperature. Experimentally this was found to be true for many solids. Once C_V is known, C_p can be obtained using the general expression $C_p - C_V = TV_m\alpha^2/\kappa T$. More detail on this subject can be found in the literature [1] and texts on condensed matter. Thermodynamics of solid and liquid mixtures is discussed in Chapter 8.

Reference

1. Prigogine, I. and Defay R., *Chemical Thermodynamics*, 4th ed. 1967, London: Longman. p. 542.

Data Sources

[A] NBS table of chemical and thermodynamic properties. *J. Phys. Chem. Reference Data*, **11**, Suppl. 2 (1982).
[B] G. W. C. Kaye and Laby T. H., (eds.) *Tables of Physical and Chemical Constants*. 1986, London: Longman.
[C] I. Prigogine and Defay R., *Chemical Thermodynamics*, 4th ed., 1967, London: Longman.
[D] J. Emsley, *The Elements*, 1989, Oxford: Oxford University Press.
[E] L. Pauling, *The Nature of the Chemical Bond*. 1960, Ithaca NY: Cornell University Press.
[F] D. R. Lide (ed.), *CRC Handbook of Chemistry and Physics*, 75th ed. 1994, Ann Arbor MI: CRC Press.

Examples

Example 6.1   Show that C_V for a van der Waals gas is the same as for an ideal gas.

Solution: The relation between C_V for real and ideal gases (6.2.9) is given by

$$C_{V,\text{real}} = C_{V,\text{ideal}} + \int_{\infty}^{V} T\left(\frac{\partial^2 p}{\partial T^2}\right)_V dV$$

For one mole of a van der Waals gas, we have

$$p = \frac{RT}{(V-b)} - a\frac{1}{V^2}$$

Since this is a linear function of T, the derivative $(\partial^2 p/\partial T^2)_V = 0$ so the integral in the expression relating $C_{V,\text{real}}$ and $C_{V,\text{ideal}}$ is zero. Hence $C_{V,\text{real}} = C_{V,\text{ideal}}$.

Example 6.2 Calculate the total internal energy of a real gas using the Berthelot equation (6.2.2a).

Solution: The internal energy of a real gas can be calculated using relation (6.2.7):

$$U_{\text{real}}(T,\ V,\ N) = U_{\text{ideal}}(T,\ N) + \int_{\infty}^{V} T^2 \left(\frac{\partial}{\partial T}\frac{p}{T}\right)_V dV$$

For the Berthelot equation

$$p = \frac{RT}{V_{\text{m}} - b} - \frac{a}{T V_{\text{m}}^2}$$

In this case the integral

$$\int_{\infty}^{V} T^2 \left(\frac{\partial}{\partial T}\frac{p}{T}\right)_V dV = -\int_{\infty}^{V} \frac{aN^2}{V^2} T^2 \frac{\partial}{\partial T}\left(\frac{1}{T^2}\right) dV = \int_{\infty}^{V} \frac{2aN^2}{T}\frac{1}{V^2} dV$$

$$= -\frac{2aN^2}{TV}$$

Hence

$$U_{\text{real}}(T,\ V,\ N) = U_{\text{ideal}}(T,\ N) - \frac{2aN^2}{TV}$$

Exercises

6.1 For an *ideal gas* obtain the explicit expressions for the following:
(i) $F(V,T,N) = U - TS$ as a function of V, T and N.
(ii) $G = U - TS + pV$ as a function of p, T, and N.
(iii) Use the relation $\mu = (\partial F/\partial N)_{V,T}$ and obtain an expression for μ as a function of the mole number density (N/V) and T. Also show that $\mu = \mu^0(T) + RT \ln p$ in which $\mu^0(T)$ is a function of T.

6.2 (a) Obtain a general expression for the entropy of mixing of two non-identical gases of equal molar densities (N/V), with mole numbers N_1 and N_2, initially occupying volumes V_1 and V_2. Also show that the entropy of mixing can be written as $\Delta S_{\text{mix}} = -RN[x_1 \ln x_1 + x_2 \ln x_2]$ where x_1 and x_2 are the mole fractions and N is the total number of moles.
(b) Using Stirling's approximation $\ln N! \approx N \ln N - N$, obtain (6.1.15) from (6.1.14).

6.3. For N_2 the critical values are $p_c = 33.5$ atm, $T_c = 126.3$ K and $V_{mc} = 90.1 \times 10^{-3}$ L mol^{-1}. Using the equations (6.2.1a)–(6.2.3b), calculate the constants a and b for the van der Waals, Berthelot and Dieterici equations. Plot the $p - V_m$ curves for the three equations at $T = 300$ K, 200 K and 100 K on the same graph in the range $V_m = 0.1$–10 L and comment on the differences between the curves.

6.4 Using the van der Waals equation write the pressure as a function of the density (N/V). Assume that the quantity $b(N/V)$ is small and use the expansion

$$\frac{1}{1-x} = 1 + x + x^2 + x^3 \ldots,$$

valid for $x < 1$, to obtain an equation similar to the virial equation

$$p = RT \frac{N}{V}\left[1 + B(T)\left(\frac{N}{V}\right) + C(T)\left(\frac{N}{V}\right)^2 + \ldots\right].$$

Comparing the two series expansions for p show that van der Waals constants a and b and the virial coefficients $B(T)$ and $C(T)$ are related by:

$$B = b - \frac{a}{RT} \qquad \text{and} \qquad C = b^2.$$

6.5 Show that the Helmholtz equation

$$\left(\frac{\partial U}{\partial V}\right)_T = T^2\left(\frac{\partial}{\partial T}\frac{\partial}{T}\right)_V$$

can be written in the alternative form:

$$\left[\left(\frac{\partial U}{\partial V}\right)_T + p\right] = T\left(\frac{\partial p}{\partial T}\right)_V.$$

6.6 (i) Assume ideal gas energy $U = C_V NT$ where $C_V = 28.46$ J K^{-1} for CO_2 and calculate the difference, ΔU, between U_{ideal} and U_{VW} for $N = 1, T = 30$ K at $V = 0.5$ L. What percent of U_{ideal} is ΔU?
(ii) Use Maple/Mathematica to obtain a 3D plot of $(\Delta U/U_{ideal})$ for *one mole* of CO_2, in the volume range $V = 22.00$ L to 0.50 L for $T = 200$ to 500 K.

6.7 Obtain (6.2.9) from (6.2.7) and the definition

$$C_{V,\text{real}} = \left(\frac{\partial U_{m,\text{real}}}{\partial T}\right)_V.$$

6.8 For CO_2 using the van der Waals equation:
(i) Obtain an expression for the compressibility factor Z. At $T = 300\,K$, and $N = 1$, using Mathematica/Maple, plot Z as a function of V from $V = 22.0$ to $V = 0.5\,L$.
(ii) Obtain an explicit expression for $(F_{VW} - F_{ideal})$ for one mole of CO_2 as a function of T and V in which, if T is in K and V is in L, then $(F_{VW} - F_{ideal})$ is in J.

6.9. Using the relation

$$\mu = \left(\frac{\partial F}{\partial N}\right)_{V,T}$$

show that for a van der Waals gas

$$\mu(n, T) = (U_0 - 2an) + \left[\frac{C_V}{R} + \frac{1}{1 - nb}\right] RT$$
$$- T\left[s_0 + C_V \ln T - R \ln\left(\frac{n}{1 - bn}\right)\right]$$

in which $n = (N/V)$.

6.10. Obtain (6.2.24) from (6.2.22) and (6.2.26) from (6.2.24) and (6.2.25).

7 PHASE CHANGE

Introduction

The transformation from the liquid to the vapor phase or from the solid to the liquid phase is caused by heat. The eighteenth century investigations of Joseph Black revealed that these transformations take place at a definite temperature—the boiling point or the melting point—and that there is a latent heat associated with them. Under suitable conditions, phases of a compound can coexist in a state of thermal equilibrium. The nature of this state of thermal equilibrium and how it changes with pressure and temperature can be understood using the laws of thermodynamics.

In addition, at the point where the phase transition takes place, some thermodynamic quantities, such as entropy, change discontinuously. Based on these and other properties, phase transitions in various materials can be classified into different groups or "orders". There are general theories that describe phase transitions of different orders. Today, the study of phase transitions has grown to be a large and interesting subject and some very important developments have occurred during the 1960s and the 1970s. In this chapter we will only present some of the basic results. For further understanding of phase transitions, we refer the reader to books devoted to this subject [1–3]

7.1 Phase Equilibrium and Phase Diagrams

The conditions of temperature and pressure under which a substance exists in different phases, gas, liquid and solid, are summarized in a phase diagram. A simple **phase diagram** is shown in Fig. 7.1. Under suitable conditions of pressure and temperature, two phases may coexist in thermodynamic equilibrium. Thermodynamic study of phase equilibrium leads to many interesting and useful results. For example, it tells us how the boiling point or freezing point of a liquid changes with the applied pressure.

We begin by looking at the equilibrium between liquid and gas phases as shown in Fig. 7.1(b). The system under consideration is closed and it consists only of the liquid in equilibrium with its vapor at a fixed temperature. In Fig. 7.2 the $p-V$ isotherms of a vapor–liquid system are shown. The region of coexistence of the liquid and vapor phases corresponds to the flat portion BC of the isotherm. When $T > T_c$, the flat portion does not exist; there is no distinction

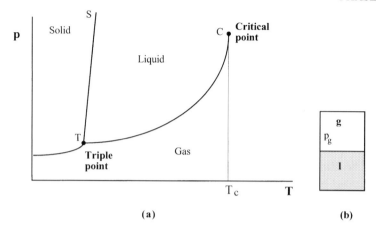

Figure 7.1 (a) Phase diagram for a one-component system showing equilibrium p–T curves (defined by the equality of the chemical potentials), the triple point and the critical point. T_c is the critical temperature above which the gas cannot be liquefied by increasing the pressure. (b) Liquid in equilibrium with its vapor.

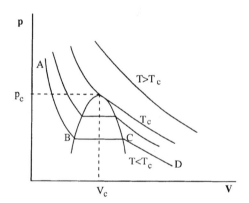

Figure 7.2 Isotherms of a gas showing critical behavior. T_c is the critical temperature above which the gas cannot be liquefied by increasing the pressure. In the flat region such as BC the liquid and the vapor phases coexist.

between the vapor and the liquid phases. The flat portion of each isotherm in Fig. 7.2 corresponds to a point on the curve TC in Fig. 7.1. As the temperature approaches T_c, we approach the critical point C.

We consider a **heterogeneous systems** in which the two phases occupy separate volumes. Under these conditions, the liquid converts irreversibly to vapor, or vice versa, until equilibrium between the two phases is attained. The exchange of matter between the two phases may be considered a "chemical reaction" which we may represent as

$$g \rightleftharpoons l \qquad (7.1.1)$$

Let the chemical potentials of the substance k in the two phases be μ_k^g and μ_k^l, with the superscripts g for gas and l for liquid. *At equilibrium, the entropy production due to every irreversible process must vanish*. This implies that the affinity corresponding to liquid–vapor conversion must vanish, i.e.

$$A = \mu_k^l(p, T) - \mu_k^g(p, T) = 0$$
$$\text{i.e.} \quad \mu_k^l(p, T) = \mu_k^g(p, T) \qquad (7.1.2)$$

in which we have made explicit that the two chemical potentials are functions of pressure and temperature. The pressure of the vapor phase in equilibrium with the liquid phase is called the **saturated vapor pressure**. The equality of the chemical potentials implies that, when a liquid is in equilibrium with its vapor, the pressure and temperature are not independent. This relationship between p and T gives the **coexistence curve** TC in the phase diagram of Fig. 7.1(a).

Clearly, equality of chemical potentials as in (7.1.2) must be valid between any two phases that are in equilibrium. If there are P phases, we have the general equilibrium condition

$$\mu_k^1(p, T) = \mu_k^2(p, T) = \mu_k^3(p, T) = \ldots \mu_k^P(p, T) \qquad (7.1.3)$$

Figure 7.1(a) also shows another interesting feature, the **critical point** C at which the coexistence curve TC (between the liquid and vapor phase) terminates. If the temperature of the gas is above T_c, the gas cannot be liquefied by increasing the pressure. As the pressure increases, the density increases but there is no *transition* to a condensed phase. In contrast, there is no critical point for the transition between solid and liquid; this is because a solid has a definite crystal structure but a liquid hasn't. Due to the definite change in symmetry, the transition between a solid and liquid is always well defined.

A change in phase of a solid is not necessarily a transformation to a liquid. A solid may exist in different phases. In the thermodynamic sense, a phase change is identified by a sharp change in properties such as specific heats. In molecular terms, these changes correspond to different arrangements of the atoms, i.e. different crystal structures. For example, at very high pressures, ice exists in different structures, they are the different solid phases of water. Figure 7.3 shows the phase diagram of water.

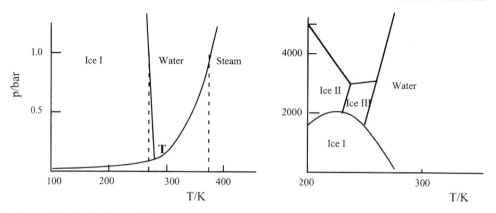

Figure 7.3   The phase diagram of water at ordinary and high pressures (not strictly to scale). At high pressures, the solid phase (ice) can exist in different phases, as shown on the right. The triple point of water is at $p = 0.006$ bar, $T = 273.16$ K. The critical point is at $p_c = 218$ bar, $T_c = 647.3$ K

THE CLAPEYRON EQUATION

The temperature on the coexistence curves corresponds to the temperature at which the transition from one phase to another takes place at a given pressure. Thus, if we obtain an explicit relation between the pressure and the temperature that defines the coexistence curve, we can know how the boiling point or freezing point changes with pressure. Using the condition for equilibrium (7.1.2), we can arrive at a more explicit expression for the coexistence curve. Let us consider two phases denoted by 1 and 2. Using the Gibbs–Duhem equation, $d\mu = -S_m\,dT + V_m\,dp$, one can derive a differential relation between p and T of the system as follows. From (7.1.3) it is clear that for a component k, $d\mu_k^1 = d\mu_k^2$. Therefore we have the equality

$$-S_{m1}\,dT + V_{m1}\,dp = -S_{m2}\,dT + V_{m2}\,dp \qquad (7.1.4)$$

in which the molar quantities for the two phases are indicated by the subscript m1 and m2. From this it follows that:

$$\frac{dp}{dT} = \frac{(S_{m1} - S_{m2})}{(V_{m1} - V_{m2})} = \frac{\Delta H_{trans}}{T(V_{m1} - V_{m2})} \qquad (7.1.5)$$

in which we have expressed the difference in the molar entropy between the two phases in terms of the enthalpy of transition, $S_{m1} - S_{m2} = (\Delta H_{trans}/T)$ where ΔH_{trans} is the *molar enthalpy* of the transition (vaporization, fusion or

Table 7.1   Enthalpies of fusion and vaporization at $p = 1$ bar $= 10^5$Pa $= 0.987$ atm*

Compound	T_{melt} (K)	ΔH_{fus} (kJ mol^{-1})	T_{boil} (K)	ΔH_{vap} (kJ mol^{-1})
He	0.95[†]	0.021	4.22	0.082
H_2	14.01	0.12	20.28	0.46
O_2	54.36	0.444	90.18	6.820
N_2	63.15	0.719	77.35	5.586
Ar	83.81	1.188	87.29	6.51
CH_2	90.68	0.941	111.7	8.18
C_2H_5OH	156	4.60	351.4	38.56
CS_2	161.2	4.39	319.4	26.74
CH_3OH	175.2	3.16	337.2	35.27
NH_3	195.4	5.652	239.7	23.35
CO_2	217.0	8.33	194.6	25.23
Hg_2	234.3	2.292	629.7	59.30
CCl_4	250.3	2.5	350	30.0
H_2O	273.15	6.008	373.15	40.656
Ga	302.93	5.59	2676	270.3
Ag	1 235.08	11.3	2485	257.7
Cu	1 356.6	13.0	2840	306.7

* Tabulated in order of increasing melting point; more extensive data may be found in sources [B], [F], and [G].
† Under pressure.

sublimation) (Table 7.1). Thus, we have the **Clapeyron equation**.

$$\boxed{\frac{dp}{dT} = \frac{\Delta H_{trans}}{T \Delta V_m}} \qquad (7.1.6)$$

Here ΔV_m is the difference in the *molar volumes* of the two phases. The temperature T in this equation is the transition temperature, the boiling point, melting point, etc. This equation tells us how the transition temperature changes with pressure. For a transition from a solid to a liquid in which there is an increase in the molar volume ($\Delta V > 0$), the freezing point will increase ($dT > 0$) when the pressure is increased ($dp > 0$); if there is a decrease in the molar volume, the opposite will happen.

THE CLAUSIUS–CLAPEYRON EQUATION

For liquid–vapor transitions, the Clapeyron equation can be further simplified. In this transition $V_{ml} \ll V_{mg}$. Therefore we may approximate $(V_{mg} - V_{ml})$ by V_{mg}. In this case the Clapeyron equation (7.1.6) simplifies to

$$\frac{dp}{dT} = \frac{\Delta H_{vap}}{T V_{mg}} \qquad (7.1.7)$$

As a first approximation we may use the molar volume of an ideal gas, $V_{mg} = RT/p$. Substituting this expression in the place of V_{mg}, and noting that $dp/p = d(\ln p)$, we arrive at the the **Clausius–Clapeyron equation**:

$$\boxed{\frac{d \ln p}{dT} = \frac{\Delta H_{vap}}{RT^2}} \qquad (7.1.8)$$

This equation is also applicable to a solid in equilibrium with its vapor (e.g iodine), since the molar volume of the vapor phase is much larger than the molar volume of the solid phase. *For a solid in equilibrium with its vapor, ΔH_{sub} takes the place of ΔH_{vap}.* At times, this equation is also written in its integrated form:

$$\boxed{\ln p_2 - \ln p_1 = \frac{\Delta H_{vap}}{R}\left(\frac{1}{T_1} - \frac{1}{T_2}\right)} \qquad (7.1.9)$$

As illustrated in Fig. 7.4, equations (7.1.8) or (7.1.9) tell us how the boiling point of a liquid changes with pressure. When a liquid subjected to an external pressure, p_{ext}, is heated, bubbles containing the vapor (in equilibrium with the liquid) can form, provided the vapor pressure $p_g \geq p_{ext}$. The liquid then begins to "boil." If the vapor pressure p is less than p_{ext}, the bubbles cannot form—they "collapse." The temperature at which $p = p_{ext}$ is what we call the boiling point T_{boil}. Hence, in equations (7.1.8) and (7.1.9) we may interpret p as the pressure to which the liquid is subjected, and T as the corresponding boiling point. It tells us that the boiling point of a liquid decreases with a decrease in pressure, p_{ext}.

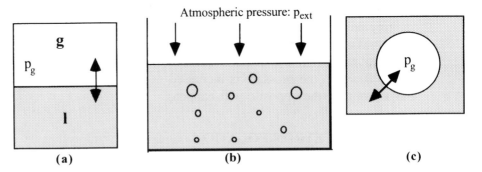

Figure 7.4 Equilibrium between liquid and vapor phases. (a) An isolated system which contains a liquid in equilibrium with its vapor; the pressure of the vapor, p_g, is called the saturated vapor pressure. (b) When the liquid is subjected to a pressure p_{ext} (atmospheric pressure) is heated, bubbles of its vapor can form when $p_g \geq p_{ext}$ and the liquid begins to "boil." (c) The vapor in the bubble is the saturated vapor in equilibrium with the liquid, as in the case of an isolated system (a)

7.2   The Gibbs Phase Rule and Duhem's Theorem

Thus far we have considered the equilibrium between two phases of a single compound. When many compounds or components and more than two phases are in equilibrium, the chemical potential of each component should be the same in every phase in which it exists. When we have a single phase, such as a gas, its intensive variables, pressure and temperature, can be varied independently. However, when we consider equilibrium between two phases, such as a gas and a liquid, p and T are no longer independent. Since the chemical potentials of the two phases must be equal $(\mu^1(p, T) = \mu^2(p, T))$, only one of the two intensive variables is independent. In the case of liquid–vapor equilibrium of a single component, p and T are related by (7.1.9). The number of independent intensive variables depends on the number of phases in equilibrium and the number of components in the system.

The *independent intensive variables* that specify a state are called its **degrees of freedom**. Gibbs observed that there is a general relationship between the number of degrees of freedom, f, the number of phases, P, and the number of components C:

$$\boxed{f = C - P + 2} \tag{7.2.1}$$

This can be seen as follows. At a given T, specifying p is equivalent to specifying the density in terms of mole number per unit volume (through the equation of state). For a given density, the mole fractions specify the composition of the system. Thus, for each phase, p, T and the C mole fractions x_k^i (in which the superscript indicates the phase and the subscript the component) are the intensive variables that specify the state. Of the C mole fractions in each phase i, there are $(C - 1)$ independent mole fractions x_k^i because $\sum_{k=1}^{C} x_k^i = 1$. In a system with C components and P phases, there are a total of $P(C - 1)$ independent mole fractions x_k^i. These together with p and T make a total of $P(C - 1) + 2$ independent variables. On the other hand, equilibrium between the P phases of a component k requires the equality of chemical potentials in all the phases:

$$\mu_k^1(p, T) = \mu_k^2(p, T) = \mu_k^3(p, T) = \ldots = \mu_k^P(p, T) \tag{7.2.2}$$

in which, as before, the superscript indicates the phase and the subscript the component. These constitute $(P - 1)$ constraining equations for each component. For the C components we then have a total of $C(P - 1)$ equations between the chemical potentials, which reduces the number of independent intensive variables by $C(P - 1)$ Thus, the total number of independent degrees of freedom is

$$f = P(C - 1) + 2 - C(P - 1) = C - P + 2$$

If a component "a" does not exist in one of the phases "b", then the corresponding mole fraction $x_a^b = 0$, reducing the number of independent variables by one. However, this also decreases the number of constraining equations (7.2.2) by one. Hence there is no overall change in the number of degrees of freedom.

As an illustration of the Gibbs phase rule, let us consider the equilibrium between solid, liquid and gas phases of a pure substances, i.e. one component. In this case we have $C = 1$ and $P = 3$, which gives $f = 0$. Hence for this equilibrium, there are no free intensive variables; there is only one pressure and temperature at which they can coexist. This point is called the **triple point** (Fig. 7.1). At the triple point of H_2O, $T = 273.16\,K = 0.01\,°C$ and $p = 611\,pa = 6.11 \times 10^{-3}$ bar. (This unique coexistence between the three phases of water may be used in defining the Kelvin scale.)

If the various components of the system also chemically react through R independent reactions, then besides (7.2.2) we also have R equations for the chemical equilibrium, represented by the vanishing of the corresponding affinities:

$$A_1 = A_2 = A_3 = \ldots = A_R = 0 \qquad (7.2.3)$$

Consequently, the number of degrees of freedom is further decreased by R and we have

$$\boxed{f = C - R - P + 2} \qquad (7.2.4)$$

In older statements of the phase rule, the term "number of independent components" is used to represent $(C - R)$. In a reaction such as $A \rightleftharpoons B + 2C$, if the amounts of B and C are entirely a result of decomposition of A, the amounts of B and C are determined by the amount of A that has converted to B and C; in this case the mole fractions of B and C are related, $x_C = 2x_B$. This additional constraint, which depends on the initial preparation of the system, decreases the degrees of freedom by one.

In addition to the phase rule discovered by Gibbs, there is another general observation which Pierre Duhem made in his treatise *Traité élémentaire de Mécanique Chimique*, which is referred to as **Duhem's theorem**. It states:

"Whatever the number of phases, components and chemical reactions, if the initial mole numbers N_k of all the components are specified, the equilibrium state of a closed system is completely specified by two independent variables."

The proof of this theorem is as follows. The state of the entire system is specified by the pressure p, the temperature T and the mole numbers N_k^i in which the superscript indicates the P phases and the subscript the C component—a total of CP mole numbers in P phases. Thus the total number of variables is $CP + 2$. Considering the constraints on these variables, for the equilibrium of each component k between the phases we have

$$\mu_k^1 = \mu_k^2 = \mu_k^3 = \ldots = \mu_k^P \qquad (7.2.5)$$

a total of $(P-1)$ equations for each component, a total of $C(P-1)$ equations. In addition, since the total number of moles, say $N_{k,\text{total}}$, of each component is specified, we have $\sum_{i=1}^{P} N_k^i = N_{k,\text{total}}$ for each component, a total of C equations. Thus the total number of constraints is $C(P-1)+C$. Hence the total number of independent equations is: $CP + 2 - C(P-1) - C = 2$.

The addition of chemical reactions does not change this conclusion, because each chemical reaction α adds a new independent variable ξ_α, its extent of reaction, to each phase and at the same time adds the constraint for the corresponding chemical equilibrium $A_\alpha = 0$. Hence there is no net change in the number of independent variables.

Let us compare the Gibbs phase rule and the Duhem equation: the Gibbs phase rule specifies the total number of independent intensive variables regardless of the extensive variables in the system, while the Duhem equation specifies the total number of independent variables, intensive or extensive, in a closed system.

7.3 Binary and Ternary Systems

Figure 7.1 shows the phase diagram for a single-component system. The phase diagrams for systems with two and three components are more complex. In this section we shall consider examples of two- and three-component systems.

BINARY LIQUID MIXTURES IN EQUILIBRIUM WITH THEIR VAPORS

Consider a liquid mixture of two components, A and B, in equilibrium with their vapors. This system contains two phases and two components. The Gibbs phase rule tells us that such a system has two degrees of freedom. We may take these degrees of freedom to be the pressure and the mole fraction, x_A, of component A. Thus, if we consider a system subjected to a constant pressure, for each value of the mole fraction x_A there is a corresponding temperature at which the two phases are in equilibrium. For example, if the applied pressure is 0.5 bar, for the liquid to be in equilibrium with its vapor the temperature must be set at an appropriate value T.

If the applied pressure is the atmospheric pressure, the temperature corresponds to the boiling point. In Fig. 7.5 curve I is the boiling point as a function of the mole fraction x_A; the boiling points of the two components A and B are T_A and T_B respectively. Curve II shows the composition of the vapor at each boiling temperature. If a mixture with composition corresponding to the

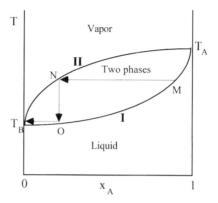

Figgure 7.5 Boiling point versus
composition for a mixture of two
similar liquids such as benzene and
toluene

point M is boiled, the vapor will have the composition corresponding to the
point N; if this vapor is now collected and condensed, its boiling point and
composition will correspond to the point O. This process enriches the mixture in
component B. For such systems, by continuing this process, a mixture can be
enriched in the more volatile component.

AZEOTROPES

The relation between the boiling point and the compositions of the liquid and
vapor phases shown in Fig. 7.5 is not valid for all binary mixtures. For many
liquid mixtures the boiling point curve is as shown in Fig 7.6. In this case there

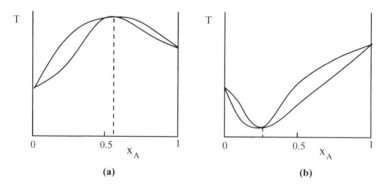

Figure 7.6 Boiling point versus composition for liquid and
vapor phases of binary mixtures called azeotropes. Azeotropes
reach a point at which the vapor and the liquid phases have the
same composition, then the boiling point is either a maximum (a)
or a minimum (b)

Table 7.2 Examples of azeotropes*

	Boiling Point (°C)		Azeotropic weight percent
	Pure compound	Azeotrope	
Azeotropes formed with water at $p = 1$ bar Boiling point of water $= 100$ °C			
Hydrogen chloride (HCl)	−85	108.58	20.22
Nitric acid (HNO$_3$)	86	120.7	67.7
Ethanol (C$_2$H$_5$OH)	78.32	78.17	96
Azeotropes formed with acetone at $p = 1$ bar Boiling point of acetone ((CH$_3$)$_2$CO) $= 56.15$ °C			
Cyclohexane (C$_6$H$_{12}$)	80.75	53.0	32.5
Methyl acetate (CH$_3$COOCH$_3$)	57.0	55.8	51.7
n-hexane (C$_6$H$_{14}$)	68.95	49.8	41
Azeotropes formed with methanol at $p = 1$ bar Boiling point of methanol (CH$_3$OH) $= 64.7$ °C			
Acetone ((CH$_3$)$_2$CO)	56.15	55.5	88
Benzene (C$_6$H$_6$)	80.1	57.5	60.9
Cyclohexane (C$_6$H$_{12}$)	80.75	53.9	63.9

* Extensive data on azeotropes can be found in data source [F].

is a value of x_A at which the composition of the liquid and the vapor remains the same. Such systems are called **azeotropes**. The components of an azeotrope cannot be separated by distillation. For example, in the case of Fig. 7.6(a), starting from the left of the maximum, if the mixture is boiled and the vapor collected, the vapor becomes richer in component B while the remaining liquid becomes richer in A and moves towards the azeotropic composition. Thus successive boiling and condensation results in pure B and the azeotrope, not pure A and pure B. We leave it as an exercise for the reader to analyze for Fig. 7.6(b). The azeotropic composition and the corresponding boiling points for binary mixtures are tabulated in data source [F]. Notice how the boiling point corresponding to the azeotropic composition occurs at an extremum (maximum or minimum). That this must be so for thermodynamic reasons has been noted by Gibbs and later by Konovolow and Duhem. This observation is called the **Gibbs–Konovalow theorem** [4]:

"At constant pressure, in an equilibrium displacement of a binary system, the temperature of coexistence passes through an extremum if the composition of the two phases is the same".

We shall not discuss the proof of this theorem here. An extensive discussion of this and other related theorems may be found in [4]. Azeotropes are an important class of solutions whose thermodynamic properties we shall discuss in more detail in Chapter 8. Some examples of azeotropes are given in Table 7.2.

SOLUTIONS IN EQUILIBRIUM WITH PURE SOLIDS: EUTECTICS

The next example we consider is a solid–liquid equilibrium of two components, A and B, which are miscible in the liquid state but not in the solid state. This system has three phases in all, the liquid with A + B, the solid A and solid B.

We can understand the equilibrium of such a system by first considering the equilibrium of two-phase systems, the liquid and one of the two solids, A or B, then extending it to three phases. In this case the Gibbs phase rule tells us that, with two components and two phases, the number of degrees of freedom equals two. We can take these two degrees of freedom to be the pressure and the composition. Thus, if the mole fraction x_A and the pressure are fixed then the equilibrium temperature is also fixed. By fixing the pressure at a given value (say the atmospheric pressure) one can obtain an equilibrium curve relating T and x_A. The two curves corresponding to solid A in equilibrium with the liquid and solid

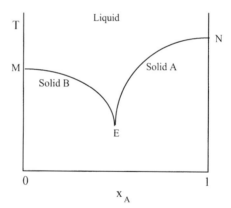

Figure 7.7 The phase diagram of a two-component system with three phases. The system has only one degree of freedom. For a fixed pressure, the three phases—the liquid, solid A and solid B—are in equilibrium at the **eutectic point** E. The curve ME gives the equilibrium between solid B and the liquid, and the curve NE gives the equilibrium between solid A and the liquid. The point of intersection, E, specifies the equilibrium composition and temperature when all three phases are in equilibrium

B in equilibrium with the liquid are shown in Fig. 7.7. In this figure, along the curve EN, the solid A is in equilibrium with the liquid; along the curve EM, solid B is in equilibrium with the liquid. The point of intersection of the two curves, E, is called the **eutectic point** and the corresponding composition and temperature are called the **eutectic composition** and the **eutectic temperature**.

Now, if we consider a three-phase system—the liquid, solid A and solid B, all in equilibrium—then the Gibbs phase rule tells us that there is only one degree of freedom. If we take this degree of freedom to be the pressure and fix it at a particular value, then there is only one point (T, x_A) at which equilibrium can exist between the three phases. This is the eutectic point, the point at which the chemical potentials, of solid A, solid B and the liquid are equal. Since the chemical potentials of solids and liquids do not change much with pressure changes, the eutectic composition and temperature do not change much.

TERNARY SYSTEMS

As was noted by Gibbs, the composition of a solution containing three components may be represented by points within an equilateral triangle of unit side length. Let us consider a system with components A, B and C. As shown in Fig. 7.8, a point P may be used to specify the mole fractions x_A, x_B and x_C. From the point P, we draw lines parallel to the sides of the equilateral triangle.

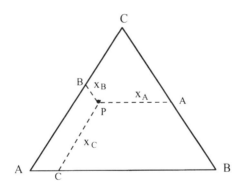

Figure 7.8 The composition of a ternary system consisting of components A, B and C can be represented on a triangular graph because $x_A + x_B + x_C = 1$. The composition is represented as a point P inside an equilateral triangle of unit side length. The mole fractions are the lengths of the lines drawn parallel to the sides of the triangle. It is left as an exercise to show that $PA + PB + PC = 1$ for any point P

The length of these lines can be used to represent the mole fractions x_A, x_B and x_C. It is left as an exercise to show how this construction ensures that $x_A + x_B + x_C = 1$. In this representation of the composition, we can see three things:

(i) The vertices A, B and C correspond to pure substances.
(ii) A line parallel to a side of the triangle corresponds to a series of ternary systems in which one of the mole fractions remains fixed.
(iii) A line drawn through an apex to the opposite side represents a series of systems in which the mole fractions of two components have a fixed ratio. As the apex is approached along this line, the system becomes increasingly richer in the component represented by the apex.

The variation of some property of a three-component solution can be shown in a three-dimensional graph in which the base is the composition triangle (Fig. 7.8); the height will then represent the property.

As an example, let us consider components A, B and C in two phases: a solution that contains A, B and C, and a solid phase of B in equilibrium with the solution. This system has three components and two phases, hence three degrees of freedom, which may be taken as the pressure, x_A and x_B. At constant pressure, each value of x_A and x_B has a corresponding equilibrium temperature. In Fig. 7.9 the point P shows the composition of the solution at a temperature T. As the temperature decreases, the relative values of x_A and x_C remain the same while more of B turns into a solid. According to the observations (iii) above, this

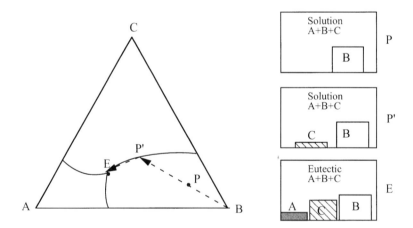

Figure 7.9 The phase diagram of a ternary system showing the composition of the solution as the system is cooled. At point P the system consists of two phases, the solution (A + B + C) in equilibrium with solid B. As the temperature decreases, the composition moves along PP'. At P' the component C begins to crystallize and the composition moves along P'E until it reaches the ternary eutectic point E at which all components begin to crystallize

means the point moves along the line BP as shown by the arrow. As the temperature decreases, a point P' is reached at which the component C begins to crystallize. The system now has two solid phases and one solution phase, with two degrees of freedom. The composition of the system is now confined to the line $P'E$. With further decrease in the temperature, the component A also begins to crystallize at the point E, which corresponds to the eutectic temperature. The system now has only one degree of freedom. At the eutectic temperature and composition, all three components will crystallize in the eutectic proportions.

7.4 Maxwell's Construction and the Lever Rule

The reader might have noticed that the isotherms obtained from an equation of state, such as the van der Waals equation, do not coincide with the isotherms shown in Fig 7.2 at the part of the curve that is flat, i.e. where the liquid and vapor states coexist. The flat part of the curve represents what is physically realized when a gas is compressed at a temperature below the critical temperature. Using the condition that the chemical potential of the liquid and the vapor phases must be equal at equilibrium, Maxwell was able to specify where the flat part of the curve would occur.

Let us consider a van der Waals isotherm for $T < T_c$ (Fig. 7.10). Imagine a steady decrease in volume starting at the point Q. Let the point P be such that, at this pressure, the chemical potentials of the liquid and vapor phases are equal. At this point the vapor will begin to condense and the volume can be decreased with no change in the pressure. This decrease in volume can continue until all the vapor condenses to a liquid at the point L. If the volume is maintained at some value between P and L, liquid and vapor coexist. Along line PL the chemical potentials of the liquid and the vapor are equal. Thus the total change in the chemical potential along curve LMNOP must be equal to zero:

$$\int_{LMNOP} d\mu = 0 \qquad (7.4.1)$$

Now since the chemical potential is a function of T and p, and since the path is an isotherm, it follows from the Gibbs–Duhem relation that $d\mu = V_m dp$. Using this relation we may write the above integral as

$$\int_P^O V_m \, dp + \int_O^N V_m \, dp + \int_N^M V_m \, dp + \int_M^L V_m \, dp = 0 \qquad (7.4.2)$$

Now since $\int_O^N V_m \, dp = -\int_N^O V_m \, dp$, the sum of the first two integrals equals the area I shown in Fig. 7.10. Similarly, since $\int_N^M V_m \, dp = -\int_M^N V_m \, dp$, the sum of

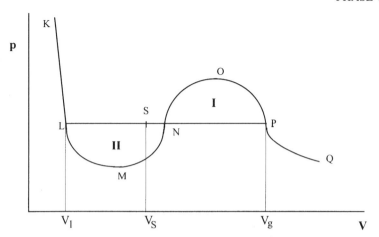

Figure 7.10 Maxwell's construction specifies the physically realized flat part LP with respect to the theoretical isotherm given by an equation of state such as the van der Waals equation. At equilibrium, the chemical potentials at the points L and P must be equal. As shown in the text, this implies that the area I must be equal to area II, specifying the line LP

the second two terms equals the negative of area II. Thus equation (7.4.2) may be interpreted as

$$(\text{Area I}) - (\text{Area II}) = 0 \qquad (7.4.3)$$

This condition specifies the flat line on which the chemical potentials of the liquid and the vapor are equal, the line that is physically realized. It is called the **Maxwell construction**.

At the point P the substance is entirely in the vapor phase with volume V_g; at the point L it is entirely in the liquid phase with volume V_1. At any point S on the line LP, if a mole fraction x of the substance is in the vapor phase, the total volume V_S of the system is

$$V_S = xV_g + (1-x)V_1 \qquad (7.4.4)$$

It follows that

$$x = \frac{V_S - V_1}{V_g - V_1} = \frac{\text{SL}}{\text{LP}} \qquad (7.4.5)$$

From this relation it is easy to show that (exc. 7.10) the mole fraction x of the vapor phase and $(1-x)$ of the liquid phase satisfy the equation:

$$\boxed{(SP)\, x = (SL)(1-x)} \qquad (7.4.6)$$

This relation is called the **lever rule** in analogy with a lever supported at S, in equilibrium with weights x and $(1 - x)$ attached to either end.

7.5 Phase Transitions

Phase transitions are associated with many interesting and general thermodynamic features. As described below, based on some of these features, phase transitions can be classified into different "orders." Thermodynamic behavior in the vicinity of the critical points has been of much interest from the viewpoint of thermodynamic stability and extremum principles discussed in Chapter 5. A classical theory of phase transitions was developed by Lev Landau but in the 1960s experiments showed that the predictions of this theory were incorrect. This led to the development of the modern theory of phase transitions during the 1960s and the 1970s. The modern theory is based on the work of C. Domb, M. Fischer, L. Kadanoff, G. S. Rushbrook, B. Widom, K. Wilson and others. In this section we will only outline some of the main results of the thermodynamics of phase transitions. A detailed description of the modern theory of phase transitions, which uses the mathematically advanced concepts of renormalization group theory, is beyond the scope of this book. For a better understanding of this rich and interesting subject we refer the reader to the literature [1–3].

GENERAL CLASSIFICATION OF PHASE TRANSITIONS

When transition from a solid to liquid or from a liquid to vapor takes place, there is a discontinuous change in the entropy. This can clearly be seen (Fig. 7.11) if we plot molar entropy $S_m = -(\partial G_m/\partial T)_p$ as a function of T, for a fixed p and N. The same is true for other derivatives of G_m such as $V_m = (\partial G/\partial p)_T$. The chemical potential changes continuously but its derivative is discontinuous. At the transition temperature, because of the existence of latent heat, the specific

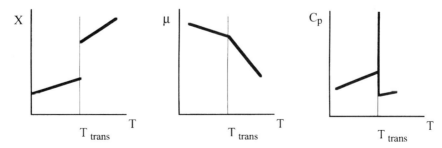

Figure 7.11 The change of thermodynamic quantities in a first-order phase transition that occurs at the temperature T_{trans}. X is an extensive quantity such as S_m or V_m that undergoes discontinuous change

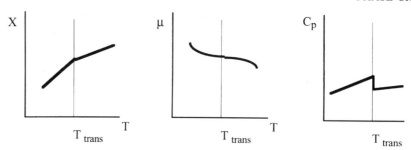

Figure 7.12 The change of thermodynamic quantities in a second-order phase transition that occurs at the temperature T_{trans}. X is an extensive quantity such as S_m or V_m whose derivative undergoes a discontinuous change

heats ($= \Delta Q/\Delta T$) have a "singularity" in the sense they become infinite, i.e. ΔQ the heat absorbed causes no change in temperature ($\Delta T = 0$). Transitions of this type are classified as **first-order** phase transitions.

The characteristic features of **second-order phase transitions** are shown in Fig. 7.12. In this case the changes in the thermodynamic quantities are not so drastic: changes in S_m and V_m are continuous but their derivatives are discontinuous. Similarly, for the chemical potential it is the second derivative that is discontinuous; the specific heat does not have a singularity but it has a discontinuity. Thus, depending on the order of the derivatives that are discontinuous, phase transitions are classified as transitions of first and second order.

BEHAVIOR NEAR THE CRITICAL POINT

Classical theory of phase transition was developed by Lev Landau to explain coexistence of phases and the critical point at which the distinction between the phases disappears. Landau's theory explains the critical behavior in terms of the minima of the Gibbs free energy. According to this theory (Fig. 7.13), in the coexistence region for a given p and T, G as a function of V has two minima. As the critical point is approached, the minima merge into a broad minimum. The classical theory of Landau makes several predictions regarding the behavior of systems near the critical point. The predictions of the theory are in fact quite general, valid for large classes of systems. Experiments did not support these predictions. We shall list below some of the discrepancies between theory and experiments using the liquid–vapor transition as an example, but the experimental values are those obtained for many similar transitions. Also, all the predictions of classical theory can be verified using the van der Waals equation of state as an example.

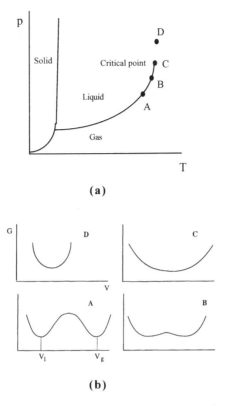

Figure 7.13 Classical theory of phase transitions is based on the shape of the Gibbs free energy. The Gibbs free energy associated with the points A, B, C and D in (a) are shown in (b). As the system moves from A to D, the Gibbs free energy changes from a curve with two minima to a curve with one minimum

- For the liquid–vapor transition, as the critical temperature was approached from below $(T < T_c)$, the theory predicted that

$$V_{mg} - V_{ml} \propto (T_c - T)^\beta, \qquad \beta = 1/2 \qquad (7.5.1)$$

However, experiments showed that β was in the range 0.3–0.4, not equal to 0.5.
- Along the critical isotherm, as the critical pressure p_c is approached from above, the theory predicted

$$V_{mg} - V_{ml} \propto (p - p_c)^{1/\delta}, \qquad \delta = 3 \qquad (7.5.2)$$

Experiments place the value of δ in the range 4.0–5.0.

- When the gas can be liquefied, it is easy to see that the isothermal compressibility $\kappa_T = -(1/V)(\partial V/\partial p)_T$ diverges during the transition (the flat part of the $p-V$ isotherm). Above the critical temperature, since there is no transition to liquid, there is no divergence. According to classical theory, as the critical temperature is approached from above, the divergence of κ_T should be given by

$$\kappa_T \propto (T - T_c)^{-\gamma}, \qquad \gamma = 1 \qquad (7.5.3)$$

Experimental values of γ were found to be in the range 1.2–1.4.

- We have seen in the last chapter that C_V values for real and ideal gases are the same if the pressure is a linear function of the temperature. This means that C_V does not change with T; it does not diverge (though C_p does diverge). Thus, according to classical theory, if

$$C_V \propto (T - T_c)^{-\alpha} \quad \text{then} \quad \alpha = 0 \qquad (7.5.4)$$

The experimental value of α was in the range -0.2 to 0.3.

The failure of the classical or Landau theory initiated a reexamination of the critical behavior. The main reason for the discrepancy was found to be the role of fluctuations. Near the critical point, due to the flat nature of the Gibbs free energy, large long–range fluctuations arise in the system. These fluctuations were successfully incorporated into the theory through the development of "renormalization theory" by Kenneth Wilson. The modern theory of critical behavior not only predicts the experimental values of the exponents α, β, γ and δ more successfully than the classical theory but it also relates these exponents. For example, the modern theory predicts that

$$\beta = \frac{2 - \alpha}{1 + \delta} \quad \text{and} \quad \gamma = \frac{(\alpha - 2)(1 - \delta)}{(1 + \delta)} \qquad (7.5.5)$$

Since a detailed presentation of the renormalization group theory is beyond the scope of this book, we will leave the reader with only this brief outline of the limitations of classical theory and the accomplishments of the modern theory.

References

1. Stanley, H. E., *Introduction to Phase Transitions and Critical Phenomena*. 1971, New York: Oxford University Press.
2. Ma, S.-K., *Modern Theory of Critical Phenomena*. 1976, New York: Addison-Wesley.
3. Pfeuty, P. and Toulouse, G., *Introduction to the Renormalization Group and Critical Phenomena*. 1977, New York: John Wiley.

4. Prigogine, I. and Defay, R., *Chemical Thermodynamics*, 4th ed. 1967, London: Longman. p. 542.

Data Sources

[A] NBS table of chemical and thermodynamic properties. *J. Phys. Chem. Reference Data*, **11**, Suppl. 2 (1982).

[B] Kaye, G. W. C. and Laby, T. H., (eds) *Tables of Physical and Chemical Constants*. 1986, London: Longman.

[C] Prigogine, I. and Defay, R., *Chemical Thermodynamics*, 4th ed. 1967, London: Longman.

[D] Emsley, J., *The Elements*. 1989, Oxford: Oxford University Press.

[E] Pauling, L., *The Nature of the Chemical Bond*. 1960, Ithaca NY: Cornell University Press.

[F] Lide, D. R., (ed.), *CRC Handbook of Chemistry and Physics*, 75th ed. 1994, Ann Arbor MI: CRC Press.

[G] The web site of the National Institute for Standards and Technology, http://webbook.nist.gov.

Examples

Example 7.1   A chemical reaction occurs in CCl_4 at room temperature but it is very slow. To increase its speed to a desired value, the temperature needs to be increased to 80 °C. Since CCl_4 boils at 77 °C at $p = 1.00$ atm, the pressure has to be increased so that CCl_4 will not boil at 77 °C. Using the data in Table 7.1, calculate the pressure at which CCl_4 will boil at 85 °C.

Solution: From the Clausius–Clapeyron equation, we have

$$\ln p - \ln(1.00\,\text{atm}) = \frac{30.0 \times 10^3}{8.314}\left(\frac{1}{350} - \frac{1}{358}\right) = 0.230$$

$$p = (1.00\,\text{atm})e^{0.23} = 1.26\,\text{atm}$$

Example 7.2   If a system contains two immiscible liquids (such as CCl_4 and CH_3OH), how many phases are there?

Solution: The system will consist of three layers. A layer rich in CCl_4, a layer rich in CH_3OH and a layer of vapor (CCl_4 and CH_3OH). Thus there are three phases in this system.

Example 7.3   Determine the number of degrees of freedom of a two-component liquid mixture in equilibrium with its vapor.

Solution: In this case $C = 2$, $P = 2$. Hence the number of degrees of freedom $f = 2 - 2 + 2 = 2$. These two degrees of freedom may be T and the mole fraction x_1 of one of the components. The pressure of the system (vapor phase in equilibrium with the liquid) is completely specified by x_1 and T.

Example 7.4   How many degrees of freedom does an aqueous solution of the weak acid, CH_3COOH have?

Solution: The acid decomposition is

$$CH_3COOH \rightleftharpoons CH_3COO^- + H^+$$

The number of components is $C = 4$, (water, CH_3COOH, CH_3COO^- and H^+). The number of phases is $P = 1$. There is one chemical reaction in equlibrium, hence $R = 1$. However, since the concentrations of CH_3COO^- and H^+ are equal, the number of degrees of freedom f is reduced by one. Hence $f = C - R - P + 2 - 1 = 4 - 1 - 1 + 2 - 1 = 3$.

Exercises

7.1   The heat of vaporization of hexane is 30.8 kJ mol^{-1}. The boiling point of hexane at a pressure of 1.00 atm is 68.9 °C. What will the boiling point be at a pressure of 0.50 atm?

7.2   The atmospheric pressure decreases with height. The pressure at a height h above sea level is given approximately by the barometric formula $p = p_0 e^{-Mgh/RT}$, in which $M = 0.0289$ kg mol^{-1}, and $g = 9.81$ ms^{-2}. Assume the enthalpy of vaporization of water is $\Delta H_{vap} = 40.6$ kJ mol^{-1} and predict the boiling temperature of water at a height of 2.5 miles. Assume a reasonable T.

7.3   At atmospheric pressure, CO_2 turns from solid to gas, i.e. it sublimates. Given that the triple point of CO_2 is at $T = 216.58$ K and $p = 518.0$ kPa, how would you obtain liquid CO_2?

7.4   In a two-component system, what is the maximum number of phases that can be in equilibrium?

7.5   Determine the number of degrees of freedom for the following systems:
(a) solid CO_2 in equilibrium with CO_2 gas
(b) an aqueous solution of fructose
(c) $Fe(s) + H_2O(g) \rightleftharpoons FeO(s) + H_2(g)$

7.6   Draw qualitative T vs x_A curves (Fig. 7.6) for the azeotropes in Table 7.2.

7.7   In Fig. 7.8 show that $PA + PB + PC = 1$ for any point P.

7.8   In the triangular representation of the mole fractions for a ternary solution, show that along the line joining an apex and a point on the opposite side, the ratio of two of the mole fractions remains constant as the mole fraction of the third component changes.

7.9 On a triangular graph, mark points representing the following compositions:
(a) $x_A = 0.2, x_B = 0.4, x_C = 0.4$
(b) $x_A = 0.5, x_B = 0, x_C = 0.5$
(c) $x_A = 0.3, x_B = 0.2, x_C = 0.5$
(d) $x_A = 0, x_B = 0, x_C = 1.0$

7.10 Obtain the lever rule (7.4.6) from (7.4.5).

7.11 When the van der Waals equation is written in terms of the reduced variables p_r, V_r and T_r (see 1.4.6), the critical pressure, temperature and volume are equal to one. Consider small deviations from the critical point, $p_r = 1 + \delta p$ and $V_r = 1 + \delta V$ on the critical isotherm. Show that δV is proportaional to $(\delta p)^{1/3}$. This corresponds to the classical prediction (7.5.2).

8 SOLUTIONS

8.1 Ideal and Nonideal Solutions

Many properties of solutions can be understood through thermodynamics. For example, we can understand how the boiling and freezing points of a solution change with composition, how the solubility of a compound changes with temperature and how the osmotic pressure depends on the concentration.

We begin by obtaining the chemical potential of a solution. The general expression for the chemical potential of a substance is: $\mu(p, T) = \mu^0(p_0, T) + RT \ln a$ in which a is the activity and μ^0 is the chemical potential of the standard state in which $a = 1$. For an ideal gas mixture, we have seen (6.1.9) that the chemical potential of a component can be written in terms of its mole fraction x_k as $\mu_k(p, T, x_k) = \mu_k^0(p, T) + RT \ln x_k$. As we shall see in this section, properties of many dilute solutions can be described by a chemical potential of the same form. This has led to the following definition of an **ideal solution** as a solution for which

$$\mu_k(p, T, x_k) = \mu_k^0(p, T) + RT \ln (x_k) \tag{8.1.1}$$

where $\mu_k^0(p, T)$ is the chemical potential of a reference state which is independent of x_k. We stress that the similarity between ideal gas mixtures and ideal solutions is only in the dependence of the chemical potential on the mole fraction; the dependence on the pressure, however, is entirely different, as can be seen from the general expression for the chemical potential of a liquid (6.3.8).

In equation (8.1.1), if the mole fraction of the "solvent" x_s, is nearly equal to one, i.e. for dilute solutions, then for the chemical potential of the solvent the reference state $\mu_k^0(p, T)$ may be taken to be $\mu_k^*(p, T)$, the chemical potential of the pure solvent. For the other components $x_k \ll 1$; for these minor components, equation (8.1.1) is still valid in a small range, but in general the reference state is not $\mu_k^*(p, T)$. Solutions for which equation (8.1.1) is valid over all values of x_k are called **perfect solutions**. When $x_k = 1$, since we must have $\mu_k(p, T) = \mu_k^*(p, T)$, it follows that for perfect solutions

$$\mu_k(p, T, x_k) = \mu_k^*(p, T) + RT \ln x_k \quad \text{for all } x_k \tag{8.1.2}$$

The activity of nonideal solutions is expressed as $a_k = \gamma_k x_k$ in which γ_k is the **activity coefficient**, a quantity introduced by G.N. Lewis. Thus the chemical

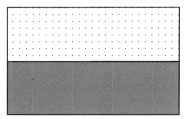

Figure 8.1 Equilibrium between a solution and its vapor

potential of nonideal solutions is written as

$$\mu_k(p, T, x_k) = \mu_k^0(p, T) + RT \ln a_k$$
$$= \mu_k^0(p, T) + RT \ln \gamma_k x_k \tag{8.1.3}$$

The activity coefficient $\gamma_k \to 1$ as $x_k \to 1$.

Let us now look at conditions under which ideal solution are realized. We consider a solution with many components, with mole fractions x_i, in equilibrium with its vapor (Fig. 8.1). At equilibrium the affinities for the conversion of liquid to gas phase are zero for each component i, i.e. for each component, the chemical potentials in the two phases are equal. If we use the ideal gas approximation for the component in the vapor phase, we have

$$\mu_{i,1}^0(p_0, T) + RT \ln a_i = \mu_{i,g}^0(p_0, T) + RT \ln (p_i/p_0) \tag{8.1.4}$$

in which the subscripts l and g indicate the liquid and gas phases. The physical meaning of the activity a_i can be seen as follows. Consider a pure liquid in equilibrium with its vapor. Then $p_i = p_i^*$, the vapor pressure of a pure liquid in equilibrium with its vapor. Since a_i is nearly equal to one for a pure liquid, $\ln a_i \approx 0$. Hence (8.1.4) can be written as

$$\mu_{i,1}^0(p_0, T) = \mu_{i,g}^0(p_0, T) + RT \ln (p_i^*/p_0) \tag{8.1.5}$$

Subtracting (8.1.5) from (8.1.4) we find that

$$\boxed{RT \ln a_i = RT \ln (p_i/p_i^*)} \quad \text{or} \quad \boxed{a_i = \frac{p_i}{p_i^*}} \tag{8.1.6}$$

that is, *the activity is the ratio of the partial vapor pressure of the component and the vapor pressure of the pure solvent*. The activity of a component can be determined by measuring its vapor pressure.

For an ideal solution, equation (8.1.4) takes the form

$$\mu^0_{i,1}(p, T) + RT \ln x_i = \mu^0_{i,g}(p_0, T) + RT \ln (p_i/p_0) \tag{8.1.7}$$

From this equation it follows that the partial pressure in the vapor phase and the mole fraction of a component can be written as

$$\boxed{p_i = K_i x_i} \tag{8.1.8}$$

in which

$$K_i(p, T) = p_0 \exp\{[\mu^0_{i,1}(p, T) - \mu^0_{i,g}(p_0, T)]/RT\} \tag{8.1.9}$$

As indicated, the term $K_i(p, T)$ is in general a function of p and T, but since the chemical potential of the liquid $\mu^0_{i,1}(p, T)$ changes little with p, it is essentially a function of T. K_i has the dimensions of pressure. For any component when $x_i = 1$, we must have $K(p^*, T) = p^*$, the vapor pressure of the pure substance (Fig. 8.2). (This is consistent with (8.1.9) because when we set $p = p_0 = p^*$, the exponent $\mu^0_1(p^*, T) - \mu^0_g(p^*, T) = 0$, because the chemical potentials of the vapor and the liquid are equal.) At a given temperature T, if $x_s \approx 1$ for a particular component, i.e., "the solvent", since the change of K_i is small for changes in pressure, we may write

$$\boxed{p_s = p^*_s x_s} \tag{8.1.10}$$

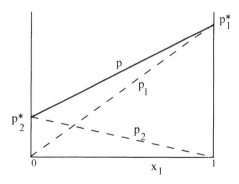

Figure 8.2 The vapor pressure diagram of a **perfect binary solution** for which (8.1.1) is valid over all values of the mole fraction x_1. p^*_1 and p^*_2 are the vapor pressures of the pure substances; p_1 and p_2 are the partial pressures of the two components in the mixture and p is the total vapor pressure

Experiments conducted by François-Marie Raoult (1830–1901) in the 1870s showed that if the mole fraction of the solvent is nearly equal to 1, i.e. for dilute solutions, equation (8.1.10) is valid. For this reason (8.1.10) is called the **Raoult's law**. The chemical potential of the vapor phase of the solvent $\mu_{s,g} = \mu_{s,g}(p_0, T) + RT \ln (p_s/p_0)$ can now be related to its mole fraction in the solution by using the Raoult's law and setting $p_0 = p^*$:

$$\mu_{s,g}(p, T, x_s) = \mu_{s,g}(p^*, T) + RT \ln x_s \qquad (8.1.11)$$

For a minor component of a solution, when its mole fraction $x_k \ll 1$, equation (8.1.10) is not valid but equation (8.1.8) is still valid. This is called **Henry's law** after William Henry (1774–1836), who studied this aspect for the solubility of gases [1]:

$$\boxed{p_i = K_i x_i \qquad (x_i \ll 1)} \qquad (8.1.12)$$

The constant K_i is called the Henry constant. Some values of the Henry constant are given in Table 8.1. In the region where Henry's law is valid, K_i is not equal to the vapor pressure of the pure substance. The graphical significance of the Henry constant is shown in Fig. 8.3. (Also, where Henry's law is valid, in general the chemical potential of the reference state μ_i^0 is not the same as the chemical potential of the pure substance.) Only for a perfect solution do we have $K_i = p_i^*$ when $x_i \ll 1$, but such solutions are very rare. Many dilute solutions obey Raoult's law and Henry's law.

When the solution is not dilute, the **nonideal** behavior is described using the activity coefficients γ_i in the chemical potential:

$$\mu_i(p, T, x_i) = \mu_i^0(p, T) + RT \ln \gamma_i x_i \qquad (8.1.13)$$

Table 8.1 Henry's law constants at 25 °C for atmospheric gases*

Gas	$K/10^4$ atm	Volume fraction in the atmosphere (ppm)
$N_2(g)$	8.5	780 900
$O_2(g)$	4.3	209 500
$Ar(g)$	4.0	9 300
$CO_2(g)$	0.16	300
$CO(g)$	5.7	–
$He(g)$	13.1	5.2
$H_2(g)$	7.8	0.5
$CH_4(g)$	4.1	1.5
$C_2H_2(g)$	0.13	–

* Henry's law constants for organic compounds can be found in data source [F].

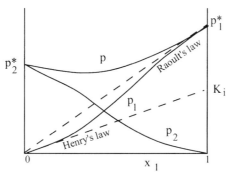

Figure 8.3 The vapor pressure diagram of a binary solution. When the mole fraction is very small or nearly equal to one, we have ideal behavior. When the component is the minor component, it obeys Henry's law; when it is the major component, it obeys Raoult's law. p_1^* and p_2^* are the vapor pressures of the pure substances; p_1 and p_2 are the partial pressures of the two components in the mixture; and p is the total vapor pressure. The deviation from the partial pressure predicted by Henry's law or Raoult's law can be used to obtain the activity coefficients

The deviation from Raoult's law or Henry's law is a measure of γ_i. For nonideal solutions, as an alternative to the activity coefficient, an **osmotic coefficient** ϕ_i is defined through

$$\mu_i(p,\, T,\, x_i) = \mu_i^0(p,\, T) + \phi_i RT \ln x_i \qquad (8.1.14)$$

As we will see in the following section, the significance of the osmotic coefficient lies in the fact that it is the ratio of the osmotic pressure to the osmotic pressure of an ideal solution. From (8.1.13) and (8.1.14) it is easy to see that

$$\phi_i - 1 = \frac{\ln \gamma_i}{\ln x_i} \qquad (8.1.15)$$

8.2 Colligative Properties

Using the chemical potential of ideal solutions, we can derive several properties of ideal solutions that depend on the *total number of solute particles** and not on

---

* Thus, a 0.2 M solution of NaCl will have a colligative concentration of 0.40 M due to the dissociation into Na^+ and Cl^-

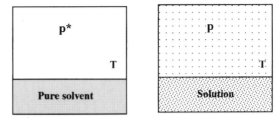

Figure 8.4 The vapor pressure of a solution with nonvolatile solute is lower than that of a pure solvent. Consequently, the boiling point of a solution increases with the solute concentration

the chemical nature of the solute. Such properties are collectively called **colligative properties**.

CHANGES IN BOILING AND FREEZING POINTS

Equation (8.1.11) could be used to obtain an expression for the increase in the boiling point and the decrease in the freezing point of solutions (Fig. 8.4). As we noted in Chapter 7, a liquid boils when its vapor pressure $p = p_{ext}$, the atmospheric or applied external pressure. Let T^* be the boiling temperature of the pure solvent and T the boiling temperature of the solution. *We assume the mole fraction of the solvent is x_2 and the mole fraction of the solute is x_1.* We assume the solute is nonvolatile, so the gas phase of the solution is pure solvent. At equilibrium, the chemical potentials of the liquid and gas phases of the solvent must be equal:

$$\mu_{2,g}^*(p_{ext}, T) = \mu_{2,1}(p_{ext}, T, x_2) \tag{8.2.1}$$

Using (8.1.11) we can now write this equation as

$$\mu_{2,g}^*(p_{ext}, T) = \mu_{2,1}^*(T) + RT \ln x_2 = \mu_{2,1}(p_{ext}, T, x_2) \tag{8.2.2}$$

where we have used $\mu_{2,1}^*(T) = \mu_{2,g}^*(p^*, T)$ valid for a pure liquid in equilibrium with its vapor.

Since the chemical potential of a pure substance $\mu = G_m$, the molar Gibbs free energy, we have

$$\frac{\mu_{2,g}^*(p_{ext}, T) - \mu_{2,1}^*(T)}{RT} = \frac{\Delta G_m}{RT} = \frac{\Delta H_m - T\Delta S_m}{RT} = \ln x_2 \tag{8.2.3}$$

in which Δ denotes the difference between the liquid and the gas phase. Generally ΔH_m does not vary much with temperature. Therefore, $\Delta H_m(T) = \Delta H_m(T^*) = \Delta H_{vap}$. Also, $\Delta S_m = \Delta H_{vap}/T^*$ and $x_2 = (1 - x_1)$, in which $x_1 \ll 1$ is the mole fraction of the solute. With these observations we can write equation (8.2.3) as

$$\ln (1 - x_1) = \frac{\Delta H_{vap}}{R} \left(\frac{1}{T} - \frac{1}{T^*} \right) \tag{8.2.4}$$

If the difference $T - T^* = \Delta T$ is small, it is easy to show that the terms containing T and T^* can be approximated to $-\Delta T/T^{*2}$. Furthermore, since $\ln (1 - x_1) \approx -x_1$ when $x_1 \ll 1$, we can approximate (8.2.4) by the relation

$$\boxed{\Delta T = \frac{RT^{*2}}{\Delta H_{vap}} x_1} \tag{8.2.5}$$

which relates the change in boiling point to the mole fraction of the solute. In a similar way, by considering a pure solid in equilibrium with the solution, one can derive the following relation for the decrease in freezing point ΔT in terms of the enthalpy of fusion, ΔH_{fus}, the mole fraction x_1 of the solute and the freezing point T^* of the pure solvent:

$$\boxed{\Delta T = \frac{RT^{*2}}{\Delta H_{fus}} x_1} \tag{8.2.6}$$

The changes in the boiling point and the freezing point are often expressed in terms of **molality** = *moles of solute per kilogram of solvent*, instead of mole fraction. For dilute solutions the conversion from mole fraction x_1 to molality m_1 is easily done. If M_s is the molar mass of the solvent in kg, the mole fraction of the solute is

$$x_1 = \frac{N_1}{N_1 + N_2} \approx \frac{N_1}{N_2} = M_s \left(\frac{N_1}{M_s N_2} \right) = M_s m_1$$

Equations (8.2.5) and (8.2.6) are often written as

$$\Delta T = K(m_1 + m_2 + \ldots m_s) \tag{8.2.7}$$

in which the molalities of all the s species of solute particles are shown explicitly. The constant K is called the **ebullioscopic constant** for changes in boiling point and the **cryoscopic constant** for changes in freezing point. Values of ebullioscopic and cryoscopic constants for some liquids are given in Table 8.2.

Table 8.2 Ebullioscopic and cryoscopic constants

Compound	$K_b\,(^\circ\mathrm{C\,kg\,mol^{-1}})$	$T_b(^\circ\mathrm{C})$	$K_f\,(^\circ\mathrm{C\,kg\,mol^{-1}})$	$T_f(^\circ\mathrm{C})$
Acetic Acid CH_3COOH	3.07	118	3.90	16.7
Acetone $(CH_3)_2CO$	1.71	56.3	2.40	-95
Benzene C_6H_6	2.53	80.10	5.12	5.53
Carbon disulfide CS_2	2.37	46.5	3.8	-111.9
Tetrachloromethane CCl_4	4.95	76.7	30	-23
Nitrobenzene $C_6H_5NO_2$	5.26	211	6.90	5.8
Phenol C_6H_5OH	3.04	181.84	7.27	40.92
Water H_2O	0.51	100.0	1.86	0.0

Source: G. W. C. Kaye and T. H. Laby (eds) *Tables of Physical and Chemical Constants*, 1986, Longman, London.

OSMOTIC PRESSURE

When a solution and a pure solvent are separated by a semipermeable membrane (Fig. 8.5(a)) that is permeable to the solvent but not to the solute molecules, the solvent flows into the chamber containing the solution until equilibrium is reached. This processes is called **osmosis** and was noticed in the mid eighteenth century. In 1877 a botanist named Pfeffer made a careful quantitative study of it. Jacobus Henricus van't Hoff (1852–1911), who was awarded the first Nobel prize for chemistry in 1901 for his contributions to thermodynamics and chemistry [1], found that a simple equation, similar to that of an ideal gas, could be used to describe the observed data.

As shown in Fig. 8.5, when the membrane separating the solution and the pure solvent is impermeable to the solute, the chemical potentials are not equal. The

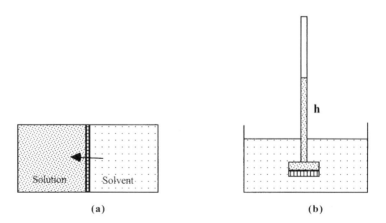

Figure 8.5 Osmosis: the pure solvent flows toward the solution through a semipermeable membrane until the chemical potentials in the two chambers are equal

Jacobus van't Hoff (1852–1911) (Reproduced, by permission, from the Emilio Segré Visual Archives of the American Institute of Physics)

affinity is

$$A = \mu^*(p,\, T) - \mu(p',\, T,\, x_2) \qquad (8.2.8)$$

in which x_2 is the mole fraction of the solvent, p' is the pressure of the solution and p the pressure of the pure solvent. A nonzero affinity is the driving force for the flow of solvent to the solution side. Equilibrium is reached when the chemical potentials become equal, and the corresponding affinity vanishes. Using equation (8.1.1) for an ideal solution, the affinity of this system can be written as

$$A = \mu^*(p,\, T) - \mu^*(p',\, T) - RT \ln x_2 \qquad (8.2.9)$$

in which $\mu^0 = \mu^*$, the chemical potential of the pure solvent.

When $p = p'$, the affinity takes the simple form

$$A = -RT \ln x_2 \tag{8.2.10}$$

The flow of solvent due to a nonzero affinity generates a pressure difference between the solvent and the solution. The flow continues until the difference between p and p' makes $A = 0$. The difference $(p - p')$ when $A = 0$, which we shall denote by π, is called the osmotic pressure. In an experimental setup (Fig. 8.5(b)) the liquid level in the solution rises to a height h above the pure solvent level when equilibrium is reached. The excess pressure in the solution $\pi = h\rho g$, in which ρ is the density of the solution and g is the acceleration due to gravity. At equilibrium, from (8.2.9) it follows that

$$A = 0 = \mu^*(p, T) - \mu^*(p + \pi, T) - RT \ln x_2 \tag{8.2.11}$$

At constant temperature the change in the chemical potential $d\mu = (\partial\mu/\partial p)_T \, dp = V_m \, dp$ where V_m is the partial molar volume. Since the partial molar volume of a liquid changes very little with pressure, we may assume it to be constant and equal to V_m^*. Hence we may write

$$\mu^*(p + \pi, T) \approx \mu^*(p, T) + \int_0^\pi V_m^* \, dp \tag{8.2.12}$$
$$= \mu^*(p, T) + V_m^* \pi$$

in which V_m^* is the molar volume of the pure solvent. (A better approximation would be to use the mean value of V_m^* in the pressure range 0 to π.) Also, for dilute solutions, $\ln x_2 = \ln(1 - x_1) \approx -x_1$. If N_1 is the number of moles of the solute and N_2 is the number of moles of the solvent, since $N_2 \gg N_1$, we see that $x_1 = N_1/(N_2 + N_1) \approx N_1/N_2$. Hence $\ln x_2 \approx -N_1/N_2$. Using (8.2.12) and the fact that $\ln x_2 \approx -N_1/N_2$, equation (8.2.11) can be written as

$$RT \frac{N_1}{N_2} = V_m^* \pi \tag{8.2.13}$$
$$RTN_1 = N_2 V_m^* \pi = V\pi$$

in which $V = N_2 V_m^*$ is nearly the volume of the solution (the correction due to solute is negligible). This shows that the osmotic pressure π obeys an ideal-gas like equation,

$$\boxed{\pi = \frac{N_{\text{solute}} RT}{V_{\text{solution}}} = [S]RT} \tag{8.2.14}$$

in which [S] is the molar concentration of the solution. This is the **van't Hoff**

Table 8.3 Comparison between theoretical osmotic pressure calculated using van't Hoff's equation and the experimentally observed osmotic pressure for an aqueous solution of sucrose at two temperatures

| | $T = 273\,\mathrm{K}$ | | | $T = 333\,\mathrm{K}$ | |
| | π (atm) | | | π (atm) | |
Concentration $(\mathrm{mol\,L^{-1}})$	Experiment	Theory	Concentration $(\mathrm{mol\,L^{-1}})$	Experiment	Theory
0.029 22	0.65	0.655	0.098	2.72	2.68
0.058 43	1.27	1.330	0.1923	5.44	5.25
0.1315	2.91	2.95	0.3701	10.87	10.11
0.2739	6.23	6.14	0.533	16.54	14.65
0.5328	14.21	11.95	0.6855	22.33	18.8
0.8766	26.80	19.70	0.8273	28.37	22.7

Source: I. Prigogine and R. Defay, Chemical Thermodynamics, 1967, Longman, London.

equation for the osmotic pressure. It is as if an ideal gas consisting of the solute particles is occupying a volume equal to the solution volume. By measuring the osmotic pressure one can determine the number of moles of a solute. Thus, if the mass of the solute is known, its molecular weight can be calculated. The measurement of osmotic pressure is used to determine the molecular weight of large biomolecules for which semipermeable membranes can be easily found (exc. 8.10).

Table 8.3 shows a comparison between experimentally measured osmotic pressures and those calculated using the van't Hoff equation (8.2.14) for an aqueous solution of sucrose. We see that for concentrations up to about 0.2 M van't Hoff equation agrees with experimental values. Deviation from the van't Hoff equation is not necessarily due to deviation from ideality. In deriving the van't Hoff equation, we also assumed a dilute solution. Using (8.1.11) and (8.2.12) it is easy to see that the osmotic pressure can also be written as

$$\pi_{\mathrm{id}} = \frac{-RT \ln x_2}{V_{\mathrm{m}}^*} \tag{8.2.15}$$

where x_2 is the mole fraction of the solvent. Here we have indicated explicitly that the osmotic pressure in this expression is valid for ideal solutions. This formula was obtained by J.J. van Laar in 1894.

For nonideal solutions, instead of using the activity coefficient γ, an osmotic coefficient ϕ is defined as in (8.1.14) through

$$\mu(p, T, x_2) = \mu^*(p, T) + \phi RT \ln x_2 \tag{8.2.16}$$

in which μ^* is the chemical potential of the pure solvent. At equilibrium, when the affinity vanishes and the osmotic pressure is π, we have the equation

$$\mu^*(p, T) = \mu^*(p + \pi, T) + \phi RT \ln x_2 \qquad (8.2.17)$$

Following the same procedure as above, we arrive at the following expression for the osmotic pressure of a nonideal solution:

$$\boxed{\pi = \frac{-\phi RT \ln x_2}{V_m^*}} \qquad (8.2.18)$$

Equation (8.2.18) was proposed by Donnan and Guggenheim in 1932. From eqns. (8.2.15) and (8.2.18) it follows that $\phi = \pi/\pi_{\rm id}$. Hence the name "osmotic coefficient" for ϕ. Also the affinity can be related to the osmotic pressure. When the solution and the pure solvent are at the same pressure, the affinity is $A = \mu^*(p, T) - \mu^*(p, T) - \phi RT \ln x_2 = -\phi RT \ln x_2$. Using this in (8.2.18) we see that

$$\pi = \frac{A}{V_m^*} \qquad \text{when} \qquad p_{\rm solution} = p_{\rm solvent} \qquad (8.2.19)$$

Another approach for nonideal solutions is similar to the virial equation description of real gases. In this case the osmotic pressure is written as

$$\pi = [S]RT\{1 + B(T)[S] + \ldots\} \qquad (8.2.20)$$

in which $B(T)$ is a constant that depends on the temperature. The experimental data on the osmotic pressure of polymer solutions (such as polyvinyl chloride in cyclohexanone) shows a fairly linear relation between $\pi/[S]$ and $[S]$. Also, the value of B changes sign from negative to positive as the temperature increases. The temperature at which B equals zero is called the **theta temperature**. If the concentration is expressed in g/L, which we shall denote by $[C]$, then (8.2.20) can be written as

$$\boxed{\pi = \frac{[C]RT}{M_s}\left\{1 + B(T)\frac{[C]}{M_s} + \right\}} \qquad (8.2.21)$$

in which M_s is the molecular weight of the solute. With this equation, a plot of $\pi/[C]$ versus $[C]$ is expected to be linear with an intercept equal to (RT/M_s). From the intercept, the molecular weight can be determined. From the slope, equal to (BRT/M_s^2), the "virial constant" B can be obtained.

8.3   Solubility Equilibrium

The solubility of a solid in a solvent depends on the temperature. **Solubility** is the concentration when the solid solute is in equilibrium with the solution; it is the concentration at saturation. Thermodynamics gives us a quantitative relation between solubility and temperature. In considering the solubilities of solids, one must distinguish between ionic solutions and nonionic solutions. When an ionic solid, such as NaCl, is dissolved in a polar solvent, such as water, the solution contains ions, Na^+ and Cl^-. Since ions interact strongly, even in dilute solutions, the activities cannot be approximated well by mole fractions. For nonionic solutions—e.g. sugar in water, or naphthalene in acetone—the activity of a dilute solution can be approximated by the mole fraction.

NONIONIC SOLUTIONS

For dilute nonionic solutions, we may assume ideality and use the expression (8.1.1) for the chemical potential to analyze the conditions for thermodynamic equilibrium. Solutions of higher concentration require a more detailed theory [2]. Recall that, like liquids, the chemical potential for a solid varies very little with pressure, so it is essentially a function only of the temperature. If $\mu_s^*(T)$ is the chemical potential of the pure solid in equilibrium with the liquid, equation (8.1.1) gives

$$\mu_s^*(T) = \mu_1(T) = \mu_1^*(T) + RT \ln x_1 \qquad (8.3.1)$$

in which μ_1 is the chemical potential of the solute in the solution phase (liquid phase), μ_1^* is the chemical potential of the pure solute in the liquid phase, and x_1 is the mole fraction of the solute. If $\Delta G_{fus}(T) = \mu_1^* - \mu_s^*$ is the molar Gibbs free energy of fusion at temperature T, equation (8.3.1) can be written in the form

$$\ln x_1 = -\frac{1}{R}\frac{\Delta G_{fus}}{T} \qquad (8.3.2)$$

In this form the temperature dependence of the solubility is not explicit because ΔG_{fus} is itself a function of T. This expression can also be written in terms of the enthalpy of fusion, ΔH_{fus}, by differentiating it with respect to T and using the Gibbs–Helmholtz equation $d(\Delta G/T)/dT = -\Delta H/T^2$ (5.2.14):

$$\boxed{\frac{d \ln x_1}{dT} = \frac{1}{R}\frac{\Delta H_{fus}}{T^2}} \qquad (8.3.3)$$

Since ΔH_{fus} does not change much with T, this expression can be integrated to obtain a more explicit dependence of solubility with temperature.

IONIC SOLUTIONS

Ionic solutions, also called **electrolytes**, are dominated by electrical forces which can be very strong. To get an idea of the strength of electrical forces, it is instructive to calculate the force of repulsion between two cubes of copper of side 1 cm in which one in a million Cu atoms is a Cu^+ ion, when the two cubes are 10 cm apart (exc 8.13). The force is sufficient to lift a 16×16^6 kg weight!. Due to the strength of electrical forces, there is almost no separation between positive and negative ions in a solution; positive and negative charges aggregate to make the net charge in every macroscopic volume nearly zero, i.e. electrically neutral. Solutions, and indeed most matter, maintain *electroneutrality* to a high degree. Thus if c_k represents the concentrations (mol/L) of positive and negative ions with *ion numbers* (number of electronic charges) z_k, the total charge carried by an ion per unit volume is Fz_kc_k, in which $F = eN_A$ is the Faraday constant, equal to the product of the electronic charge $e = 1.609 \times 10^{-19}$ C and the Avogadro number N_A. Since electroneutrality implies that the net charge is zero, we have

$$\sum_k Fz_kc_k = 0 \qquad (8.3.4)$$

Let us consider the solubility equilibrium of a sparingly soluble electrolyte AgCl in water:

$$AgCl(s) \rightleftharpoons Ag^+ + Cl^- \qquad (8.3.5)$$

At equilibrium

$$\mu_{AgCl} = \mu_{Ag^+} + \mu_{Cl^-} \qquad (8.3.6)$$

In ionic systems, since the positive and negative ions always come in pairs, physically it is not possible to measure the chemical potentials μ_{Ag^+} and μ_{Cl^-} separately; only the sum can be measured. A similar problem arises for the definition of enthalpy and Gibbs free energy of formation. For ions, these two quantities are defined with respect to a new reference state based on the H^+ ions, as described in Box 8.1. For the chemical potential, a **mean chemical potential** is defined by

$$\mu_\pm = \tfrac{1}{2}(\mu_{Ag^+} + \mu_{Cl^-}) \qquad (8.3.7)$$

so that (8.3.6) becomes

$$\mu_{AgCl} = 2\mu_\pm \qquad (8.3.8)$$

In general, for the decomposition of a neutral compound W into positive and negative ions, A^{z+} and B^{z-} (with ion numbers z^+ and z^-), we have

$$W \rightleftharpoons \nu_+ A^{z+} + \nu_- B^{z-} \qquad (8.3.9)$$

Box 8.1 Enthalpy and Gibbs Free Energy of Formation for Ions

When ionic solutions form, the ions occur in pairs, so it is not possible to isolate the enthalpy of formation of a positive or negative ion. Hence it is not possible to obtain the heats of formation of ions with the usual reference state of elements in their standard state. The enthalpy of formation for ions is tabulated by defining ΔH_f for H^+ as zero at all temperatures. Thus

$$\Delta H_f^0[H^+(aq)] = 0 \quad \text{at all temperatures}$$

With this definition it is now possible to obtain ΔH_f of all other ions. For example, to obtain the heat of formation of $Cl^-(aq)$ at a temperature T, the enthalpy of solution of HCl is measured. Thus, $\Delta H_f^0[Cl^-(aq)]$ is the heat of solution at temperature T:

$$HCl \rightarrow H^+(aq) + Cl^-(aq)$$

The tabulated values of enthalpy are based on this convention. Similarly, for the Gibbs free energy

$$\Delta G_f^0[H^+(aq)] = 0 \quad \text{at all temperatures}$$

For ionic systems it has become customary to use the **molality scale** (mol/kg solvent). This scale has the advantage that the addition of another solute does not change the molality of a given solute. Values of ΔG_f^0 and ΔH_f^0 for the formation of ions in water at $T = 298.15\,K$ are tabulated for the standard state of an ideal solution at a concentration of 1 mol/kg. This standard state is given the subscript ao. Thus the chemical potential or the activity of an ion is indicated by **ao**. The chemical potential of an ionized salt, $\mu_{salt} \equiv \nu_+\mu_+ + \nu_-\mu_-$, and the corresponding activity is denoted by the subscript **ai.**

---

in which ν_+ and ν_- are the stoichiometric coefficients. The mean chemical potential in this case is defined as

$$\mu_{\pm} = \frac{(\nu_+\mu_+ + \nu_-\mu_-)}{\nu_+ + \nu_-} = \frac{\mu_{salt}}{\nu_+ + \nu_-}$$

$$\mu_{salt} \equiv \nu_+\mu_+ + \nu_-\mu_- \tag{8.3.10}$$

in which we have written the chemical potential of the positive ion A^{z+} as μ_+ and the chemical potential of the negative ion, B^{z-}, as μ_-.

The activity coefficients γ of electrolytes are defined with respect to ideal solutions. For example, the mean chemical potential for AgCl is written as

$$\mu_{\pm} = \tfrac{1}{2}\left(\mu_{Ag^+}^0 + RT\ln\left(\gamma_{Ag^+}x_{Ag^+}\right) + \mu_{Cl^-}^0 + RT\ln\left(\gamma_{Cl^-}x_{Cl^-}\right)\right)$$
$$= \mu_{\pm}^0 + RT\ln\sqrt{\gamma_{Ag^+}\gamma_{Cl^-}x_{Ag^+}x_{Cl^-}} \tag{8.3.11}$$

where $\mu_\pm^0 = \frac{1}{2}(\mu_{Ag^+}^0 + \mu_{Cl^-}^0)$. Once again, since the activity coefficients of the positive and negative ions cannot be measured individually, a mean activity coefficient $\gamma_\pm$ is defined by

$$\gamma_\pm = (\gamma_{Ag^+}\gamma_{Cl^-})^{1/2} \tag{8.3.12}$$

In the more general case of (8.3.9), the **mean ionic activity coefficient** is defined as

$$\boxed{\gamma_\pm = (\gamma_+^{\nu_+}\gamma_-^{\nu_-})^{1/(\nu_+ + \nu_-)}} \tag{8.3.13}$$

where we have used γ_+ and γ_- for the activity coefficients of the positive and negative ions.

The chemical potentials of dilute solutions may be expressed in terms of molality m_k (moles of solute per kilogram of solvent) or molarities c_k (moles of solute per liter of solution) instead of mole fractions x_k. In electrochemistry it is more common to use molality m_k. For dilute solutions, since $x_k = (N_k/N_{solvent})$, we have the following conversion formulas for the different units

$$x_k = m_k M_s \quad \text{and} \quad x_k = V_{ms}c_k \tag{8.3.14}$$

in which M_s is the molar mass of the solvent in kilograms and V_{ms} the molar volume of the solvent in liters. The corresponding chemical potentials are then written as

$$\mu_k^x = \mu_k^{x0} + RT \ln \gamma_k x_k \tag{8.3.15}$$

$$\mu_k^m = \mu_k^{x0} + RT \ln M_s + RT \ln \gamma_k m_k$$
$$= \mu_k^{m0} + RT \ln (\gamma_k m_k/m^0) \tag{8.3.16}$$

$$\mu_k^c = \mu_k^{x0} + RT \ln V_{ms} + RT \ln \gamma_k c_k$$
$$= \mu_k^{c0} + RT \ln (\gamma_k c_k/c^0) \tag{8.3.17}$$

in which the definition of the reference chemical potentials μ_k^{m0} and μ_k^{c0} in each concentration scale is self-evident. The acitivity in the molality scale is written in the dimensionless form as $a_k = \gamma_k m_k/m^0$ in which m^0 is the standard value of molality equal to 1 mole of solute per kilogram of solvent. Similarly, the activity in the molarity scale is written as $a_k = \gamma_k c_k/c^0$ in which c^0 equals 1 mole per liter of solution. For electrolytes the mean chemical potential $\mu_\pm$ is usually expressed in the molality scale; the tabulation of ΔG_f^0 and ΔH_f^0 for the formation of ions in water at $T = 298.15$ K is usually for the standard state of an ideal solution at a concentration of 1 mol/kg. This standard state is given the subscript ao.

In the commonly used molality scale, the solution equilibrium of AgCl expressed in (8.3.8) can now be written as

$$\mu^0_{AgCl} + RT \ln a_{AgCl} = 2\mu^{m0}_{\pm} + RT \ln \left(\frac{\gamma^2_{\pm} m_{Ag^+} m_{Cl^-}}{(m^0)^2} \right) \tag{8.3.18}$$

Since the activity of a solid is nearly equal to one, $a_{AgCl} \approx 1$. Hence we obtain the following expression for the equilibrium constant[*] for solubility in the molality scale

$$K_m(T) \equiv \frac{\gamma^2_{\pm} m_{Ag^+} m_{Cl^-}}{(m^0)^2} = a_{Ag^+} a_{Cl^-} = \exp \left[\frac{\mu^0_{AgCl} - 2\mu^{m0}_{\pm}}{RT} \right] \tag{8.3.19}$$

The equilibrium constant for electrolytes is also called the **solubility product** K_{sp}. For sparingly soluble electrolytes such as AgCl, even at saturation, the solution is very dilute and $\gamma_{\pm} \approx 1$. In this limiting case the solubility product

$$K_{sp} \approx m_{Ag^+} m_{Cl^-} \tag{8.3.20}$$

in which we have not explicitly included m^0 which has a value of one.

ACTIVITY, IONIC STRENGTH AND SOLUBILITY

A theory of ionic solutions developed by Peter Debye and Erich Hückel in 1923 (which is based on statistical mechanics and beyond the scope of this text) provides an expression for the activity. We shall only state the main result of this theory, which works well for dilute electrolytes. The activity depends on a quantity called the ionic strength I, defined by

$$I = \frac{1}{2} \sum_k z^2_k m_k \tag{8.3.21}$$

The activity coefficient of an ion k in the molality scale is given by

$$\log_{10} \gamma_k = -Az^2_k \sqrt{I} \tag{8.3.22}$$

in which

$$A = \frac{N^2_A}{2.3026} \left(\frac{2\pi \rho_s}{R^3 T^3} \right)^{1/2} \left(\frac{e^2}{4\pi \varepsilon_0 \varepsilon_r} \right)^{3/2} \tag{8.3.23}$$

where N_A is the Avogadro number, ρ_s is the density of the solvent, e is the electronic charge, $\varepsilon_0 = 8.854 \times 10^{-12} C^2 N^{-1} m^{-2}$ is the permittivity of a vacuum, and ε_r is the relative permitivity of the solvent ($\varepsilon_r = 78.54$ for water).

---

[*] The general definition of equilibrium constant is given in chapter 9.

For ions in water at $T = 298.15$ K, we find $A = 0.509$ kg$^{1/2}$ mol$^{-1/2}$. Thus at 25 °C the activity of ions in dilute solutions can be approximated well by

$$\log_{10}(\gamma_k) = -0.509 z_k^2 \sqrt{I} \qquad (8.3.24)$$

Debye–Hückel theory tells us how solubility is influenced by ionic strength. For example, let us look at the solubility of AgCl. If $m_{Ag^+} = m_{Cl^-} = S$, the *solubility*, we may write the equilibrium constant K_m as

$$K_m(T) \equiv \gamma_\pm^2 m_{Ag^+} m_{Cl^-} = \gamma_\pm^2 S^2 \qquad (8.3.25)$$

The ionic strength depends not only on the concentration of Ag^+ and Cl^- ions but also on all the other ions. Thus, for example, the addition of nitric acid HNO_3, which adds H^+ and NO_3^- ions to the system, will change the activity coefficient. But the equilibrium constant, which is a function of T only (8.3.19), remains constant if T is constant. As a result, the value of m (or solubility in molal) will change with the ionic strength I. If the concentration of nitric acid (which dissociates completely) is m_{HNO_3}, the ionic strength will be

$$I = \tfrac{1}{2}(m_{Ag^+} + m_{Cl^-} + m_{H^+} + m_{NO_3^-})$$
$$= S + m_{HNO_3} \qquad (8.3.26)$$

Using (8.3.12) for $\gamma_\pm$ of AgCl and (8.3.24) in (8.3.25), we can obtain the following relation between the solubility S of AgCl and the concentration of HNO_3:

$$\log_{10} S = \tfrac{1}{2}\log_{10} K_m(T) + 0.509\sqrt{S + m_{HNO_3}} \qquad (8.3.27)$$

If $S \ll m_{HNO_3}$ then the above relation can be approximated by

$$\log_{10} S = \tfrac{1}{2}\log_{10} K_m(T) + 0.509\sqrt{m_{HNO_3}} \qquad (8.3.28)$$

Thus, a plot of $\log S$ versus $\sqrt{m_{HNO_3}}$ should yield a straight line, which is indeed found to be the case experimentally. In fact, such plots can be used to determine the equilibrium constant K_m and the activity coefficients.

8.4 Mixing and Excess Functions

PERFECT SOLUTIONS

A perfect solution is one for which the chemical potential of the form $\mu_k(p, T, x_k) = \mu_k^*(p, T) + RT \ln x_k$ is valid for all values of the mole fraction

x_k. The molar Gibbs free energy of such a solution is

$$G_m = \sum_k x_k \mu_k = \sum_k x_k \mu_k^* + RT \sum_k x_k \ln x_k \qquad (8.4.1)$$

If each of the components were separated, the total Gibbs free energy for the components would be the sum $G_m^* = \sum_k x_k G_{mk}^* = \sum_k x_k \mu_k^*$ in which we have used the fact that for a pure substance the molar Gibbs free energy of k, G_{mk}^*, is equal to the chemical potential μ_k^*. Hence the change (decrease) in the *molar* Gibbs free energy due to the *mixing* of the components in the solution is

$$\Delta G_{mix} = RT \sum_k x_k \ln x_k \qquad (8.4.2)$$

and

$$G_m = \sum_k x_k G_{mk}^* + \Delta G_{mix} \qquad (8.4.3)$$

Since the molar entropy $S_m = -(\partial G_m / \partial T)_p$, it follows from (8.4.2) and (8.4.3) that

$$S_m = \sum_k x_k S_{mk}^* + \Delta S_{mix} \qquad (8.4.4)$$

$$\Delta S_{mix} = -R \sum_k x_k \ln x_k \qquad (8.4.5)$$

where ΔS_{mix} is the molar entropy of mixing. This shows that during the formation of a perfect solution from pure components at a fixed temperature, the decrease in G is $\Delta G_{mix} = -T \Delta S_{mix}$. Since $\Delta G = \Delta H - T \Delta S$ we can conclude that $\Delta H = 0$ for the formation of a perfect solution at a fixed temperature. This can be verified explicitly by noting that the Gibbs–Helmholtz equation (5.2.13) can be used to evaluate the enthalpy. For G given by (8.4.2) and (8.4.3) we find

$$H_m = -T^2 \left(\frac{\partial}{\partial T} \frac{G_m}{T} \right) = \sum_k x_k H_{mk}^* \qquad (8.4.6)$$

Thus, the enthalpy of the solution is the same as the enthalpy of the pure components; *there is no change in the enthalpy of a perfect solution due to mixing*. Similarly, by noting that $V_m = (\partial G_m / \partial p)_T$, it is easy to see (exc. 8.16) that there is no change in the molar volume due to mixing, i.e. $\Delta V_{mix} = 0$. Furthermore, since $\Delta U = \Delta H - p \Delta V$, we see also that $\Delta U_{mix} = 0$. Thus, for a perfect solution, the *molar quantities* for mixing are

$$\Delta G_{\text{mix}} = RT \sum_k x_k \ln x_k \qquad (8.4.7)$$

$$\Delta S_{\text{mix}} = -R \sum_k x_k \ln x_k \qquad (8.4.8)$$

$$\Delta H_{\text{mix}} = 0 \qquad (8.4.9)$$

$$\Delta V_{\text{mix}} = 0 \qquad (8.4.10)$$

$$\Delta U_{\text{mix}} = 0 \qquad (8.4.11)$$

In a perfect solution the irreversible process of mixing of the components at constant p and T is entirely due to the increase in entropy; no heat is evolved or absorbed.

IDEAL SOLUTIONS

Dilute solution may be ideal over a small range of mole fractions x_i. In this case the molar enthalpy H_{m} and the molar volume V_{m} may be linear functions of the partial molar enthalpies $H_{\text{m}i}$ and the partial molar volumes $H_{\text{m}i}$. Thus

$$H_{\text{m}} = \sum_k x_i H_{\text{m}i} \quad \text{and} \quad V_{\text{m}} = \sum_i x_i V_{\text{m}i} \qquad (8.4.12)$$

However, the partial molar enthalpies $H_{\text{m}i}$ may not be equal to the molar enthalpies of pure substances if the corresponding mole fractions are small. The same is true for the partial molar volumes. However, if the x_i are nearly equal to one, then the $H_{\text{m}i}$ will be nearly equal to the molar enthalpy of the pure substance. A dilute solution for which (8.4.12) is valid will exhibit ideal behavior, but it may have a nonzero enthalpy of mixing. To see this more explicitly, consider a dilute binary solution $(x_1 \gg x_2)$ for which $H_{\text{m}1}^*$ and $H_{\text{m}2}^*$ are the molar enthalpies of the two pure components. Then, before mixing, the molar enthalpy is

$$H_{\text{m}}^* = x_1 H_{\text{m}1}^* + x_2 H_{\text{m}2}^* \qquad (8.4.13)$$

After mixing, since the major component (for which $x_1 \approx 1$) has $H_{\text{m}1}^* = H_{\text{m}1}$, the molar enthalpy will be

$$H_{\text{m}} = x_1 H_{\text{m}1}^* + x_2 H_{\text{m}2} \qquad (8.4.14)$$

The molar enthalpy of mixing is then the difference between the above two enthalpies

$$\Delta H_{\text{mix}} = H_{\text{m}} - H_{\text{m}}^* = x_2 (H_{\text{m}2} - H_{\text{m}2}^*) \qquad (8.4.15)$$

In this way, an ideal solutions may have a nonzero enthalpy of mixing. The same may be true for the volume of mixing.

EXCESS FUNCTIONS

For nonideal solutions, the molar Gibbs free energy of mixing is

$$\Delta G_{\text{mix}} = RT \sum_i x_i \ln \gamma_i x_i \qquad (8.4.16)$$

The difference between the Gibbs free energies of mixing of ideal and nonideal solutions is called the **excess Gibbs free energy**, which we shall denote by ΔG_E. From (8.4.7) and (8.4.16) it follows that

$$\Delta G_E = RT \sum_i x_i \ln \gamma_i \qquad (8.4.17)$$

Other excess functions, such as excess entropy and enthalpy, can be obtained from ΔG_E. For example, we have

$$\Delta S_E = -\left(\frac{\partial \Delta G_E}{\partial T}\right)_p = -RT \sum_i x_i \frac{\partial \ln \gamma_i}{\partial T} - R \sum_i x_i \ln \gamma_i \qquad (8.4.18)$$

Similarly ΔH_E can be obtained from the relation

$$\Delta H_E = -T^2 \left(\frac{\partial}{\partial T} \frac{\Delta G_E}{T}\right)$$

These excess functions can be obtained experimentally through measurements of vapor pressure and heats of reaction (Fig. 8.6).

REGULAR AND ATHERMAL SOLUTIONS

Nonideal solutions may be classified into two limiting cases. In one limiting case, called **regular solutions**, $\Delta G_E \approx \Delta H_E$, i.e. most of the deviation from ideality is due to the excess enthalpy of mixing. Since $\Delta G_E = \Delta H_E - T\Delta S_E$, it follows that for regular solutions $\Delta S_E \approx 0$. Furthermore, since $\Delta S_E = -(\partial \Delta G_E/\partial T)_p$, from (8.4.18) it follows that the activity coefficients are given by

$$\ln \gamma_i \propto \frac{1}{T} \qquad (8.4.19)$$

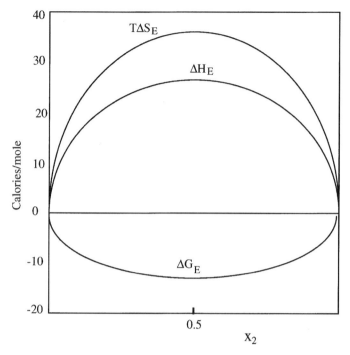

Figure 8.6 Thermodynamic excess function for a solution of *n*-heptane (component 1) and *n*-hexadecane (component 2) at 20 °C. The graph shows molar excess functions as a function of the mole fraction x_2 of *n*-hexadecane

For a special class of solutions called regular binary solutions the activities may be approximated by the function $\ln\gamma_k = \alpha x_k^2/RT$.

The other limiting case of nonideal solutions is when $\Delta G_E \approx -T\Delta S_E$, in which case the deviation from ideality is mostly due to the excess entropy of mixing; $\Delta H_E \approx 0$. In this case, using (8.4.17) in

$$\Delta H_E = -T^2\left(\frac{\partial}{\partial T}\frac{\Delta G_E}{T}\right)$$

we see that the $\ln\gamma_i$ are independent of *T*. Such solutions are called **athermal solutions**. Solutions in which the component molecules are of nearly the same size but differ in intermolecular forces generally behave like regular solutions. If the component molecules have very different sizes but do not differ significantly in their intermolecular forces, as in the case of some monomers and polymers, the solution behaves like an athermal solution.

8.5 Azeotropy

We shall now apply the ideas of this chapter to the azeotropes introduced
in Chapter 7. For an azeotrope in equilibrium with its vapor, the composition of
the liquid and the vapor phases are the same. At a fixed pressure, a liquid
mixture is an azeotrope at a particular composition called the **azeotropic
composition**. An **azeotropic transformation** is one in which there is an
exchange of matter between two phases without change in composition. In
this regard, an azeotrope is similar to the vaporization of a pure substance.
This enables us to obtain the activity coefficients of azeotropes just as for a
pure substance.

Let us consider a binary azeotrope. As we have seen in secion 8.1, the
chemical potentials of the components can be written in the form
$\mu_k(T, p, x_k) = \mu_k^0(T, p) + RT \ln \gamma_k x_k$, in which activity coefficient γ_k is a
measure of the deviation from ideality. If $\gamma_{k,1}$ and $\gamma_{k,g}$ are the activity
coefficients of component k in the liquid and gas phases respectively, then by
considering an azeotropic transformation, we can show (exc. 8.17) that

$$\ln \left(\frac{\gamma_{k,g}}{\gamma_{k,1}} \right) = \int_{T_k^*}^{T} \frac{\Delta H_{\text{vap},k}}{RT^2} dT - \frac{1}{RT} \int_{p^*}^{p} \Delta V_{\text{m},k}^* \, dp \qquad (8.5.1)$$

in which $\Delta H_{\text{vap},k}$ is the heat of vaporization of component k, and $\Delta V_{\text{m},k}^*$ is
the change in the molar volume of the pure component between the liquid
and the vapor phases. T^* is the boiling point of a pure component at pressure p^*.
If we consider an azeotropic transformation at a fixed pressure e.g.
$p = p^* = 1$ atm, then since ΔH_{vap} generally does not change much with T,
we obtain

$$\ln \left(\frac{\gamma_{k,g}}{\gamma_{k,1}} \right) = \frac{-\Delta H_{\text{vap},k}}{R} \left(\frac{1}{T} - \frac{1}{T^*} \right) \qquad (8.5.2)$$

For the activity coefficient of the vapor phase, if we use the ideal gas
approximation $\gamma_{k,g} = 1$. This gives us an explicit expression for the activity
coefficient of the liquid phase:

$$\ln \gamma_{k,1} = \frac{\Delta H_{\text{vap},k}}{R} \left(\frac{1}{T} - \frac{1}{T^*} \right) \qquad (8.5.3)$$

This expression allows the activity coefficient to be calculated for a component
of an azeotrope, and it gives a simple physical meaning to the activity
coefficient. More on azeotropes can be found in [3].

References

1. Laidler, K. J., *The World of Physical Chemistry.* 1993, Oxford: Oxford University Press.
2. Prigogine, I. and Defay, R., *Chemical Thermodynamics*, 4th ed. 1967, London: Longman.
3. Prigogine, I. *Molecular Theory of Solutions.* 1957, New York: Interscience Publishers.

Data Sources

[A] NBS Table of chemical and thermodynamic properties. *J. Phys. Chem. Reference Data*, **11**, suppl. 2 (1982).
[B] Kaye, G. W. C. and Laby, T. H. (eds.), *Tables of Physical and Chemical Constants*, 1986, London: Longman.
[C] Prigogine, I. and Defay, R., *Chemical Thermodynamics*, 1967, London: Longman.
[D] Emsley, J., *The Elements* 1989, Oxford: Oxford University Press.
[E] Pauling, L., *The Nature of the Chemical Bond*, 1960, Ithaca NY: Cornell University Press.
[F] Lide, D. R., (ed.): *CRC Handbook of Chemistry and Physics*, 75th ed. 1994, Ann Arbor MI: CRC Press.
[G] The web site of the National Institute for Standards and Technology, http://webbook.nist.gov.

Examples

Example 8.1   In the oceans, to a depth of about $100\,m$ the concentration of O_2 is about $0.25 \times 10^{-3}\,mol/L$. Compare this with the value obtained using Henry's law, assuming equilibrium between the atmospheric oxygen and the dissolved oxygen.

Solution: The partial pressure p_{O_2} of O_2 in the atmosphere is about 0.2 atm. Using the Henry's law constant in Table 8.1, we have for the mole fraction x_{O_2} of the dissolved oxygen

$$p_{O_2} = K_{O_2} x_{O_2}$$

Hence

$$x_{O_2} = \frac{p_{O_2}}{K_{O_2}} = \frac{0.21\,\text{atm}}{4.3 \times 10^4\,\text{atm}} = 4.8 \times 10^{-6}$$

i.e. there is 4.6×10^{-6} mole of O_2 per mole of H_2O. Noting that $1\,L$ of H_2O is equal to 55.55 moles, this mole fraction of O_2 can be converted to a concentration in mol/L:

$$[O_2] = 4.6 \times 10^{-6} \times 55.5 = 2.7 \times 10^{-4} \, \text{mol/L}$$

which is nearly equal to the measured concentration of O_2 in the oceans.

Example 8.2   In an aqueous solution of NH_3 at $25.0\,°C$, the mole fraction of NH_3 is 0.05. For this solution, assuming ideality, calculate the partial pressure of water vapor. If the vapor pressure is found to be $3.40\,\text{kPa}$, what is the activity, a, of water, and what is its activity coefficient γ?

Solution: If p^* is the vapor pressure of pure water at $25.0\,°C$, then according to Raoult's law (8.1.10), the vapor pressure of the ammonia solution is given by $p = x_{H_2O}p^* = 0.95p^*$. The value of p^* can be obtained as follows. Since water boils at $373.0\,K$ under $p = 1.0\,\text{atm}\,(101.3\,\text{kPa})$, we know that its vapor pressure is $101.3\,\text{kPa}$ at $373.0\,K$. Using the Clausius–Clapeyron equation, we can calculate the vapor pressure at $25\,°C\,(298.0\,K)$:

$$\ln p_1 - \ln p_2 = \frac{\Delta H_{\text{vap}}}{R}\left(\frac{1}{T_2} - \frac{1}{T_1}\right)$$

With $p_2 = 1\,\text{atm}$, $T_2 = 373.0$, $T_1 = 298.0$, $\Delta H_{\text{vap}} = 40.66\,\text{kJ/mol}$ (Table 7.1) the vapor pressure p_1 (in atm) can be computed

$$\ln p_1 = -3.299$$

i.e. $p_1 = \text{Exp}(-3.299) = 0.0369\,\text{atm} = 101.3 \times 0.0369\,\text{kPa}$
$$= 3.73\,\text{kPa} = p^*$$

Hence the vapor pressure p^* of pure water at $25°\,C$ is $3.73\,\text{kPa}$. For the ammonia solution in which the mole fraction of water is 0.95, according to Raoult's law, the vapor pressure for an ideal solution should be

$$p = 0.95 \times 3.73\,\text{kPa} = 3.54\,\text{kPa}$$

For an ideal solution the activity a is the same as the mole fraction x_1. As shown in (8.1.6), in the general case the activity $a = p/p^*$. Hence, if the measured vapor pressure is $3.40\,\text{kPa}$, the activity is

$$a_1 = 3.40/3.73 = 0.909$$

The activity coefficient is defined by $a_k = \gamma_k x_k$. Hence $\gamma_1 = a_1/x_1 = 0.909/0.95 = 0.96$.

Example 8.3 Living cells contain water with many ions. The osmotic pressure corresponds to that of an NaCl solution of about 0.15 M. Calculate the osmotic pressure at $T = 27\,°C$.

Solution: Osmotic pressure depends on the "colligative concentration," i.e. the number of particles per unit volume. Since NaCl dissociates into Na^+ and Cl^- ions, the colligative molality of the above solution is 0.30 M. Using the van't Hoff's equation (8.2.14), we can calculate the osmotic pressure π:

$$\pi = RT[S] = (0.0821\,L\,atm\,L^{-1})(300.0\,K)(0.30) = 7.40\,atm$$

If an animal cell is immersed in water, the water flowing into the cell due to osmosis will exert a pressure of about 7.4 atm, causing the cell to burst. Plant cell walls are strong enough to withstand this pressure

Example 8.4 At $p = 1\,atm$, the boiling point of an azeotropic mixture of C_2H_5OH and CCl_4 is 338.1 K. The heat of vaporization of C_2H_5OH is 38.56 kJ/mol and its boiling point is 351.4 K. Calculate the activity coefficient of ethanol in the azeotrope.

Solution: This can be done by direct application of (8.5.3) where $\Delta H_{1,vap} = 38.56\,kJ$, $T = 338.1\,K$ and $T^* = 351.4\,K$:

$$\ln(\gamma_{1,k}) = \frac{38.56 \times 10^3}{8.314}\left(\frac{1}{338.1} - \frac{1}{351.4}\right) = 0.519$$

$$\text{i.e. } \gamma_{1,k} = 1.68$$

Exercises

8.1   Obtain equation (8.1.8) from (8.1.7)

8.2   14.0 g of NaOH is dissolved in 84.0 g of H_2O and the solution has a density of $1.114 \times 10^3\,kg/m^3$. For the two components, NaOH and H_2O, in this solution, obtain (a) the mole fractions, (b) the molality and (c) molarity

8.3   The composition of the atmosphere is shown in Table 8.1.
(a) Calculate the partial pressures of N_2, O_2 and CO_2.
(b) Using the Henry's law constants calculate the concentrations of N_2, O_2 and CO_2 in a lake.

8.4   Obtain (8.2.5) from (8.2.4) for small changes in the boiling point of a solution.

8.5 (a) The solubility of $N_2(g)$ in water is about the same as in blood serum. Calculate the concentration (in $mol\,L^{-1}$) of N_2 in the blood.
(b) The density of seawater is 1.01 g/mL. What is the pressure at a depth of 100 m? What will be the blood serum concentration (in $mol\,L^{-1}$) of N_2 at this depth? If divers rise too fast, the excess N_2 can form bubbles in the blood, causing pain, paralysis and distress in breathing.

8.6 Assuming Raoult's law, predict the boiling point of a 0.5 M aqueous solution of sugar. Do the same for NaCl but by noting that the number of particles (ions) per molecule is twice that of a nonionic solution. Raoult's law is a colligative property that depends on the number of solute particles.

8.7 Ethylene glycol ($OH - CH_2 - CH_2 - OH$) is used as an antifreeze. (Its boiling point is $197\,°C$ and its freezing point is $-17.4\,°C$.)
(a) Look up the density of ethylene glycol in the CRC handbook or other tables and write a general formula for the freezing point of a mixture of X mL of ethylene glycol in 1.00 L of water for X in the range 0–100 mL.
(b) If the lowest expected temperature is about $-10.0\,°C$, what is the minimum amount (in mL/liter H_2O) of ethylene glycol that you need in your coolant?
(c) What is the boiling point of the coolant that contains 300 mL of ethylene glycol per liter of water?

8.8 What will be the boiling point of a solution of 20.0 g of urea $((NH_2)_2CO)$ in 1.25 kg of nitrobenzene? Use Table 8.2.

8.9 A 1.89 g pellet of an unknown compound was dissolved in ·50 mL of acetone. The change in the boiling point was $0.64°\,C$. Calculate the molar mass of the unknown compound. The density of acetone is 0.7851 g/mL and the value of K_b may be found in Table 8.2.

8.10 A solution of 4.00 g hemoglobin in 100 mL water was prepared and its osmotic pressure was measured. The osmotic pressure was 0.0130 atm at 280 K. (a) Estimate the molar mass of hemoglobin. (b) If 4.00 g of NaCl were dissolved in 100 mL of water, what would the osmotic pressure be? (Molecular weights of some proteins: ferricytochrome c 12 744; myoglobin 16 951; lysozyme 14 314; immunoglobulin G 156 000; myosin 570 000)

8.11 The concentration of the ionic constituents of seawater are as follows:

Ion	Cl^-	Na^+	SO_4^{2-}	Mg^{2+}	Ca^{2+}	K^+	HCO_3^-
Concentration(M)	0.55	0.46	0.028	0.054	0.010	0.010	0.0023

Many other ions are present in much lower concentrations. Estimate the osmotic pressure of seawater due to its ionic constituents.

8.12 The concentration of NaCl in seawater is approximately 0.5 M. In the process of reverse osmosis, seawater is forced through a membrane impermeable to the ions to obtain pure water. The applied pressure has to be larger than the osmotic pressure.

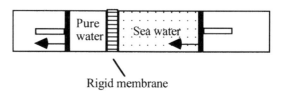

Rigid membrane

(a) At 25 °C what is the minimum pressure needed to achieve reverse osmosis? What is the work done in obtaining 1.0 L of pure water from seawater?

(b) If the cost of 1 kWh of electrical energy is about $0.15, what would be the energy cost of producing 100 L of water from seawater through reverse osmosis if the process is 50% efficient in using the electric power to obtain pure water?

(c) Suggest another process to obtain pure water from seawater.

8.13 Consider two cubes of copper of side 1 cm. In each cube, assume that one out of a million Cu atoms is Cu^+. Using Coulomb's law, calculate the force between the two cubes if they are placed at a distance of 10 cm.

8.14 Calculate the ionic strength and the activity coefficients of Ca^{2+} and Cl^- in a 0.02 M solution of $CaCl_2$.

8.15 The solubility product of AgCl is 1.77×10^{-10}. Calculate the concentration of Ag^+ ions in equlibrium with solid AgCl.

8.16 Show that for a perfect solution the molar volume of mixing $\Delta V_{mix} = 0$.

8.17 Consider a binary azeotrope. The chemical potentials of a component, say 2, in the gas and the liquid phases can be written as:

$$\mu_{2,g}(T,p,x) = \mu_{2,g}^*(T,p) + RT \ln (\gamma_{2,g} x_2)$$

and

$$\mu_{2,l}(T,p,x) = \mu_{2,l}^*(T,p) + RT \ln (\gamma_{2,l} x_2)$$

in which μ^* is the chemical potential of the pure substance. Note that the mole fraction is the same in the two phases. Use equation (5.3.7) to derive the relation (8.5.1).

9 CHEMICAL TRANSFORMATIONS

9.1 Transformations of Matter

Transformations of matter take place through chemical, nuclear and elementary particle reactions. We shall speak of "chemical transformations" in this broader sense. Though thermodynamics was founded in our daily experience, its reach is vast, ranging from the simplest changes, like the melting of ice, to the state of matter during the first few minutes after the Big Bang to the radiation that fills the entire universe today.

Let us begin by looking at the transformation that matter undergoes at various temperatures. Box 9.1 gives an overview of the reactions that take place at various temperatures ranging from those during the first few minutes after the Big Bang [1] to terrestrial and interstellar temperatures. All these transformations or reactions can be associated with affinities of reaction and an equilibrium characterized by the vanishing of the corresponding affinities.

In the present state of the universe, only a very small part of the energy is in the form of protons, neutrons and electrons that make up ordinary matter in all the galaxies. The rest consists of thermal radiation at a temperature of about 2.8 K and particles called neutrinos that interact very weakly with other particles. The small amount of matter which is in the form of stars and galaxies, however, is not in thermodynamic equilibrium. The affinities for the reactions that are currently occurring in the stars are not zero. The nuclear reactions in the stars produce all the known elements from hydrogen [2–4]. Hence the observed properties such as the abundance of elements in stars and planets cannot be described using the theory of chemical equilibrium. A knowledge of the rates of reaction and the history of the star or planet are necessary to understand the abundance of elements.

When a system reaches thermodynamic equilibrium, however, its history is of no importance. Regardless of the path leading to equilibrium, the state of equilibrium can be described by general laws. In this chapter we shall first look at the nature of chemical reactions; then we study the relation between entropy production and the rates of chemical reactions that drive the system to equilibrium.

Box 9.1 Transformation of Matter at Various Temperatures

Temperature $> 10^{10}$ K. This was the temperature during the first few minutes after the Big Bang. At this temperature the thermal motion of the protons and neutrons is so violent that even the strong nuclear forces cannot hold them together. Electron–positron pairs appear and disappear spontaneously and are in thermal equilibrium with radiation. (The threshold for electron–positron pair production is about 6×10^9 K.)

Temperature range 10^9 to 10^7 K. At about 10^9 K, nuclei begin to form and nuclear reactions occur. Temperatures of around 10^9 K occur in stars and supernova where heavier elements are synthesized from H and He. The *binding energy per nucleon* (proton or neutron) is in the range $(1.0-1.5)$ 10^{-12} J $\approx (6.0-9.0) \times 10^6$ eV, which corresponds to $(6.0-9.0) \times 10^8$ kJ/mol.

Temperature range 10^6 to 10^4 K. In this range, electrons bind with nuclei to form atoms, but the bonding forces between atoms are not strong enough to form stable molecules. At a temperature of about 1.5×10^5 K, hydrogen atoms begin to ionize. The ionization energy of 13.6 eV corresponds to 1310 kJ/mol. Heavier atoms require larger energies for complete ionization. To completely ionize a carbon atom, for example, 490 eV of energy is required, which corresponds to 47187 kJ.mol.* Carbon atoms will be completely dissociated at $T \approx 5 \times 10^6$ K into electrons and nuclei. In this temperature range, matter exists as free electrons and nuclei, a state of matter called *plasma*.

Temperature range 10 to 10^4 K. Chemical reactions take place in this range. The chemical bond energies are of the order of 10^2 kJ/mol. The C–H bond energy is about 412 kJ/mol. At a temperature of about 5×10^4 K, chemical bonds will begin to break. The intermolecular forces such as hydrogen bonds are of the order 10 kJ/mol. The enthalpy of vaporization of water, which is essentially the breaking of hydrogen bonds, is about 40 kJ/mol.

*1 eV $= 1.6 \times 10^{-19}$ J $= 96.3$ kJ/mol; T = (energy in J/mol)/R = (energy in J)/k_B

9.2 Chemical Reaction Rates

In studying chemical reactions and their approach to equilibrium, it is our purpose to look explicitly at the entropy production while the reactions are in progress. In other words, we would like to obtain explicit expressions for the entropy production (d_iS/dt) in terms of the rates of reaction. The introduction of reaction rates takes us beyond classical thermodynamics of equilibrium states, as formulated by Gibbs and others.

In general, thermodynamics cannot specify reaction rates (which depend on many factors such as the presence of catalysts) but, as we shall see in later chapters, close to thermodynamic equilibrium — called the "linear regime" —

thermodynamic formalism can be used to show that rates are linearly related to the affinities. But the general problem of specifying the rates of chemical reactions has become a subject in itself and it goes by the name of "chemical kinetics." Some basic aspects of chemical kinetics will be discussed in this section.

We have already seen (4.1.17) that the entropy production due to a chemical reaction may be written in the form

$$\frac{d_i S}{dt} = \frac{A}{T}\frac{d\xi}{dt} \qquad (9.2.1)$$

in which ξ is the extent of reaction introduced in section 2.5, and A is the affinity. The time derivative of ξ is related to the rate of reaction. The precise definition of the rate of reaction is given in Box 9.2. For the simple reaction[*]

$$Cl(g) + H_2(g) \rightleftharpoons HCl(g) + H(g) \qquad (9.2.2)$$

Box 9.2 Reaction Rate and Reaction Velocity

The reaction rate is defined as the number of reactive events per second per unit volume. It is usually expressed as $mol\,L^{-1}\,s^{-1}$. Chemical reactions depend on collisions. In most reactions, only a very small fraction of the collisions result in a chemical reaction. For each reacting species, since the number of collisions per unit volume is proportional to its concentration, the rates are proportional to the product of the concentrations. A **reaction rate** refers to conversion of the reactants to the products or vice versa. Thus, for the reaction

$$Cl(g) + H_2(g) \rightleftharpoons HCl(g) + H(g)$$

the **forward rate** $R_f = k_f[Cl][H_2]$ and the **reverse rate** $R_r = k_r[HCl][H]$ in which k_f and k_r are the forward and reverse rate constants and [H], etc., are concentrations. In a reaction, both forward and reverse reactions take place simultaneously. For thermodynamic considerations, we define the **velocity** of a reaction as the net conversion of the reactants to products. Thus

$$\text{Reaction velocity } v = \text{Forward rate} - \text{reverse rate}$$
$$= k_f[Cl][H_2] - k_r[HCl][H]$$
$$= R_f - R_r$$

In a homogeneous system the reaction velocity in terms of the extent of reaction is given by

$$\boxed{\text{Reaction velocity } v = \frac{d\xi}{V\,dt} = R_f - R_r}$$

in which V is the volume of the system. In practice, monitoring the progress of a reaction by noting the change in some property of the system (such as refractive index or spectral absorption) generally amounts to monitoring the change in the extent of reaction ξ.

---

[*] A detailed study of this reaction can be found in *Science* **273** (1996) 1519.

The affinity A and the extent of reaction ξ are defined by

$$A = \mu_{Cl} + \mu_{H_2} - \mu_{HCl} - \mu_H \qquad (9.2.3)$$

$$d\xi = \frac{dN_{Cl}}{-1} = \frac{dN_{H_2}}{-1} = \frac{dN_{HCl}}{1} = \frac{dN_H}{1} \qquad (9.2.4)$$

As explained in Box 9.2, the forward reaction rate is $k_f[Cl][H_2]$, in which the square brackets indicate concentrations and k_f is the forward rate constant that depends on temperature. Similarly the reverse reaction rate is $k_r[HCl][H]$. The time derivative of ξ is the *net rate of change* due to forward and reverse reactions. Since the reaction rates are generally expressed as functions of concentrations, it is more convenient to define this net rate per unit volume. Accordingly, we define a **reaction velocity** as

$$\text{Reaction velocity } v = \frac{d\xi}{V\,dt} = k_f[Cl][H_2] - k_r[HCl][H] \qquad (9.2.5)$$

Note that this equation follows from (9.2.4) and the definition of the forward and reverse rates. For example, in a homogeneous system the rate of change of the concentration of Cl is $(1/V)dN_{Cl}/dt = -k_f[Cl][H_2] + k_r[HCl][H]$. More generally, if R_f and R_r are the forward and reverse reaction rates, we have

$$\boxed{\text{Reaction velocity } v = \frac{d\xi}{V\,dt} = R_f - R_r} \qquad (9.2.6)$$

The velocity of reaction has the dimensions of $\text{mol L}^{-1}\,\text{s}^{-1}$.

In the above example, the rate of reaction bears a direct relation to the stoichiometry of the reactants, but this is not always true. In general, for a reaction such as

$$2X + Y \rightarrow \text{products we have Rate} = k[X]^a[Y]^b \qquad (9.2.7)$$

in which k is a temperature-dependent **rate constant**, and the exponents a and b, are not necessarily integers. The rate is said to be of *order a in* [X] and *of order b in* [Y]. The sum of all the orders of the reactants $(a + b)$ is called the **order of the reaction**. Reaction rates can take complex forms because they may be the result of many intermediate steps with widely differing rates that depend on the presence of catalysts. If all the intermediate steps are known, then each step is called an **elementary step**. Rates of elementary steps do bear a simple relation to the stoichiometry: the exponents equal the stoichiometric coefficients. If reaction (9.2.7) were an elementary step, for example, its rate would be $k[X]^2[Y]$.

Box 9.3 The Arrhenius Equation and Transition State Theory

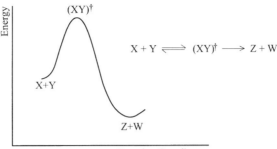

$$X + Y \rightleftharpoons (XY)^\dagger \longrightarrow Z + W$$

According to the Arrhenius equation, the rate constant of a chemical reaction is of the form

$$k = k_0 e^{-E_a/RT}$$

The rate constant k has this form because, for the reactants to convert to products, the collisions must have sufficient energy to overcome an energy barrier. As illustrated, the transformation from the reactants to the products is schematically represented with a "reaction coordinate" and the energy of the molecules undergoing the reaction.

According to transition state theory, the reactants X and Y form a **transition state** $(XY)^\dagger$. The transition state then irreversibly transforms to the products. The difference in the enthalpy and entropy between the free molecules X and Y and the transition state are denoted by $\Delta H^\dagger$ and $\Delta S^\dagger$ respectively. The main result of transition state theory (obtained using the principles of statistical mechanics) is that the rate constant has the form

$$k = \kappa (k_B T/h) \exp\left[-(\Delta H^\dagger - T\Delta S^\dagger)/RT\right] = \kappa (k_B T/h) \exp\left(-\Delta G^\dagger/RT\right)$$

in which $k_B = 1.381 \times 10^{-23}\,\mathrm{J\,K^{-1}}$ is Boltzmann's constant and $h = 6.626 \times 10^{-34}\,\mathrm{J\,s}$ is Planck's constant; κ is a small term of the order of 1 that is characteristic of the reaction. *A catalyst increases the rate of reaction by altering the transition state such that* $(\Delta H^\dagger - T\Delta S^\dagger) = \Delta G^\dagger$ *decreases.*

In many cases the temperature dependence of the rate constant is given by the **Arrhenius equation**:

$$k = k_0\, e^{-E_a/RT} \tag{9.2.8}$$

Svante Arrhenius (1859–1927) proposed it in 1889 and showed its validity for a large number of reactions [5, 6]. The term k_0 is called the "preexponential factor" and E_a the **activation energy**. For the forward reaction of (9.2.2), $Cl + H_2 \rightarrow HCl + H$, the values of k_0 and E_a are $k_0 = 7.9 \times 10^{10}\,\mathrm{L\,mol^{-1}\,s^{-1}}$ and $E_a = 23\,\mathrm{kJ\,mol^{-1}}$. For large variations in temperature, the Arrhenius equation was found to be inaccurate in predicting the variation of the rate constant, though it is quite useful for many reactions.

A more recent theory that is based on statistical mechanics and quantum theory was developed in the 1930s by Wigner, Pelzer, Eyring, Polanyi and Evans. According to this theory, the reaction occurs through a **transition state** (Box 9.3). The concept of a transition state leads to the following expression for the rate constant:

$$\boxed{k = \kappa(k_B T/h) \exp\left[-(\Delta H^\dagger - T\Delta S^\dagger)/RT\right] = \kappa(k_B T/h) \exp\left(-\Delta G^\dagger/RT\right)}$$

$$(9.2.9)$$

in which $k_B = 1.38 \times 10^{-23}\,\mathrm{J\,K^{-1}}$ is Boltzmann's constant and $h = 6.626 \times 10^{-34}\,\mathrm{J\,s}$ is Planck's constant. The terms $\Delta H^\dagger$ and $\Delta S^\dagger$ are the transition state enthalpy and entropy, as defined in Box 9.3. The term κ is small, of order 1, characteristic of the reaction. *A catalyst increases the rate of reaction by altering the transition state such that* $(\Delta H^\dagger - T\Delta S^\dagger) = \Delta G^\dagger$ *decreases.*

RATE EQUATIONS USING THE EXTENT OF REACTIONS

Reaction rates are generally determined empirically. The mechanisms of reaction, which detail all the elementary steps, are usually a result of long and detailed study. Once the reaction rate laws are known, the time variation of the concentration can be obtained by integrating the rate equations, which are coupled differential equations in general. For example, if we have an elementary reaction of the form

$$X \underset{k_r}{\overset{k_f}{\rightleftharpoons}} 2Y \qquad (9.2.10)$$

the concentrations are governed by the following differential equations:

$$-\frac{1}{V}\frac{d\xi}{dt} = \frac{d[X]}{dt} = -k_f[X] + k_r[Y]^2 \qquad (9.2.11)$$

$$2\frac{1}{V}\frac{d\xi}{dt} = \frac{d[Y]}{dt} = 2k_f[X] - 2k_r[Y]^2 \qquad (9.2.12)$$

Without loss of generality, we may assume $V = 1$, and simplify the notation. These two equations are not independent. *In fact, there is only one independent variable ξ for every independent reaction.* If $[X]_0$ and $[Y]_0$ are the values of the concentrations at $t = 0$, by assigning $\xi(0) = 0$, and using $d\xi = -d[X]$ and $2d\xi = d[Y]$ it is easy to see that $[X] = [X]_0 - \xi$ and $[Y] = [Y]_0 + 2\xi$.

Substituting these values into (9.2.11) we obtain the equation

$$\frac{d\xi}{dt} = k_f \left([X]_0 - \xi\right) - k_r \left([Y]_0 + 2\xi\right)^2 \tag{9.2.13}$$

In this equation the initial concentrations $[X]_0$ and $[Y]_0$ appear explicitly and $\xi(0) = 0$ for all initial concentrations. The solutions of such equations, $\xi(t)$, can be used to obtain the entropy production, as will be shown explicitly in section 9.5. Differential equations such as these, and more complicated systems, can be numerically solved on a computer, using software such as Mathematica or Maple. Sample Mathematica codes are provided in Appendix 9.1.

When many reactions are to be considered simultaneously, we will have one ξ for each independent reaction, denoted by ξ_k, and the entire system will be described by a set of coupled differential equations in ξ_k. Only in simple cases can we find analytical solutions to such equations while solutions for complex reactions can be readily found numerically using computers.

REACTION RATES AND ACTIVITIES

Though reaction rates are generally expressed in terms of concentrations, one could equally well express them in terms of activities. In fact, we shall see in the following sections that the connection between affinities and reaction rates can be made more easily if the reaction rates are expressed in terms of activities. For the elementary reaction

$$X + Y \rightleftharpoons 2W \tag{9.2.14}$$

the forward rate R_f and the reverse rate R_f may be written as

$$R_f = k_f\, a_X a_Y \quad \text{and} \quad R_r = k_r\, a_W^2 \tag{9.2.15}$$

The rate constants k_f and k_r in (9.2.15) have dimensions of mol L^{-1} s^{-1}; their numerical values and dimensions differ from those of the rate constants when R_f and R_r are expressed in terms of concentrations (exc. 9.8).

Experimentally we know that reaction rates do depend on the activities; they are not specified by concentrations alone. For example, at fixed values of temperature and fixed reactants concentrations, it is well known that rates of ionic reactions can be altered by changing the ionic strength of the solution (usually called salt effects). This change in the rate is due to a change in the activities. It has become general practice, however, to express the reaction rates in terms of the concentrations and to include the effects of changing activities in the rate constants. Thus, the rates constants are considered functions of the ionic strength when rates are expressed in terms of concentrations. Alternatively, if the rates were expressed in terms of activities, the rate constant would be independent of the ionic strength; the change in rates due to changes in ionic strength would be because the activities depended on ionic strength.

Box 9.4 Elementary Reactions

To obtain an explicit analytic expression for the concentrations of the reactants and products as functions of time, we must solve differential equations such as (9.2.11) and (9.2.12). Generally this is possible only in the case of simple reactions. For more complex reactions, one can obtain numerical solutions using a computer. Two elementary reactions for which we can obtain explicit expressions for the concentrations as functions of time are given below.

First-order reaction. For a decomposition reaction $X \rightarrow$ products, in which the reverse reaction rate is so small that it can be neglected, we have the differential equation

$$\frac{d[X]}{dt} = -k_f[X]$$

It is easy to see that the solution of this equation is

$$[X](t) = [X]_0 e^{-k_f t}$$

in which $[X]_0$ is the concentration at time $t = 0$. This is the well-known exponential decay; in a given amount of time, $[X]$ decreases by the same fraction. In particular, the time it takes for any initial value of $[X]$ to decrease by a factor of $1/2$ is called the **half-life**. It is usually denoted by $t_{1/2}$. The half-life can be computed by noting that $\exp[-k_f t_{1/2}] = 1/2$. i.e.

$$t_{1/2} = \frac{\ln 2}{k_f} = \frac{0.6931}{k_f}$$

Second-order reaction. For the elementary reaction $2X \rightarrow$ products, if the reverse reaction can be neglected, the rate equation is:

$$\frac{d[X]}{dt} = -2k_f[X]^2$$

The solution is obtained by evaluating

$$\int_{[X]_0}^{[X]} \frac{d[X]}{[X]^2} = -\int_0^t 2k_f \, dt,$$

which give us

$$\frac{1}{[X]} - \frac{1}{[X]_0} = 2k_f t$$

Given k_f and $[X]_0$ at $t = 0$, this expression yields $[X]$ at any time t.

9.3   Chemical Equilibrium and The Law of Mass Action

In this section we shall study chemical equilibrium in detail. At equilibrium the pressures and temperatures are uniform; moreover affinities and the corresponding velocities of reaction vanish. For a reaction such as

$$X + Y \rightleftharpoons 2Z \tag{9.3.1}$$

at equilibrium, we have

$$A = (\mu_X + \mu_Y - 2\mu_Z) = 0 \quad \text{and} \quad \frac{d\xi}{dt} = 0 \tag{9.3.2}$$

or

$$\mu_X + \mu_Y = 2\mu_Z \tag{9.3.3}$$

The condition that the "thermodynamic force," affinity A equals zero implies that the corresponding "thermodynamic flow," i.e. the reaction velocity $d\xi/dt$, also equals zero. The condition $A = 0$ means that at equilibrium the "stoichiometric sums" for the chemical potentials of the reactants and products are equal, as in (9.3.3). It is easy to generalize this result to an arbitrary chemical reaction of the form

$$a_1A_1 + a_2A_2 + a_3A_3 + \ldots + a_nA_n \rightleftharpoons b_1B_1 + b_2B_2 + b_3B_3 + \ldots + b_mB_m \tag{9.3.4}$$

in which the a's are the stoichiometric coefficients of the reactants A_k and the b's are the stoichiometric coefficients of the products B_k. The corresponding condition for chemical equilibrium will then be

$$a_1\mu_{A1} + a_2\mu_{A2} + a_3\mu_{A3} + \ldots + a_n\mu_{An} = b_1\mu_{B1} + b_2\mu_{B2} + b_3\mu_{B3} + \ldots + b_m\mu_{Bm} \tag{9.3.5}$$

Such equalities of chemical potentials are valid for all reactions, changes of phase, chemical, nuclear and elementary particle reactions. Just as a difference in temperature drives the flow of heat until the temperatures difference vanishes, a nonzero affinity drives a chemical reaction until the affinity vanishes.

To understand the physical meaning of the mathematical conditions such as (9.3.3) or (9.3.5), we express the chemical potential in terms of experimentally measurable quantities. We have seen in (5.3.5) that a general chemical potential can be expressed as

$$\mu_k(p, T) = \mu_{k0}(T) + RT \ln a_k \tag{9.3.6}$$

in which a_k is the activity and $\mu_{k0} = \Delta G_f^0[k]$ is the standard molar Gibbs free energy of formation (Box 5.1) which is tabulated. For gases, liquids and solids, we have the following explicit expressions:

- Ideal gas: $a_k = (p_k/p_0)$, where $p_k =$ partial pressure.
- Real gases: expressions for activity can be derived using (6.2.31), as was shown in section 6.2.
- Pure solids and liquids: $a_k \approx 1$.
- Solutions: $a_k \approx \gamma_k x_k$, where $\gamma_k =$ activity coefficient and $x_k =$ mole fraction.

For ideal solutions $\gamma_k = 1$. For nonideal solutions γ_k is obtained by various means, depending on the type of solution. The chemical potential can also be written in terms of the concentrations by appropriately redefining μ_{k0}.

We can now use (9.3.6) to express the condition for equilibrium (9.3.3) in terms of the activities, which are experimentally measurable quantities:

$$\mu_{x0}(T) + RT \ln a_{x,\text{eq}} + \mu_{y0}(T) + RT \ln a_{y,\text{eq}} = 2\left[\mu_{z0}(T) + RT \ln(a_{z,\text{eq}})\right]$$

$$(9.3.7)$$

where the equilibrium values of the activities are indicated by the subscript equation. This equation can be rewritten as

$$\boxed{\frac{a_{z,\text{eq}}^2}{a_{x,\text{eq}}\, a_{y,\text{eq}}} = \exp\left[\frac{\mu_{x0}(T) + \mu_{y0}(T) - 2\mu_{z0}(T)}{RT}\right] \equiv K(T)} \qquad (9.3.8)$$

$K(T)$, as defined above, is called the **equilibrium constant**. The equilibrium constant as defined above is a function of T only, and this is an important thermodynamic result. It is called the **law of mass action**. In terms of the tabulated molar Gibbs free energies of formation $\Delta G_f^0[k] = \mu_{k0}$, it is convenient to define a **Gibbs free energy of reaction** ΔG_{rxn} as

$$\Delta G_{\text{rxn}}^0 = 2\Delta G_f^0[\text{Z}] - \Delta G_f^0[\text{X}] - \Delta G_f^0[\text{Y}] \qquad (9.3.9)$$

The equilibrium constant is then written as

$$\boxed{\begin{aligned} K(T) &= \exp[-\Delta G_{\text{rxn}}^0/RT] \\ &= \exp[-(\Delta H_{\text{rxn}}^0 - T\Delta S_{\text{rxn}}^0)/RT] \end{aligned}} \qquad (9.3.10)$$

in which ΔH_{rxn}^0 and ΔS_{rxn}^0 are enthalpy and the entropy of the reaction. The activities in (9.3.8) can be written in terms of partial pressures p_k or mole fractions x_k. If (9.3.1) were an ideal gas reaction, $a_k = p_k/p_0$. With $p_0 = 1$ bar and p_k measured in bars, the equilibrium constant takes the form

$$\frac{p_{z,eq}^2}{p_{x,eq}\, p_{y,eq}} = K_p(T) = \exp[-\Delta G_{rxn}^0/RT] \qquad (9.3.11)$$

At a given temperature, regardless of the initial partial pressures, the chemical reaction (9.3.1) will irreversibly proceed towards the state of equilibrium in which the partial pressures will satisfy equation (9.3.11). This is one form of the *law of mass action*; K_p is the *equilibrium constant* expressed in terms of the partial pressures. In an ideal gas mixture, since $p_k = (N_k/V)RT = [k]RT$ (in which R is in the units of bar L mol^{-1} K^{-1}), the law of mass action can also be expressed in terms of the concentrations of the reactants and products:

$$\frac{[Z]_{eq}^2}{[X]_{eq}\,[Y]_{eq}} = K_c(T) \qquad (9.3.12)$$

in which K_c is the equilibrium constant expressed in terms of the concentrations. In general, for a reaction of the form $aX + bY \rightleftharpoons cZ$ it is easy to obtain the relation $K_c = (RT)^\alpha K_p$ where $\alpha = a + b - c$ (exc. 9.11). In the particular case of reaction (9.3.1) α happens to be zero.

If one of the reactants were a pure liquid or a solid, then the equilibrium constant would not contain corresponding "concentration" terms. Let us consider the reaction

$$O_2(g) + 2C(s) \rightleftharpoons 2CO(g) \qquad (9.3.13)$$

Since $a_{C(s)} \approx 1$ for the solid phase, the equilibrium constant in this case is written as

$$\frac{a_{co,\,eq}^2}{a_{o_2,eq}\, a_{c,\,eq}^2} = \frac{p_{co,\,eq}^2}{p_{o_2,eq}} = K_p(T) \qquad (9.3.14)$$

Equations (9.3.9) and (9.3.10) provide us with a means of calculating the equilibrium constant $K(T)$ using the tabulated values of $\Delta G_f^0[k]$. If the activities are expressed in terms of partial pressures, we have K_p. Some examples are shown in Box 9.5.

RELATION BETWEEN EQUILIBRIUM CONSTANTS AND RATE CONSTANTS

Chemical equilibrium can also be described as a state in which the forward rate of every reaction equals its reverse rate. If the reaction $X + Y \rightleftharpoons 2Z$ is an *elementary reaction*, and if we express the reaction rates in terms of the

---

Box 9.5 The Equilibrium Constant

A basic result of equilibrium chemical thermodynamics is that the equilibrium constant $K(T)$ is a function of temperature only. It can be expressed in terms of the standard Gibbs free energy of reaction ΔG_{rxn}^0 (Equations 9.3.9 and 9.3.10)

$$K(T) = \exp[-\Delta G_{rxn}^0/RT]$$

For a reaction such as $O_2(g) + 2C(s) \rightleftharpoons 2CO(g)$, the equilibrium constant can be calculated using the tabulated values of ΔG_f^0:

$$\Delta G_{rxn}^0 = 2\Delta G_f^0[CO] - 2\Delta G_f^0[C] - \Delta G_f^0[O_2]$$
$$= -2(137.2)\,kJ\,mol^{-1} - 2(0) - (0) = -274.4\,kJ\,mol^{-1}$$

Using this value in the expression $K(T) = \exp[-\Delta G_{rxn}^0/RT]$, we can calculate $K(T)$ at $T = 298.15\,K$:

$$K(T) = \exp[-\Delta G_{rxn}^0/RT] = \exp[274.4 \times 10^3/(8.314 \times 298.15)] = 1.18 \times 10^{48}$$

Similarly, for the reaction $CO(g) + 2H_2(g) \rightleftharpoons CH_3OH(g)$, we have

$$\Delta G_{rxn}^0 = \Delta G_f^0[CH_3OH] - \Delta G_f^0[CO] - 2\Delta G_f^0[H_2]$$
$$= -161.96\,kJ\,mol^{-1} - (-137.2\,kJ\,mol^{-1}) - 2(0) = -24.76\,kJ\,mol^{-1}$$

The equilibrium constant at $T = 298.15\,K$ is:

$$K(T) = \exp[-\Delta G_{rxn}^0/RT] = \exp[24.76 \times 10^3/(8.314 \times 298.15)] = 2.18 \times 10^4$$

---

activities, when the velocity of the reaction is zero, we have

$$k_f\,a_X\,a_Y = k_r\,a_Z^2 \tag{9.3.15}$$

From a theoretical viewpoint, writing the reaction rates in terms of activities rather than concentrations is better because the state of equilibrium is directly related to activities, not concentrations.

Comparing (9.3.15) and the equilibrium constant (9.3.8), we see that

$$K(T) = \frac{a_{Z,\,eq}^2}{a_{X,\,eq}\,a_{Y,\,eq}} = \frac{k_f}{k_r} \tag{9.3.16}$$

Thus, the equilibrium constant can also be related to the rate constants k_r and k_f when the rates are expressed in terms of the activities. But note that $K(T) = a_{Z,\,eq}^2/(a_{X,\,eq}\,a_{Y,\,eq})$ is valid even if the reaction rates do not have the form (9.3.15) while $K(T) = (k_f/k_r)$ is valid only when (9.3.15) is valid. *The relation between the activities and the equilibrium constant is purely a consequence of the laws of thermodynamics; it is independent of the rates of the forward and reverse reactions.*

THE VAN'T HOFF EQUATION

Using (9.3.10) the temperature variation of the equilibrium constant $K(T)$ can be related to the enthalpy of reaction ΔH_{rxn}. From (9.3.10) it follows that

$$\frac{d \ln K(T)}{dT} = -\frac{d}{dT} \frac{\Delta G^0_{rxn}}{RT} \tag{9.3.17}$$

But according to the Helmholtz equation (5.2.14), the variation of ΔG with temperature is related to ΔH by $(\partial/\partial T)(\Delta G/T) = -\Delta H/T^2$. Using this in equation (9.3.17) gives

$$\boxed{\frac{d \ln K(T)}{dT} = \frac{\Delta H^0_{rxn}}{RT^2}} \tag{9.3.18}$$

This relation is called the **van't Hoff equation**. Generally, the heat of reaction ΔH_{rxn} changes very little with temperature and may be assumed constant. Thus we may integrate (9.3.18) and obtain

$$\ln K(T) = \frac{-\Delta H^0_{rxn}}{RT} + C \tag{9.3.19}$$

Experimentally, the equilibrium constant $K(T)$ can be obtained at various temperatures. According to equation (9.3.19), a plot of $\ln K(T)$ versus $(1/T)$ should result in a straight line with a slope equal to $(-\Delta H^0_{rxn}/R)$. This method can be used to obtain the values of ΔH^0_{rxn}.

RESPONSE TO PERTURBATION FROM EQUILIBRIUM: THE LE CHATELIER–BRAUN PRINCIPLE

When a system is perturbed from its state of equilibrium, it will relax to a new state of equilibrium. Le Chatelier and Braun noted in 1888 that a simple principle may be used to predict the direction of the response to a perturbation from equilibrium. Le Chatelier stated it like this:

"Any system in chemical equilibrium undergoes, as a result of a variation in one of the factors governing the equilibrium, a compensating change in a direction such that, had this change occurred alone it would have produced a variation of the factors considered in the *opposite* direction."

To illustrate this principle, let us consider the reaction

$$N_2 + 3H_2 \rightleftharpoons 2NH_3$$

in equilibrium. In this reaction there is a decrease in the number of moles when the reactants convert to products, which leads to a reduction in the pressure (at a fixed temperature). If the pressure of this system in equilibrium is suddenly increased, it will respond by producing more NH_3, which tends to decrease the pressure. The compensating change in the system is in a direction opposite to the perturbation. The new state of equilibrium will contain more NH_3. Similarly, if a reaction is exothermic, if heat is supplied to the system, the product will be converted to reactants, thus opposing the increase in temperature. This principle is a special case of a more general approach under the name "theorems of moderation". These theorems may be found in the literature [7].

9.4 The Principle of Detailed Balance

There are a few important aspects of the state of chemical equilibrium, and the state of thermodynamic equilibrium in general, that must be noted. The **principle of detailed balance** is one of them.

We have noted earlier that, for a given reaction, the state of equilibrium depends only on the stoichiometry of the reaction, not its actual mechanism. For example, in the reaction $X + Y \rightleftharpoons 2Z$ considered above, if the forward and reverse reaction rates were given by

$$R_f = k_f a_X a_Y \qquad R_r = k_r a_Z^2 \tag{9.4.1}$$

respectively, then the relation $a_Z^2/a_X a_Y = K(T)$ at equilibrium* can be interpreted as the balance between forward and reverse reactions:

$$R_f = k_f a_X a_Y = R_r = k_r a_Z^2$$

so that

$$\frac{a_Z^2}{a_X a_Y} = K(T) = \frac{k_f}{k_r} \tag{9.4.2}$$

However, the equilibrium relation $a_Z^2/a_X a_Y = K(T)$ was not obtained using any assumption regarding the kinetic mechanism of the reaction. It remains valid even if there is a complex set of intermediate reactions that result in the overall

---

* We drop the subscript "eq" for notational simplicity.

reaction $X + Y \rightleftharpoons 2Z$. This feature could be understood through the **principle of detailed balance**, according to which:

In the state of equilibrium, every elementary transformation is balanced by its exact opposite or reverse.

The principle of detailed balance implies $a_Z^2/a_X a_Y = K(T)$, regardless of the mechanism; this can be seen through the following example. Assume the reaction really consists of two steps:

$$(a) \quad X + X \rightleftharpoons W \qquad\qquad\qquad (9.4.3)$$

$$(b) \quad W + Y \rightleftharpoons 2Z + X \qquad\qquad (9.4.4)$$

which ultimately achieve $X + Y \rightleftharpoons 2Z$. According to the principle of detailed balance, at equilibrium we must have

$$\frac{a_W}{a_X^2} = \frac{k_{fa}}{k_{ra}} \equiv K_a \qquad \frac{a_Z^2 a_X}{a_W a_Y} = \frac{k_{fb}}{k_{rb}} \equiv K_b \qquad (9.4.5)$$

in which the subscripts a and b stand for reactions (9.4.3) and (9.4.4). The thermodynamic equation for equilibrium $a_Z^2/a_X a_Y = K(T)$ can now be obtained as the product of these two equations:

$$\frac{a_W a_Z^2 a_X}{a_X^2 a_W a_Y} = \frac{a_Z^2}{a_X a_Y} = K_a K_b = K \qquad (9.4.6)$$

From this derivation it is clear that equation (9.4.6) will be valid for an arbitrary set of intermediate reactions.

The principle of detailed balance is even more general. It is in fact valid for the exchange of matter and energy between any two volume elements of a system in equilibrium. The amount of matter and energy transferred from volume element X to volume element Y exactly balances the energy and matter transferred from volume element Y to volume element X (Fig. 9.1). The same can be said of the interaction between the volume elements Y and Z, and X and Z. One important consequence of this type of balance is that the removal or isolation of one of the volume elements from the system, say Z, does not alter the states of X or Y or the interaction between them. This is another way of saying that there is no long-range correlation between the various volume elements. As we shall see in the later chapters, in nonequilibrium systems that make a transition to organized dissipative structures, the principle of detailed balance is no longer valid. Consequently, the removal or isolation of a volume element at one part will alter the state of a volume element located elsewhere. It is then said to have long-range correlations. We can see this clearly if we compare a droplet of water that contains carbon compounds in thermal

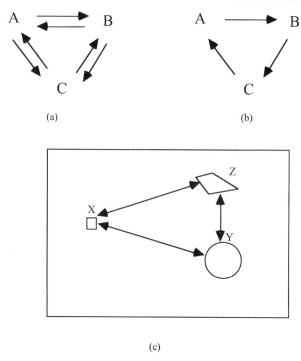

Figure 9.1 The principle of detailed balance. (a) The equilibrium between three interconverting compounds A, B and C is a result of "detailed balance" between each pair of compounds. (b) Although a conversion from one compound to another can also produce concentrations that remain constant in time, this is not the equilibrium state. (c) The principle of detailed balance has a more general validity. The exchange of matter (or energy) between any two regions of a system is balanced in detail; the amount of matter going from X to Y is balanced by exactly the reverse process, etc.

equilibrium and a living cell that is in an organized state far from thermodynamic equilibrium. Removal of a small part of the water droplet does not change the state of other parts of the droplet, but removing a small part of a living cell may have a drastic influence on other parts of the cell.

9.5   Entropy Production due to Chemical Reactions

The formalism of the previous sections can now be used to relate entropy production to reaction rates more explicitly. In Chapter 4 we have seen from

(4.1.16) that the entropy production due to a chemical reaction is given by

$$\frac{d_iS}{dt} = \frac{A}{T}\frac{d\xi}{dt} \geq 0 \tag{9.5.1}$$

Our objective is to relate the affinity A to the reaction rates, so that the entropy production is written in terms of the reaction rates. Let us consider the reaction

$$X + Y \rightleftharpoons 2Z \tag{9.5.2}$$

that we have considered before. Assuming this is an elementary step, we have for the forward and reverse rates

$$R_f = k_f a_X a_Y \qquad R_r = k_r a_Z^2 \tag{9.5.3}$$

Since the forward and reverse rates must be equal at equilibrium, we have seen from (9.4.2) that

$$K(T) = \frac{k_f}{k_r} \tag{9.5.4}$$

We have already seen that the velocity of a reaction is simply the difference between the forward and reverse reaction rates and that the reaction rates can be expressed in terms of ξ (equation 9.2.6):

$$\frac{1}{V}\frac{d\xi}{dt} = [R_f(\xi) - R_r(\xi)] \tag{9.5.5}$$

To obtain the velocity of reaction as a function of time, this differential equation has to be solved. An example is given below.

We can also relate the affinity to the reaction rates in the following manner. By definition, the affinity of the reaction (9.5.2) is

$$\begin{aligned}
A &= \mu_X + \mu_Y - 2\mu_Z \\
&= \mu_{X0}(T) + RT \ln a_X + \mu_{Y0}(T) + RT \ln a_Y - 2[\mu_{Z0}(T) + RT \ln a_Z] \\
&= [\mu_{X0}(T) + \mu_{Y0}(T) - 2\mu_{Z0}(T)] + RT \ln a_X + RT \ln a_Y - 2RT \ln a_Z
\end{aligned} \tag{9.5.6}$$

Since $[\mu_{X0}(T) + \mu_{Y0}(T) - 2\mu_{Z0}(T)] = -\Delta G^0_{rxn} = RT \ln K(T)$ (equation 9.3.10) this equation can be written as

$$A = RT \ln K(T) + RT \ln\left(\frac{a_X a_Y}{a_Z^2}\right) \tag{9.5.7}$$

This is an alternative way of writing the affinity. At equilibrium $A = 0$. To relate A to the reaction rates, we use (9.5.4) and combine the two logarithmic terms:

$$A = RT \ln \left(\frac{k_f}{k_r} \right) + RT \ln \left(\frac{a_X a_Y}{a_Z^2} \right) = RT \ln \left(\frac{k_f a_X a_Y}{k_r a_Z^2} \right) \qquad (9.5.8)$$

This leads us to the relations we are seeking if we use (9.5.3) to write this expression in terms of the reaction rates:

$$A = RT \ln \left(\frac{R_f}{R_r} \right) \qquad (9.5.9)$$

Clearly this equation is valid for any elementary step because the rates of elementary steps are directly related to the stoichiometry. Now we can substitute (9.5.5) and (9.5.9) into the expression for the entropy production (9.5.1) and obtain

$$\frac{1}{V} \frac{d_i S}{dt} = \frac{1}{V} \frac{A}{T} \frac{d\xi}{dt} = R(R_f - R_r) \ln (R_f/R_r) \geq 0 \qquad (9.5.10)$$

which is an expression that relates *entropy production per unit volume* to the reaction rates. (Note that R is the gas constant.) Also, as required by the Second Law, the right-hand side of this equation is positive, whether $R_f > R_r$ or $R_f < R_r$. Another point to note is that in equation (9.5.10) the forward and reverse rates, R_f and R_r, can be expressed in terms of concentrations, partial pressures or any other convenient variables of the reactants; the reaction rates need not be expressed only in terms of activities as in (9.5.3).

The above equation can be generalized to several simultaneous reactions, each indexed by the subscript k. The entropy production per unit volume is the sum of the entropies produced by each reaction:

$$\frac{1}{V} \frac{d_i S}{dt} = \frac{1}{V} \sum_k \frac{A_k}{T} \frac{d\xi_k}{dt} = R \sum_k (R_{kf} - R_{kr}) \ln(R_{kf}/R_{kr}) \qquad (9.5.11)$$

in which R_{kf} and R_{kr} are the forward and reverse reaction rates of the kth reaction. This expression is useful for computing the entropy production in terms of the reaction rates but *it is valid only for elementary steps whose reaction rates are specified by the stoichiometry*. This, however, is not a serious limitation because every reaction is ultimately the result of many elementary steps. If the detailed mechanism of a reaction is known, an expression for the entropy production can be written for any chemical reaction.

AN EXAMPLE

As an example of entropy production due to an irreversible chemical reaction, consider the simple reaction

$$L \rightleftharpoons D \qquad (9.5.12)$$

which is the interconversion or "racemization" of molecules with mirror-image structures. Molecules that are not identical to their mirror image are said to be *chiral* and the two mirror-image forms are called *enantiomers*. Let [L] and [D] be the concentrations of the enantiomers of a chiral molecule. If at time $t = 0$, the concentrations are $[L] = L_0$ and $[D] = D_0$, and $\xi(0) = 0$, then we have the following relations:

$$\frac{d[L]}{-1} = \frac{d[D]}{+1} = \frac{d\xi}{V} \qquad (9.5.13)$$

$$[L] = L_0 - (\xi/V) \qquad [D] = D_0 + (\xi/V) \qquad (9.5.14)$$

Relations (9.5.14) are obtained by integrating (9.5.13) and using the initial conditions. For notational convenience we shall assume $V = 1$. At the end of the calculation we shall reintroduce the factor V. Racemization can be an elementary first-order reaction for which the forward and reverse reactions are

$$R_f = k[L] = k(L_0 - \xi) \qquad R_r = k[D] = k(D_0 + \xi) \qquad (9.5.15)$$

Note that the rate constants for the forward and reverse reaction are the same due to symmetry: L must convert to D with the same rate constant as D to L. Also, from (9.5.15) and (9.5.9), one can see that the affinity is a function of the state variable ξ for a given set of initial concentrations.

To obtain the entropy production as an explicit function of time, we must obtain R_f and R_r as functions of time. This can be done by solving the differential equation defining the velocity of this reaction:

$$\frac{d\xi}{dt} = R_f - R_r = k(L_0 - \xi) - k(D_0 + \xi)$$

$$\text{i.e.} \quad \frac{d\xi}{dt} = 2k \left[\frac{(L_0 - D_0)}{2} - \xi \right] \qquad (9.5.16)$$

This first-order differential equation can easily be solved by defining $x = \left[\frac{1}{2}(L_0 - D_0) - \xi \right]$ so that the equation reduces to $dx/dt = -2kx$. The solution is

$$\xi(t) = \frac{(L_0 - D_0)}{2} \left[1 - e^{-2kt} \right] \qquad (9.5.17)$$

The rates (9.5.15) can now be written as explicit functions of time using (9.5.17):

$$R_f = \frac{k(L_0 + D_0)}{2} + \frac{k(L_0 - D_0)}{2} e^{-2kt} \tag{9.5.18}$$

$$R_r = \frac{k(L_0 + D_0)}{2} - \frac{k(L_0 - D_0)}{2} e^{-2kt} \tag{9.5.19}$$

Now the entropy production (9.5.10) can also be written as an explicit function of time:

$$\frac{1}{V}\frac{d_iS}{dt} = R(R_f - R_r)\ln(R_f/R_r)$$

$$\frac{1}{V}\frac{d_iS}{dt} = R\left[k(L_0 - D_0)e^{-2kt}\right]\ln\left\{\frac{(L_0 + D_0) + (L_0 - D_0)e^{-2kt}}{(L_0 + D_0) - (L_0 - D_0)e^{-2kt}}\right\} \tag{9.5.20}$$

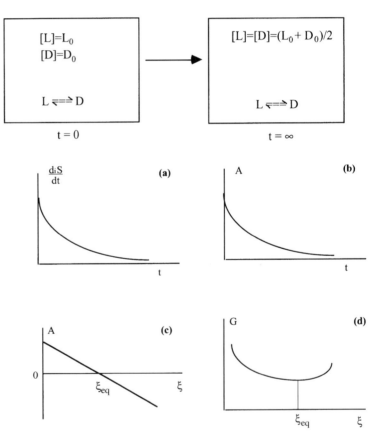

Figure 9.2 Racemization of enantiomers as an example of a chemical reaction. The associated entropy production and the time variation of A are shown in (a) and (b). State functions A and G as functions of ξ are shown in (c) and (d)

As $t \rightarrow \infty$ the system reaches equilibrium, at which

$$\xi_{eq} = \frac{L_0 - D_0}{2} \quad \text{and} \quad [L]_{eq} = [D]_{eq} = \frac{L_0 + D_0}{2} \tag{9.5.21}$$

The Gibbs free energy and the affinity are state functions. In Chapter 5 (5.1.12) we noted that $A = -(\partial G/\partial \xi)_{p,T}$. As ξ goes to its equilibrium value ξ_{eq}, the Gibbs free energy reaches its minimum value and the affinity goes to zero (Fig 9.2). The volume can be reintroduced by replacing ξ_{eq} by (ξ_{eq}/v).

   The entropy production for more complex reaction can be obtained numerically using modern computational software. A Mathematica code for the above example is given in Appendix 9.1. The student is encouraged to alter it to develop codes for more complex reactions.

Appendix 9.1: Mathematica Codes

Numerical solutions to rate equations can be obtained in Mathematica using the `NDSolve` command. Examples in solving simple rate equations are given below.

CODE A: MATHEMATICA CODE FOR LINEAR KINETICS $X \rightarrow$ (PRODUCTS)

```
(*Linear Kinetics*)
k = 0.12;
Soln1 = NDSolve[{X'[t] == -k*X[t], X[0] == 2.0}, X, {t, 0, 10}]
```

The output indicates that the solution as an interpolating function has been generated. The solution can be plotted using the following command:

```
Plot[Evaluate[X[t]/.Soln1], {t, 0, 10}]
```

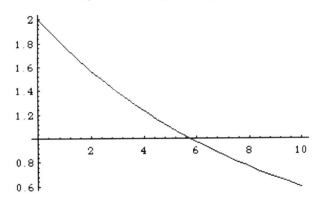

CODE B: MATHEMATICA CODE FOR THE REACTION $X + 2Y \rightleftharpoons 2Z$

```
(*Reaction X + 2Y → 2Z*)
kf = 0.5; kr = 0.05;
Soln2 = NDSolve[{X'[t] == -kf*X[t]*(Y[t]∧2) + kr*Z[t]∧2,
                Y'[t] == 2*(-kf*X[t]*(Y[t]∧2) + kr*Z[t]∧2),
                Z'[t] == 2*(kf*X[t]*(Y[t]∧2) - kr*Z[t]∧2),
                X[0] == 2.0, Y[0] == 3.0, Z[0] == 0.0},
                {X, Y, Z}, {t, 0, 3}]
```

The output indicates that the solution as an interpolating function has been generated. The solution can be plotted using the following command:

```
Plot[Evaluate[{X[t],Y[t],Z[t]}/.Soln2],{t,0,3}]
```

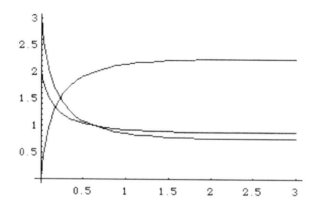

CODE C: MATHEMATICA CODE FOR RACEMIZATION REACTION $L \rightleftharpoons D$ AND CONSEQUENT ENTROPY PRODUCTION

```
(*Racemization Kinetics : L === D*)
kf = 1.0; kr = 1.0;
Soln3 = NDSolve[{XL'[t] == -kf*XL[t] + kr*XD[t],
                XD'[t] == -kr*XD[t] + kf*XL[t],
                XL[0] == 2.0, XD[0] == 0.001},
                {XL, XD}, {t, 0, 3}]
```

The output indicates that an interpolating function has been generated. The entropy production as a function of time can be obtained from the numerical solution using the expression $(1/V)(d_iS/dt) = R(R_f - R_r) \ln (R_f/R_r)$. Note that in Mathematica the function Log is used for ln.

```
(*Calculation of entropy production ''Sigma''*)
R = 8.314;
```

```
sigma = R*(kf*XD[t] – kf*XL[t])*Log[(kf*XD[t])/(kf*XL[t])];
Plot[Evaluate[sigma/.Soln3],{t,0,0.5}]
```

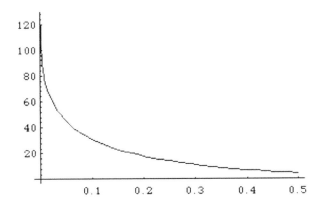

References

1. Weinberg, S., *The First Three Minutes.* New York: Bantam.
2. Taylor, R. J., *The Origin of the Chemical Elements.* 1975, London: Wykeham Publications.
3. Norman, E. B., *J. Chem. Ed.*, **71** (1994) 813–820.
4. Clayton, D. D., *Principles of Stellar Evolution and Nucleosynthesis.* 1983, Chicago: University of Chicago Press.
5. Laidler, K. J., *The World of Physical Chemistry.* 1993, Oxford: Oxford University Press.
6. Laidler, K. J., *J. Chem. Ed.*, **61** (1984) 494–498.
7. Prigogine, I. and Defay, R., *Chemical Thermodynamics*, 4th ed. 1967, London: Longman.

Data Sources

[A] NBS table of chemical and thermodynamic properties. *J. Phys. Chem. Reference Data*, **11**, suppl. 2 (1982).
[B] G. W. C. Kaye and T. H. Laby (eds) *Tables of Physical and Chemical Constants*. 1986, London: Longman.
[C] I. Prigogine and R. Defay, *Chemical Thermodynamics*, 4th ed. 1967, London: Longman.
[D] J. Emsley, *The Elements*, 1989, Oxford: Oxford University Press.
[E] L. Pauling, *The Nature of the Chemical Bond.* 1960, Ithaca NY: Cornell University Press.
[F] D. R. Lide (ed.), *CRC Handbook of Chemistry and Physics* 75th ed. 1994, Ann Arbor MI: CRC Press.
[G] The web site of the National Institute for Standards and Technology, http://webbook.nist.gov.

Examples

Example 9.1 At a temperature T, the average energy $h\nu$ of a thermal photon is roughly equal to kT. As discussed in Chapter 2, at high temperatures electron–positron pairs will be spontaneously produced when the energy of photons is larger the than the rest energy $2mc^2$ of an electron–positron pair (where m is the mass of the electron). Calculate the temperature at which electron–positron pair production occurs.

Solution: For pair production, we have

$$h\nu = kT = 2mc^2 = (2 \times 9.10 \times 10^{-31}\,\mathrm{kg})(3.0 \times 10^8\,\mathrm{m\,s^{-1}})^2 = 1.6 \times 10^{-13}\,\mathrm{J}$$

Hence the corresponding temperature is

$$T = (1.64 \times 10^{-13}\,\mathrm{J})/(1.38 \times 10^{-23}\,\mathrm{J\,K^{-1}}) = 1.19 \times 10^{10}\,\mathrm{K}$$

Example 9.2 Consider a second-order reaction $2X \rightarrow$ (products), whose rate equation is $d[X]/dt = -k[X]^2$ ($k = 2k_f$).

(a) Show that the half-life $t_{1/2}$ for this reaction depends on the initial value of $[X]$ and is equal to $1/([X]_0\,k)$.

(b) Assume that $k = 2.3 \times 10^{-1}\,\mathrm{mol^{-1}\,L\,s^{-1}}$ and obtain the value of $[X]$ at a time $t = 60.0\,\mathrm{s}$ if the initial concentration $[X]_0 = 0.50\,\mathrm{mol\,L^{-1}}$.

Solution: (a) As shown in Box 9.4, the solution to the rate equation is

$$\frac{1}{[X]} - \frac{1}{[X]_0} = k\,t$$

Multiplying both sides by $[X]_0$, we obtain

$$\frac{[X]_0}{[X]} = 1 + [X]_0\,k\,t$$

Since at $t = t_{1/2}$ the half-life $[X]_0/[X] = 2$, we must have $[X]_0\,k\,t_{1/2} = 1$ or $t_{1/2} = 1/([X]_0\,k)$.

(b) If the initial concentration $[X]_0 = 0.50\,\mathrm{mol\,L^{-1}}$, $k = 0.23\,\mathrm{mol^{-1}\,L\,s^{-1}}$ and $t = 60.0\,\mathrm{s}$, we have

$$\frac{1}{[X]} - \frac{1}{0.50} = 0.23 \times 60\,\mathrm{mol^{-1}\,L}$$

Solving for $[X]$ gives $[X] = 0.063\,\mathrm{Mol\,L^{-1}}$.

Example 9.3 For the water dissociation reaction $H_2O \rightleftharpoons OH^- + H^+$ the enthalpy of reaction $\Delta H_{rxn} = 55.84\,\mathrm{kJ}$. At $25\,^\circ\mathrm{C}$ the value of the equilibrium constant $K = 1.00 \times 10^{-14}$ and the pH $= 7.0$. What will be the pH at $50\,^\circ\mathrm{C}$?

Solution: Given $K(T)$ at one temperature T_1, its value at another temperature T_2 can be obtained using the van't Hoff equation (9.3.19):

$$\ln K(T_1) - \ln K(T_2) = \frac{-\Delta H_{rxn}}{R}\left[\frac{1}{T_1} - \frac{1}{T_2}\right]$$

For this example, we have for K at $50\,^\circ$C

$$\ln K = \ln(1.0 \times 10^{-14}) + \frac{55.84 \times 10^3}{8.314}\left[\frac{1}{298} - \frac{1}{323}\right] = -30.49$$

Hence K at $50\,^\circ$C is equal to $\exp(-30.49) = 5.73 \times 10^{-14}$. Since the equilibrium constant $K = [OH^-][H^+]$ and because $[OH^-] = [H^+]$, we have

$$pH = -\log[H^+] = -\log[\sqrt{K}] = -\frac{1}{2}\log[5.73 \times 10^{-14}] = 6.62$$

Exercises

9.1 When the average kinetic energy of molecules is nearly equal to the bond energy, molecular collisions will begin to break the bonds. The average kinetic energy of molecules is equal to $3RT/2$. (a) The C–H bond energy is about $414\,\text{kJ mol}^{-1}$. At what temperature will the C–H bonds in methane begin to break? (b) The average binding energy per ·nucleon (neutron or proton) is in the range $(6.0\text{–}9.0) \times 10^6\,\text{eV}$ or $(6.0\text{–}9.0) \times 10^8\,\text{kJ mol}^{-1}$. At what temperature do you expect nuclear reactions to take place?

9.2 For the reaction $Cl + H_2 \rightarrow HCl + H$, the activation energy $E_a = 23.0\,\text{kJ mol}^{-1}$ and $k_0 = 7.9 \times 10^{10}\,\text{mol}^{-1}\,\text{L}\,\text{s}^{-1}$. What is the value of rate constant at $T = 300.0\,\text{K}$? If $[Cl] = 1.5 \times 10^{-4}\,\text{mol L}^{-1}$ and $[H_2] = 1.0 \times 10^{-5}\,mol\,L^{-1}$, what is the forward reaction rate at $T = 350.0\,\text{K}$?

9.3 For the decomposition of urea in an acidic medium, the following data was obtained for rate constants at various temperature:

Temperature (°C)	50	55	60	65	70
Rate constant $k\,(\text{s}^{-1})$	2.29×10^{-8}	4.63×10^{-8}	9.52×10^{-8}	1.87×10^{-7}	3.72×10^{-7}

(a) Using an Arrhenius plot, obtain the activation energy E_a and the preexponential factor k_0.
(b) Apply transition state theory to the same data, plot $\ln(k/T)$ versus $1/T$ and obtain ΔH^+ and ΔS^+ for the transition state.

9.4 Consider the dimerization of the triphenylmethyl radical $Ph_3C\cdot$, which can be written as the reaction

$$A \rightleftharpoons 2B$$

The forward and reverse rate constants (at $300\,k$) for this reaction are $k_f = 0.406\,s^{-1}$ and $k_r = 3.83 \times 10^2\,mol^{-1}\,L\,s^{-1}$. Assume this reaction is an elementary step. At $t = 0$ the initial concentrations of A and B are: $[A]_0 = 0.041\,mol\,L^{-1}$ and $[B]_0 = 0.015\,mol\,L^{-1}$.
(a) What is the velocity of the reaction at $t = 0$?
(b) If ξ_{eq} is the extent of reaction at equilibrium, ($\xi = 0$ at $t = 0$), write the equilibrium concentrations of A and B in terms of $[A]_0$, $[B]_0$ and ξ_{eq}.
(c) Obtain the value of ξ_{eq} by solving the appropriate quadratic equation (you may use Maple) and obtain the equilibrium concentrations of $[A]$ and $[B]$.

9.5 (a) Write the rate equations for the concentrations of X, Y and Z in the following reaction:

$$X + Y \rightleftharpoons 2Z$$

(b) Write the rate equation for the extent of reaction ξ.
(c) At thermal equilibrium, $\xi = \xi_{eq}$. If $[X]_0$, $[Y]_0$ and $[Z]_0$ are the initial concentrations, write the equilibrium concentrations in terms of the initial concentrations and ξ_{eq}.

9.6 Radioactive decay is a first-order reaction. If N is the number of radioactive nuclei at any time t, then $dN/dt = -kN$. Carbon-14 is radioactive with a half-life of 5730 years. What is the value of k? For this process, do you expect k to change with temperature?

9.7 The chirping rate of crickets depends on temperature. When the chirping rate is plotted against $(1/T)$, it is observed to follow the Arrhenius law (K.J. Laidler, *J. Chem. Ed.*, **49** (1972) 343). How would you explain this observation?

9.8 Consider the reaction $X + Y \rightleftharpoons 2Z$ in the gas phase. Write the reaction rates in terms of the concentrations $[X]$, $[Y]$ and $[Z]$ as well as in terms of the activities. Find the relation between the rate constants in the two ways of writing the reaction rates.

9.9 When atmospheric CO_2 dissolves in water, it produces carbonic acid H_2CO_3 (which causes natural rain to be slightly acidic). At 25.0 °C the equilibrium constant K_a for the reaction $H_2CO_3 \rightleftharpoons HCO_3^- + H^+$ is specified by $pK_a = 6.63$; the enthalpy of reaction is $\Delta H_{rxn} = 7.66 \, kJ \, mol^{-1}$. Calculate the pH at 25 °C and at 35 °C. Use Henry's law to obtain $[H_2CO_3]$.

9.10 Obtain equilibrium constants for the following reactions at $T = 298.15 \, K$, using tables for $\mu_0(p_0, T_0) = \Delta G_f^0$:
(i) $2NO_2(g) \rightleftharpoons N_2O_4(g)$
(ii) $2CO(g) + O_2(g) \rightleftharpoons 2CO_2(g)$
(iii) $N_2(g) + O_2(g) \rightleftharpoons 2NO(g)$
Equilibrium constants can vary over an extraordinary range as these examples show.

9.11 For a reaction of the form $aX + bY \rightleftharpoons cZ$, show that the equilibrium constants K_c and K_p are related by $K_c = (RT)^\alpha K_p$ where $\alpha = a + b - c$.

9.12 Ammonia may be produced through the following reaction:

$$N_2(g) + 3H_2(g) \rightleftharpoons 2NH_3(g)$$

(i) Calculate the equilibrium constant of this reaction at 25 °C using thermodynamic tables.
(ii) Assuming there is no significant change in the enthalpy of reaction ΔH_{rxn}, use the van't Hoff equation to obtain approximate values for ΔG_{rxn} and the equilibrium constant at 400 °C.

9.13 The gas 2-butene has two isomeric forms, *cis* and *trans*. For the reaction

$$cis \text{ - 2 - butene} \rightleftharpoons trans \text{ - 2 - butene}$$

$\Delta G_{rxn}^0 = -2.4 \, kJ \, mol^{-1}$. Calculate the equilibrium constant for this reaction at $T = 298.15 \, K$. If the total amount of butene is 2.5 mol, assuming ideal gas behavior, determine the number of moles of each isomer.

9.14 Determine whether or not the introduction of a catalyst will alter the affinity of a reaction.

9.15 For the reaction $X + 2Y \rightleftharpoons 2Z$ write an explicit expression for the entropy production in terms of the rates, and as a function of ξ.

10 FIELDS AND INTERNAL DEGREES OF FREEDOM

The Many Faces of Chemical Potential

The concept of chemical potential is very general, applicable to almost any transformation of matter as long as there is a well-defined temperature. We have already seen how the condition for thermodynamic equilibrium for chemical reactions leads to the law of mass action. We shall now see how diffusion, electrochemical reactions and relaxation of polar molecules in the presence of an electric field, can all be viewed as "chemical transformations" with associated chemical potential and affinity.

10.1 Chemical Potential in a Field

The formalism for the chemical potential presented in the previous chapter can be extended to electrochemical reactions and to systems in an external field such as a gravitational field. In the presence of a field, the energy due to a field must be included while considering changes in energy. As a result, the energy of a constituent depends on its location.

We start with a simple system: the transport of chemical species which carry electrical charge from a location where the potential is ϕ_1 to a location where the potential is ϕ_2. For simplicity, we shall assume that our system consists of two parts, each with a well-defined potential, but the system as a whole is closed (Fig. 10.1). It is as if the system consisted of two phases and transport of particles dN_k were a "chemical reaction." For the corresponding degrees of advancement $d\xi_k$ we have

$$-dN_{1k} = dN_{2k} = d\xi_k \tag{10.1.1}$$

in which dN_{1k} and dN_{2k} are the changes in the mole numbers in each part. The change in energy due to the transport of the ions is given by

$$
dU = T\,dS - p\,dV + F\phi_1 \sum_k z_k\,dN_{1k} + F\phi_2 \sum_k z_k\,dN_{2k}
$$
$$
+ \sum_k \mu_{1k}\,dN_{1k} + \sum_k \mu_{2k}\,dN_{2k} \tag{10.1.2}
$$

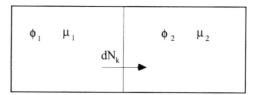

Figure 10.1 A simple situation illustrating the thermodynamics of a system in the presence of an electric field. We consider two compartments: one associated with potential ϕ_1 and the other with potential ϕ_2. It is as if there were two phases; ions will migrate from one compartment to the other until the electrochemical potentials are equal

in which z_k is the ion number of ion k and F is the Faraday constant (the product of the electronic charge e and the Avogadro number N_A : $F = eN_A = 9.6485 \times 10^4$ C mol^{-1}). Using (10.1.1) the change in the entropy dS can now be written as

$$T\,dS = dU + p\,dV - \sum_k [(F\phi_2 z_k + \mu_{2k}) - (F\phi_1 z_k + \mu_{1k})]d\xi_k \quad (10.1.3)$$

Thus we see that the introduction of a potential ϕ associated with a field is equivalent to adding a term to the chemical potential. This makes it possible to extend the definition of the chemical potential to include the field. The **electrochemical potential** $\tilde{\mu}$, introduced by Guggenheim in 1929 [1], is defined as

$$\boxed{\tilde{\mu}_k = \mu_k + Fz_k\phi} \quad (10.1.4)$$

Clearly this formalism can be extended to any field with which a potential may be associated. If ψ is the potential associated with the field, the energy of interaction *per mole* of the component k may be written in the form $\tau_k\psi$. For the electric field $\tau_k = Fz_k$; for the gravitational field $\tau_k = M_k$, the molar mass. The corresponding chemical potential which includes the potential is

$$\boxed{\tilde{\mu}_k = \mu_k + \tau_k\psi} \quad (10.1.5)$$

The affinity $\tilde{A}_k$ for the electrochemical reactions can now be defined, just as for other chemical reactions

$$\tilde{A}_k = \tilde{\mu}_{1k} - \tilde{\mu}_{2k} = [(F\phi_1 z_k + \mu_{1k}) - (F\phi_2 z_k + \mu_{2k})] \quad (10.1.6)$$

The increase in the entropy due to transfer of charged particles from one potential to another can now be written as

$$d_iS = \sum_k \frac{\tilde{A}_k}{T} d\xi_k \tag{10.1.7}$$

At equilibrium,

$$\tilde{A}_k = 0 \quad \text{or} \quad \mu_{1k} - \mu_{2k} = -z_k F(\phi_1 - \phi_2) \tag{10.1.8}$$

The basic equations of equilibrium electrochemistry follow from (10.1.8)

Because electrical forces are very strong, in ionic solutions the electric field produced by even small changes in charge density produce very strong forces between the ions. Consequently, in most cases the concentrations of positive and negative ions are such that net charge density is virtually zero, i.e. **electroneutrality** is maintained to a high degree. In a typical electrochemical cell, most of the potential difference applied to the electrodes appears in the vicinity of the electrodes, and only a small fraction of the total potential difference occurs across the bulk of the solution. The solution is electrically neutral to an excellent approximation. As a result, an applied electric field does not separate positive and negative charges; it does not create an appreciable concentration gradient.

When we consider the much weaker gravitational field, however, an external field can produce a concentration gradient. For a gravitational field, the coupling constant τ_k is the molar mass M_k. For a gas in a uniform gravitational field, $\phi = gh$, where g is the strength of the field and h is the height; from (10.1.8) we see that

$$\mu_k(h) = \mu_k(0) - M_k gh \tag{10.1.9}$$

For an ideal gas mixture, using $\mu_k(h) = \mu_k(T) + RT \ln[p_k(h)/p_0]$ in (10.1.9) we obtain the well-known **barometric formula**:

$$\boxed{p_k(h) = p(0)e^{-M_k gh/RT}} \tag{10.1.10}$$

Note how this formula is derived by assuming the temperature T is uniform, i.e. the system is in equilibrium. The temperature of the earth's atmosphere is not uniform; in fact, it varies roughly between 220 K and 300 K as shown in Fig. 10.2(b).

ENTROPY PRODUCTION IN A CONTINUOUS SYSTEM

In considering thermodynamic systems in a field, we often have to consider continuous variation of the thermodynamic fields. In this case $\tilde{\mu}$ is a function of

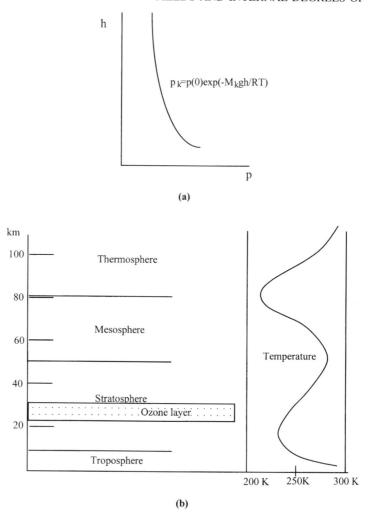

Figure 10.2 (a) The concept of a chemical potential that includes a field leads to the well-known barometric formula at thermal equilibrium in which T is uniform. (b) The actual state of the earth's atmosphere is not in thermal equilibrium; the temperature varies with height as shown

position, and entropy has to be expressed in terms of entropy density $s(\mathbf{r})$, entropy per unit volume, which depends on position $\mathbf{r}$. For simplicity, let us consider a one-dimensional system, i.e. a system in which the entropy and all other variables, such as μ, change only along one direction, say x. Let $s(x)$ be the entropy density per unit length. We shall assume the temperature is constant throughout the system. Then the entropy in a small volume element between x and $x + \delta$ is equal to $s(x)\delta$ (Fig. 10.3). An expression for affinity in this small

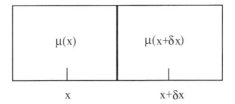

Figure 10.3 An expression for the entropy production in a continuous system can be obtained by considering two adjacent cells separated by a small distance δ. The entropy in the region between x and $x + \delta$ is equal to $s(x)\delta$. The affinity, which is the difference in the chemical potential, is given by $\tilde{A} = \tilde{\mu}(x) - \tilde{\mu}(x + \delta) = \tilde{\mu}(x) - [\tilde{\mu}(x) + (\partial\mu/\partial x)\delta] = (\partial\mu/\partial x)\delta$

volume element can be written as

$$\tilde{A} = \tilde{\mu}(x) - \tilde{\mu}(x + \delta) = \tilde{\mu}(x) - \left(\mu(x) + \frac{\partial\mu}{\partial x}\delta\right) = -\frac{\partial\mu}{\partial x}\delta \qquad (10.1.11)$$

The velocity of the reaction $d\xi_k/dt$ for this elemental volume is the flow of particles of kth component, i.e. particle current k. We shall denote this particle current by J_{NK}. Then by writing (10.1.7) for this elemental volume, we obtain

$$\frac{d_i(s(x)\delta)}{dt} = \sum_k \frac{\tilde{A}_k}{T}\frac{d\xi_k}{dt} = -\sum_k \frac{1}{T}\left(\frac{\partial\tilde{\mu}_k}{\partial x}\right)\delta\frac{d\xi_k}{dt} \qquad (10.1.12)$$

Simplifying this expression, and using the definition $J_{NK} \equiv d\xi_k/dt$ the following expression for *entropy production per unit length* due to particle flow is obtained

$$\boxed{\frac{d_i s(x)}{dt} = -\sum_k \frac{1}{T}\left(\frac{\partial\tilde{\mu}_k}{\partial x}\right) J_{Nk}} \qquad (10.1.13)$$

OHM'S LAW AND ENTROPY PRODUCTION DUE TO ELECTRICAL CONDUCTION

To understand the significance of expression (10.1.13), let us consider the flow of electrons in a conductor. In a conductor in which the electron density and temperature are uniform, the chemical potential μ_e of the electron (which is a function of the electron density and T) is constant. Therefore the derivative of the electrochemical potential becomes:

$$\frac{\partial \tilde{\mu}_e}{\partial x} = \frac{\partial}{\partial x}(\mu_e - Fe\phi) = -\frac{\partial}{\partial x}(Fe\phi) \qquad (10.1.14)$$

The electric field $E = -\partial\phi/\partial x$, and the conventional electric current $I = -eFJ_e$; using (10.1.14) in expression (10.1.13), we obtain the following expression for the entropy production

$$\frac{d_i s}{dt} = eF\left(\frac{\partial\phi}{\partial x}\right)J_e = \frac{EI}{T} \qquad (10.1.15)$$

Since the electric field is the change of potential per unit length, it follows that the integral of E over the entire length L of the conductor is the potential difference V across the entire conductor. The total entropy production from $x = 0$ to $x = L$ is

$$\frac{dS}{dt} = \int_0^L \left(\frac{d_i s}{dt}\right)dx = \int_0^L \frac{EI}{T}dx = \frac{VI}{T} \qquad (10.1.16)$$

Now it is well known that the product VI of potential difference and current is the heat generated, called the **ohmic heat**, per unit time. The flow of an electric current through a resistor is an irreversible dissipative process that converts electrical energy into heat. For this reason we may write $VI = dQ/dt$. For a flow of electric current, we have

$$\boxed{\frac{d_i S}{dt} = \frac{VI}{T} = \frac{1}{T}\frac{dQ}{dt}} \qquad (10.1.17)$$

This shows that the entropy production is equal to the dissipated heat divided by the temperature.

We have noted in Chapter 3 that the entropy production due to each irreversible process is a product of a thermodynamic force and the flow it drives (3.4.7). Here the flow is the electric current; the corresponding force is the term V/T. Now it is generally true that when a system is close to thermodynamic equilibrium, the flow is proportional to the force. Hence, based on thermodynamic reasoning, we arrive at the conclusion

$$I = L_e \frac{V}{T} \qquad (10.1.18)$$

in which L_e is a constant of proportionality for the electron current; L_e is called the **linear phenomenological coefficient**. Relations such as (10.1.18) are the basis of linear nonequilibrium thermodynamics which we shall consider in detail in Chapter 16. We see at once that this corresponds to the familiar **Ohm's law**, $V=IR$, where R is the resistance, if we identify

$$L_e = \frac{T}{R} \qquad (10.1.19)$$

This is an elementary example of how the expression for entropy production can be used to obtain linear relations between thermodynamic forces and flows, which often turn out to be empirically discovered laws such as Ohm's law. In section 10.3 we shall see that similar consideration of entropy production due to diffusion leads to another empirically discovered law called the Fick's law of diffusion. Modern thermodynamics enables us incorporate many such phenomenological laws into one unified formalism.

10.2   Membranes and Electrochemical Cells

MEMBRANE POTENTIALS

Just as equilibrium with a semipermeable membrane produced a difference in pressure (the osmotic pressure) between the two sides of the membrane, equilibrium of ions across a membrane that is permeable to one ion but not another results in an electric potential difference. As an example, consider a membrane separating two solutions of KCl of *unequal* concentrations (Fig. 10.4). We assume that the membrane is permeable to K^+ ions but is impermeable to the larger Cl^- ions. Since the concentrations of the K^+ ions on the two sides of the membrane are unequal, K^+ ions will begin to flow to the

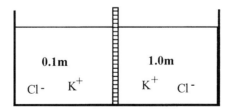

Figure 10.4   A membrane potential is generated when a membrane permeable to K^+ but not to Cl^- separates two solutions of KCl having unequal concentrations. In this case the flow of the permeable K^+ ions is counterbalanced by the membrane potential

region of lower concentration from the region of higher concentration. Such a flow of positive charge, without a counterbalancing flow of negative charge, will cause a buildup in a potential difference that will oppose the flow. Equilibrium is reached when the electrochemical potentials on the two sides become equal, at which point the flow will stop. We shall denote the two sides with superscripts α and β. Then the equilibrium of the K^+ ion is established when

$$\tilde{\mu}_{K^+}^{\alpha} = \tilde{\mu}_{K^+}^{\beta} \tag{10.2.1}$$

Since the electrochemical potential of an ion k is $\tilde{\mu}_k = \mu_k + z_k F\phi = \mu_k^0 + RT \ln a_k + z_k F\phi$, in which a_k is the activity and z_k the ion number (which is $+1$ for K^+), equation (10.2.1) can be written as

$$\mu_{K^+}^0 + RT \ln a_{K^+}^{\alpha} + F\phi^{\alpha} = \mu_{K^+}^0 + RT \ln a_{K^+}^{\beta} + F\phi^{\beta} \tag{10.2.2}$$

From this equation it follows that the potential difference, the **membrane potential** $= (\phi^{\alpha} - \phi^{\beta})$ across the membrane, can now be written as

$$(\phi^{\alpha} - \phi^{\beta}) = \frac{RT}{F} \ln \left(\frac{a_{K^+}^{\beta}}{a_{K^+}^{\alpha}} \right) \tag{10.2.3}$$

In electrochemistry the concentrations are generally measured using the molality scale, as discussed in Chapter 8. In the simplest approximation, the activities may be replaced by molalities m_{K^+}, i.e. the activity coefficients are assumed to be unity. Hence one may estimate the membrane potential using $(\phi^{\alpha} - \phi^{\beta}) = (RT/F) \ln (m_{K^+}^{\beta}/m_{K^+}^{\alpha})$.

ELECTROCHEMICAL AFFINITY AND ELECTROMOTIVE FORCE

In an electrochemical cell the reactions at the electrodes that transfer electrons can generate an electromotive force (EMF). An electrochemical cell generally has different phases that separate the two electrodes (Fig. 10.5). By considering entropy production due to the overall reaction and the electric current flowing through the system, we can derive a relationship between the electrochemical activity and the EMF. In an electrochemical cell the reactions at the two electrodes can generally be written as

$$X + ne^- \longrightarrow X_{red} \qquad \text{``Reduction''} \tag{10.2.4}$$
$$Y \longrightarrow Y_{ox} + ne^- \qquad \text{``Oxidation''} \tag{10.2.5}$$

Each is called a **half-reaction**; the overall reaction is

$$X + Y \rightarrow X_{red} + Y_{ox} \tag{10.2.6}$$

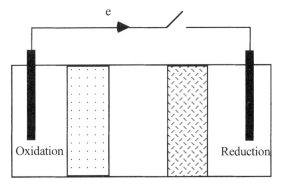

Figure 10.5 An electrochemical cell consisting of many phases that generate an EMF due to half-reactions at the electrodes. The electrode reactions are

$$X + ne^- \longrightarrow X_{red} \qquad \text{''reduction'' (Right)}$$
$$Y \longrightarrow Y_{ox} + ne^- \quad \text{''oxidation'' (Left)}$$

Upon closing the circuit, chemical reaction occurring within the cell will generate an electromotive force that will drive a current. Cells such as this are represented by a cell diagram denoting the various phases and junctions. In a cell diagram the reduction reaction is on the right

$$Electrode|Y|\ldots|\ldots|\ldots|X|Electrode$$

For example, the half-reactions

$$Cu^{2+} + 2e^- \longrightarrow Cu(s)$$
$$Zn(s) \longrightarrow Zn^{2+} + 2e^-$$

at the electrodes lead to the overall reaction

$$Cu^{2+} + Zn(s) \longrightarrow Zn^{2+} + Cu(s)$$

(Thus, if a zinc rod is placed in an aqueous solution of $CuSO_4$, the zinc rod dissolves and metallic copper is deposited.)

Reactions at the electrodes may be more complicated than the reactions above but the main idea is the same. At one electrode, electrons are transferred *from* the electrode; at the other electrode, electrons are transferred *to* the electrode. In representing electrochemical cells diagrammatically, it has become a convention to place the **reduction half-reaction on the right**. Thus the electrode on the right-hand side of the diagram supplies the electrons that reduce reactants.

Since the reactions at the electrodes may occur at different electrical potentials, we must use the electrochemical affinity to formulate the thermodynamics of an electrochemical cell. If $\tilde{A}$ is the electrochemical affinity and ξ is the extent of reaction, the entropy production is

$$\frac{d_iS}{dt} = \frac{\tilde{A}}{T}\frac{d\xi}{dt} \tag{10.2.7}$$

If each mole of reacting X transfers n moles of electrons, since $d\xi/dt$ is the velocity of reaction, the relation between the current I (which is the amount of charge transferred per second) is

$$I = nF\frac{d\xi}{dt} \tag{10.2.8}$$

in which F is the Faraday constant. Substituting (10.2.8) into (10.2.7) we find

$$\frac{d_iS}{dt} = \frac{1}{T}\frac{\tilde{A}}{nF}I \tag{10.2.9}$$

Comparing this expression with (10.1.17) gives the following relation between the electrochemical affinity and the associated voltage:

$$V = \frac{\tilde{A}}{nF} \tag{10.2.10}$$

in which n is the number of electrons transferred in the oxidation–reduction reaction. For a given $\tilde{A}$, the larger the number of electrons transferred, the smaller the potential difference.

Using the electrode reactions (10.2.4) and (10.2.5), the affinities can be more explicitly written in terms of the chemical potentials:

$$X + ne^- \rightarrow X_{red} \quad \text{(right)} \quad \tilde{A}^R = (\mu_X^R + n\mu_e^R - nF\phi^R) - \mu_{Xred}^R \tag{10.2.11}$$

$$Y \rightarrow Y_{ox} + ne^- \quad \text{(left)} \quad \tilde{A}^L = \mu_Y^L - (n\mu_e^L - nF\phi^L + \mu_{Yox}^L) \tag{10.2.12}$$

in which the superscripts indicate the reactions at the right and left electrodes. The electrochemical affinity of the electron at the left electrode, is written as $\tilde{\mu}_e = \mu_e^L - F\phi^L$. The overall electrochemical affinity $\tilde{A}$, which is the sum of the two affinities, can now be written as

$$\tilde{A} = \tilde{A}^R + \tilde{A}^L = (\mu_X^R + \mu_Y^L - \mu_{Xred}^R - \mu_{Yox}^L) + n(\mu_e^R - \mu_e^L) - nF(\phi^R - \phi^L) \tag{10.2.13}$$

If the two electrodes are identical then $\mu_e^R = \mu_e^L$; the only difference is the potential ϕ. By virtue of (10.2.10) we can now write as

$$V = \frac{\tilde{A}}{nF} = \frac{1}{nF}(\mu_X^R + \mu_Y^L - \mu_{X\,red}^R - \mu_{Y\,ox}^L) - (\phi^R - \phi^L) \qquad (10.2.14)$$

Let us now consider the potential difference or the "terminal voltage", $(\phi^R - \phi^L)$, that makes the affinity zero (open circuit, zero current). It is similar to the osmotic pressure difference at zero affinity. The terminal voltage $V_{cell} = (\phi^R - \phi^L)$, at $\tilde{A} = 0$ (zero current) is called the **EMF of the cell**. From (10.2.14) it follows that:

$$\boxed{V_{cell} = \frac{1}{nF}(\mu_X^R + \mu_Y^L - \mu_{X\,red}^R - \mu_{Y\,ox}^L)} \qquad (10.2.15)$$

Box. 10.1 Electrochemical Cells and Cell Diagrams

When there is an external flow of current, there must be a compensating current within the cell. This can be accomplished in many ways, hence there are electrochemical cells of many types. The choice of electrodes is also decided by the experimental conditions and the need to use the electrode without undesirable side reactions. Electrochemical cells often incorporate **salt bridges** and **liquid junctions**.

Liquid junctions. When two different liquids are in contact, usually through a porous separation, it is called a liquid junction. The concentrations of ions on either side of a liquid junction are generally not equal, hence there is a diffusional flow of ions. If the rates of flow of the different ions are unequal, a potential difference will be generated across the liquid junction. Such a potential is called the **liquid junction potential**. The liquid junction potential may be reduced by the use of a **salt bridge**, in which the flow of the positive and negative ions are nearly equal.

Salt bridges. A commonly used salt bridge consists of a solution of KCl in agarose jelly. In this medium, the flows of K^+ and Cl^- are nearly equal.

Cell diagrams. An electrochemical cell diagram adopts the following conventions:
• Reduction occurs at the electrode on the right.
• A phase boundary is drawn |, e.g. the boundary between a solid electrode and the solution.
• A liquid junction is drawn ⋮, e.g. the porous wall separating a solution of $CuSO_4$ and CuCl.
• A salt bridge is drawn ‖, e.g. KCl in agarose jelly.

For example, the cell in Fig. 10.6 is represented by the cell diagram

$$Zn(s)|Zn^{2+}||H^+|Pt(s)$$

For a nonzero affinity, i.e., nonzero current, the terminal voltage is less than the EMF. On the other hand, if the potentials of the two electrodes are equalized by shorting the two terminals, the flow of current $I = nF(d\xi/dt)$ is limited only by the rate of electron transfer at the electrodes. Under these conditions, the voltage V corresponding to the affinity is also given by (10.2.15).

It is more convenient to write the cell EMF (10.2.15) in terms of the activities by using the general expression $\mu_k = \mu_k^0 + RT \ln a_k$ for the reactants and products. This leads to

$$V_{cell} = V_0 - \frac{RT}{nF} \ln \left(\frac{a_{X\,red}^R\, a_{Y\,ox}^L}{a_X^R\, a_Y^L} \right)$$ (10.2.16)

where

$$V_0 = \frac{1}{nF}(\mu_{X0}^R + \mu_{Y0}^L - \mu_{X\,red\,0}^R - \mu_{Y\,ox\,0}^L) = \frac{-\Delta G_{rxn}^0}{nF}$$ (10.2.17)

Equation (10.2.16) relates the cell potential to the activities of the reactants; it is called the **Nernst Equation**. At equilibrium, as we expect, V is zero and the equilibrium constant of the electrochemical reaction can be written as

$$\ln K = \frac{-\Delta G_{rxn}^0}{RT} = \frac{nFV_0}{RT}$$ (10.2.18)

GALVANIC AND ELECTROLYTIC CELLS

A cell in which a chemical reaction generates an electric potential difference is called a **galvanic cell**; if an external source of electric voltage drives a chemical reaction, it is called an **electrolytic cell**.

Let us consider a simple reaction. When Zn reacts with an acid, H_2 is evolved (Fig. 10.6). This reaction is a simple electron-transfer reaction:

$$Zn(s) + 2H^+ \longrightarrow Zn^{2+} + H_2(g)$$ (10.2.19)

The reason why the electrons migrate from one atom to another is a difference in electrical potential, i.e. in the above reaction, when an electron moves from a Zn atom to an H^+ ion it is moving to a location of lower potential energy. Now an interesting possibility arises. If the reactants are placed in a "cell" such that the only way an electron transfer can occur is through a conducting wire, then we have a situation in which a chemical affinity drives an electric current. Such a cell would be a **galvanic cell**. Conversely, through an external electromotive force, the electron transfer can be reversed, which is the case in an **electrolytic cell**.

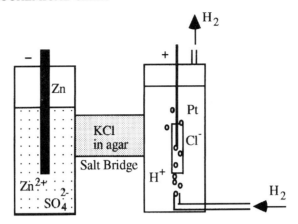

Figure 10.6 An example of a galvanic cell that is driven by the reaction

$$Zn(s) + 2H^+ \longrightarrow Zn^{2+} + H_2(g)$$

The two electrode chambers are connected through a salt bridge that allows for the flow of current without introducing a liquid junction potential

The EMF generated by a galvanic cell is given by the Nernst equation. The cell EMF in the above example is given by

$$V = V_0 - \frac{RT}{nF} \ln\left(\frac{a_{H_2}\, a_{Zn^{2+}}}{a_{Zn}\, a_{H^+}^2}\right) \tag{10.2.20}$$

CONCENTRATION CELL

The affinity generated by a concentration difference can also generate an EMF. A simple example of a $CuCl_2$ concentration cell in which a concentration-driven EMF can be realized is shown in Fig. 10.7. In this cell the membrane is permeable to Cl^- but not to Cu^{2+}. Two copper rods are placed in the two chambers, α and β, which become a source and a sink for the Cu^{2+} ions. The concentration difference drives Cl^- ions from α to β. The flow of Cl^- ions generates a potential difference (as in the case of membrane potential), but due to the presence of copper rods, this potential can drive electrons from the copper rod in β to the copper rod in α, thus generating Cu^{2+} ions in β and neutralizing Cu^{2+} ions to Cu atoms in α. The overall process is an effective transport of Cu^{2+} ions from α to β. The system finds a way to reach equilibrium through the flow of electrons. For a concentration cell, $V_0 = 0$, because the standard states for the "reactant" and the "product" are the same; the only difference between the two is in their activities (which nearly equal the molalities in dilute solutions). It

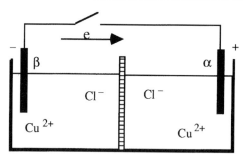

Figure 10.7 A concentration difference can generate an electromotive force. The two compartments are separated by a membrane permeable to Cl^- but not Cu^{2+}. If two copper rods are introduced into the two cells and an electrical circuit is completed, as the Cl^- ions diffuse from α to β, the flow of electrons in the opposite direction will effectively transport Cu^{2+} ions from α to β. This flow of electrons produces an electromotive force given by (10.2.21)

follows from the Nernst equation that

$$V = -\frac{RT}{nF} \ln\left(\frac{a^{\beta}_{Cu^{2+}}}{a^{\alpha}_{Cu^{2+}}}\right) \tag{10.2.21}$$

STANDARD ELECTRODE POTENTIALS

Just as the tabulation of the Gibbs free energies of formation facilitates the computation of equilibrium constants, the tabulation of "standard electrode potentials" facilitates the computation of equilibrium constants for electrochemical reactions. A voltage is assigned to each electrode reaction with the convention that the voltage of the hydrogen–platinum electrode, $H^+|Pt$, is zero. That is, the electrode reaction $H^+ + e^- \rightarrow \frac{1}{2}H_2(g)$ at a Pt electrode is taken to be the reference and the voltages associated with all other electrode reactions are measured with respect to it. *The standard electrode potentials are the potentials when the activities of all the reactants and products equal one at $T = 298.15$ K.* For any cell, the voltages of the corresponding standard potentials are added to obtain the cell potentials. Since these potential correspond to the situation when all the activities are equal to one, it follows from the Nernst equation that the standard cell voltage is equal to V_0.

Table 10.1 Standard electrode potentials*

Electrode reaction	$V_0/(V)$	Electrode
$\frac{1}{3}Au^{3+} + e^- \longrightarrow \frac{1}{3}Au$	1.50	$Au^{3+}|Au$
$\frac{1}{2}Cl_2(g) + e^- \longrightarrow Cl^-$	1.360	$Cl^-|Cl_2(g)|Pt$
$Ag^+ + e^- \longrightarrow Ag(s)$	0.799	$Ag^+|Ag$
$Cu^+ + e^- \longrightarrow Cu(s)$	0.521	Cu^+/Cu
$\frac{1}{2}Cu^{2+} + e^- \longrightarrow \frac{1}{2}Cu(s)$	0.339	$Cu^{2+}|Cu$
$AgCl + e^- \longrightarrow Ag + Cl^-$	0.222	$Cl^-|AgCl(s)|Ag$
$Cu^{2+} + e^- \longrightarrow Cu^+$	0.153	$Cu^{2+}, Cu^+|Pt$
$H^+ + e^- \longrightarrow \frac{1}{2}H_2(g)$	0.0	$H^+|H_2|Pt$
$\frac{1}{2}Pb^{2+} + e^- \rightarrow \frac{1}{2}Pb(s)$	-1.126	$Pb^{2+}|Pb(s)$
$\frac{1}{2}Sn^{2+} + e^- \longrightarrow \frac{1}{2}Sn(s)$	-0.140	$Sn^{2+}|Sn(s)$
$\frac{1}{2}Ni^{2+} + e^- \longrightarrow \frac{1}{2}Ni(s)$	-0.250	$Ni^{2+}|Ni(S)$
$\frac{1}{2}Cd^{2+} + e^- \longrightarrow \frac{1}{2}Cd(s)$	-0.402	$Cd^+|Cd(S)$
$\frac{1}{2}Zn^{2+} + e^- \longrightarrow \frac{1}{2}Zn(s)$	-0.763	$Zn^{2+}|Zn(s)$
$Na^+ + e^- \longrightarrow Na(s)$	-2.714	$Na^+|Na(s)$
$Li^+ + e^- \longrightarrow Li(s)$	-3.045	$Li^+|Li(s)$

* If the reaction is reversed, the sign of V_0 also reverses.

Example 10.3 shows how an equilibrium constant may be computed using the standard electrode potentials. Table 10.1 lists some of the commonly used standard electrode potentials. In using the standard potentials, one must note that if the reaction is reversed, the sign of V_0 also reverses.

10.3 Diffusion

We have already seen in section 4.3 that the flow of particles from a region of high concentration to a region of lower concentration is driven by unequal chemical potentials. For a discrete system consisting of two parts of equal temperature T, one with chemical potential μ_1 and mole number N_1 and the other with chemical potential μ_2 and mole number N_2, we have the following relations:

$$-dN_1 = dN_2 = d\xi \qquad (10.3.1)$$

The entropy production that results from unequal chemical potentials is

$$d_iS = -\left(\frac{\mu_2 - \mu_1}{T}\right)d\xi = \frac{A}{T}d\xi > 0 \qquad (10.3.2)$$

The positivity of this quantity, required by the Second Law, implies that particle transport is from a region of higher chemical potential to a region of lower chemical potential. This is the diffusion of particles from a region of higher chemical potential to a region of lower chemical potential. In many situations this is a flow of a component from a higher concentration to a lower concentration. At equilibrium the concentrations become uniform. But this need not be so in every case. For example, when a liquid is in equilibrium with its vapor or when a gas reaches equilibrium in a gravitational field, the chemical potentials become uniform, not the concentrations. *The tendency of the thermodynamic forces that drive matter flow is to equalize the chemical potential, not the concentrations.*

DIFFUSION IN A CONTINUOUS SYSTEM AND FICK'S LAW

Expression (10.3.2) can be generalized to describe a continuous system as was done for the general case of a field in section 10.1 (Fig. 10.3). Let us consider a system in which the variation of the chemical potential is along one direction only, say x. We shall also assume that T is uniform and does not change with position. Then, as in (10.1.13), we have for diffusion that

$$\frac{d_i s(x)}{dt} = -\sum_k \frac{1}{T}\left(\frac{\partial \mu_k}{\partial x}\right) J_{Nk} \tag{10.3.3}$$

For simplicity, let us consider the flow of a single component k:

$$\frac{d_i s(x)}{dt} = -\frac{1}{T}\left(\frac{\partial \mu_k}{\partial x}\right) J_{Nk} \tag{10.3.4}$$

We note, once again, that the entropy production is a product of a thermodynamic flow J_{Nk} and the force $-(1/T)(\partial \mu_k/\partial x)$ that drives it. The identification of a thermodynamic force and the corresponding flow enables us to relate the two. Near equilibrium, the flow is linearly proportional to the force. In the above case we can write this linear relation as

$$J_{Nk} = -L_k \frac{1}{T}\left(\frac{\partial \mu_k}{\partial x}\right) \tag{10.3.5}$$

The constant of proportionality, L_k, is the linear phenomenological coefficient for diffusional flow. We have seen earlier that in an ideal fluid mixture the chemical potential can be written as $\mu(p, T, x_k) = \mu(p, T) + RT \ln x_k$, in which x_k is the mole fraction per unit volume of k, generally a function of position. If n_{tot} is the total mole number density and n_k is the mole number density of component k, the mole fraction $x_k = n_k/n_{tot}$. We shall assume the change of n_{tot} due to diffusion is insignificant, so $\partial \ln(x_k)/\partial x = \partial \ln(n_k)/\partial x$. Then, substituting $\mu(p,T,x_k) = \mu(p,T) + RT \ln x_k$ into (10.3.5), we obtain the following thermodynamic relation between the diffusion current J_{Nk} and the concentration:

$$J_{Nk} = -L_k R \frac{1}{n_k} \frac{\partial n_k}{\partial x} \qquad (10.3.6)$$

Empirical studies of diffusion have led to what is called Fick's law. According to **Fick's law**

$$\boxed{J_{Nk} = -D_k \frac{\partial n_k}{\partial x}} \qquad (10.3.7)$$

in which D_k is the diffusion coefficient of the diffusing component k. Typical values of the diffusion coefficients in gases and liquids are given in Table 10.2. Clearly this expression is the same as (10.3.6) if we make the identification

$$\boxed{D_k = \frac{L_k R}{n_k}} \qquad (10.3.8)$$

Table 10.2 Diffusion coefficients of molecules in gases and liquids[†]

System	$D(m^2 s^{-1})$
D at $p = 101.325$ kPa and $T = 298.15$ K	
CH_4 in air	0.106×10^{-4}
Ar in air	0.148×10^{-4}
CO_2 in air	0.160×10^{-4}
CO in air	0.208×10^{-4}
H_2O in air	0.242×10^{-4}
He in air	0.580×10^{-4}
H_2 in air	0.627×10^{-4}
D at $T = 298.15$ K for aqueous solutions	
Sucrose	0.52×10^{-9}
Glucose	0.67×10^{-9}
Alanine	0.91×10^{-9}
Ethylene glycol	1.16×10^{-9}
Ethanol	1.24×10^{-9}
Acetone	1.28×10^{-9}

[*] More extensive data may be found in source [F].

This give us a relation between the thermodynamic phenomenological coefficient L_k and the diffusion coefficient. In Chapter 16 we will consider diffusion in detail and see how the diffusion of one species affects the diffusion of another species, using the modern theory of nonequilibrium thermodynamics.

An important point to note is that the thermodynamic relation (10.3.5) is valid in all cases while "Fick's law" (10.3.7) is not. For example, in the case of a liquid in equilibrium with its vapor, since the chemical potential is uniform, $(\partial \mu_k / \partial x) = 0$ and (10.3.5) correctly predicts $J_{Nk} = 0$; but (10.3.7) does not predict $J_{Nk} = 0$ because $(\partial n_k / \partial x) \neq 0$. In general, if we write (10.3.5) as $J_{Nk} = -(L_k / T) (\partial \mu_k / \partial n_k)(\partial n_k / \partial x)$ we see that, depending on the sign of $(\partial \mu_k / \partial n_k)$, J_{Nk} can be positive or negative when $(\partial n_k / \partial x) > 0$. Thus, when $(\partial \mu_k / \partial n_k) > 0$ the flow is toward the region of lower concentration but when $(\partial \mu_k / \partial n_k) < 0$, the flow can be to the region of higher concentration. The later situation arises when a mixture of two components is separating into two phases; each component flows from a region of lower concentration to a region of higher concentration. As we shall see in later chapters, when $(\partial \mu_k / \partial n_k) < 0$, the system is "unstable".

THE DIFFUSION EQUATION

In the absence of chemical reactions, the only way the mole number density $n_k(x, t)$ can change with time is due to the flow J_{Nk}. Consider a small cell of size δ at a location x (Fig. 10.8). The number of moles in this cell is equal to $(n_k(x, t)\delta)$. The rate of change of the number of moles in this cell is $\partial(n_k(x, t)\delta)/\partial t$. This change is due the net flow, i.e. the difference between the inflow and the outflow of particles in the cell. The net flow into the cell of size δ is equal to

$$J_{Nk}(x) - J_{Nk}(x + \delta x) = J_{Nk}(x) - \left(J_{Nk}(x) + \frac{\partial J_{Nk}}{\partial x} \delta \right) = -\frac{\partial J_{Nk}}{\partial x} \delta \quad (10.3.9)$$

Equating the net flow to the rate of change of the mole number, we obtain the equation

$$\frac{\partial n_k(x, t)}{\partial t} = -\frac{\partial J_{Nk}}{\partial x} \quad (10.3.10)$$

Using Fick's law (10.3.7), we can write this equation entirely in terms of $n_k(x, t)$ as

$$\boxed{\frac{\partial n_k(x, t)}{\partial t} = D_k \frac{\partial^2 n_k(x, t)}{\partial x^2}} \quad (10.3.11)$$

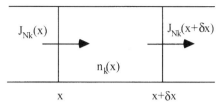

Figure 10.8 In the absence of chemical reactions, the change in the mole number in a small cell of size δ, at a location x, is due to the difference in the flow J_{NK} into and out of the cell, which is the net flow. The number of particles in the cell of size δ is $(n_k\delta)$. The net flow into the cell of size δ is equal to $J_{Nk}(x) - J_{Nk}(x + \delta x) = J_{Nk}(x) - [J_{Nk}(x) + (\partial J_{Nk}/\partial x)\delta] = -(\partial J_{Nk}/\partial x)\delta$. This difference in the flow will cause a net rate of change in the mole number $\partial(n_k(x,t)\delta)/\partial t$. Equating the net flow to the rate of change of the mole number, we obtain the equation $\partial n_k(x,t)/\partial t = -\partial J_{Nk}/\partial x$

This partial differential equation for $n_k(x)$ is the **diffusion equation** for the component k. It is valid in a homogeneous system. Diffusion tends to eliminate concentration differences and equalize the concentrations throughout the system. But it must be borne in mind that in general the thermodynamic force tends to equalize the chemical potential, not the concentrations.

THE STOKES–EINSTEIN RELATION

Fick's law gives us the diffusion current in the presence of a concentration gradient. In the presence of a field, there is also a current which is proportional to the strength of the field. For example, in the presence of an electric field **E**, an ion carrying a charge ez_k will drift at constant speed proportional to the force $ez_k|\mathbf{E}|$. This happens because the force due to the field, F_{field} ($= ez_k|\mathbf{E}|$ for ions), accelerates the ion till the viscous or frictional force, which is proportional to the velocity and acts against the direction of motion, balances F_{field}. When the ion moves at a speed v, the viscous force equals $\gamma_k v$, in which γ_k is the coefficient of friction. When the two forces balance, $\gamma_k v = F_{\text{field}}$ and the ion will drift with a **terminal velocity** v. Hence the terminal or **drift velocity** can be written as

$$v = \frac{F_{\text{field}}}{\gamma_k} \tag{10.3.12}$$

Since the number of ions that drift is proportional to the concentration n_k, the ionic drift gives rise to a current density*

$$I = v n_k = \frac{e z_k}{\gamma_k} n_k |\mathbf{E}| = -\Gamma_k n_k \frac{\partial \phi}{\partial x} \qquad (10.3.13)$$

in which the constant $\Gamma_k = (e z_k / \gamma_k)$ is called the **ionic mobility** of the ion k. Similarly, a molecule of mass m_k, falling freely in the atmosphere or any fluid, will reach a "terminal velocity" $v = m_k g / \gamma_k$, where g is the acceleration due to gravity. For any general potential ψ, the mobility a component k is defined by

$$J_{\text{field}} = -\Gamma_k n_k \frac{\partial \psi}{\partial x} \qquad (10.3.14)$$

Linear phenomenological laws of nonequilibrium thermodynamics lead to a general relation between mobility Γ_k and the diffusion coefficient D_k. This relation can be obtained as follows. The general expression for the chemical potential in a field with potential ψ is given by $\tilde{\mu}_k = \mu_k + \tau_k \psi$, in which τ_k is the interaction energy per mole due to the field (10.1.5). In the simplest approximation of an ideal system, if we write the chemical potential in terms of the mole fraction $x_k = (n_k / n_{\text{tot}})$, we have

$$\tilde{\mu}_k = \mu_k^0 + RT \ln x_k + \tau_k \psi \qquad (10.3.15)$$

A gradient in this chemical potential will result in a thermodynamic flow

$$J_{Nk} = -L_k \frac{1}{T} \left(\frac{\partial \tilde{\mu}_k}{\partial x} \right) = -\frac{L_k}{T} \left(\frac{RT}{n_k} \frac{\partial n_k}{\partial x} + \tau_k \frac{\partial \psi}{\partial x} \right) \qquad (10.3.16)$$

where we used $(\partial \ln x_k / \partial_k) = (\partial \ln n_k / \partial x)$. In this expression the first term on the right-hand side is the familiar diffusion current, and the second term is the drift current due to the field. Comparing this expression with Fick's law (10.3.7) and expression (10.3.13) that defines mobility, we see that

$$\frac{L_k R}{n_k} = D_k \qquad \frac{L_k \tau_k}{T} = \Gamma_k n_k \qquad (10.3.17)$$

From these two relations it follows that the diffusion coefficient D_k and the mobility Γ_k satisfy the following general relation:

$$\boxed{\frac{\Gamma_k}{D_k} = \frac{\tau_k}{RT}} \qquad (10.3.18)$$

---

* Note that the electric current density is $\sum_k e z_k I_k$.

This general relation was obtained by Einstein and is sometimes called the **Einstein relation**. For ionic systems, as we have seen in section 10.1 (10.1.5), $\tau_k = Fz_k = eN_A z_k$ and $\Gamma_k = ez_k/\gamma_k$. Since $R = k_B N_A$, where k_B is the Boltzmann constant $(= 1.381 \times 10^{-23} \text{J K}^{-1})$ and N_A the Avogadro number, equation (10.3.18) for the **ionic mobility** Γ_k becomes

$$\frac{\Gamma_k}{D_k} = \frac{ez_k}{\gamma_k D_k} = \frac{z_k F}{RT} = \frac{ez_k}{k_B T} \qquad (10.3.19)$$

which leads to the following general relation between the diffusion coefficient D_k and the friction coefficient γ_k of a molecule or ion k, called the **Stokes–Einstein relation**:

$$\boxed{D_k = \frac{k_B T}{\gamma_k}} \qquad (10.3.20)$$

10.4   Chemical Potential for an Internal Degree of Freedom

The notion of a chemical potential can also be extended to transformations in an internal degree of freedom of molecules, such as orientation of a polar molecule with respect to an external field (Fig 10.9), deformation of a macromolecule due to flow and similar phenomena [2]. This can be done by defining an internal coordinate θ just as we define an "external coordinate" such as the position x. In this section we shall only consider the orientation of an electric dipole with

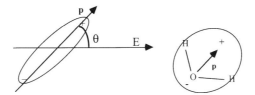

Figure 10.9   Chemical potential $\mu(\theta)$ can be defined for an internal degree of freedom such as the orientation of a polar molecule with respect to an electric field **E**. The electric dipole moment is denoted by **p**. The energy of an electric dipole in field **E** is given by $-\mathbf{p}.\mathbf{E}$. A water molecule is an example of a molecule with a dipole moment. The oxygen atom tends to accumulate negative charge, and the resulting slight charge separation gives rise to an electric dipole moment

respect to an electric field (generalization to other situations is straightforward). In this case, θ is the angle between the direction of the field and the dipole, as shown in Fig. 10.9. And just as we defined a concentration as a function of position, we can also define a concentration $n(\theta)$ as a function of θ. Just as chemical potential is a function of position, for an internal coordinate θ the chemical potential of component k is a function of θ

$$\tilde{\mu}_k(\theta, T) = \mu_k(\theta, T) + g_k\phi(\theta) \tag{10.4.1}$$

in which $g_k\phi(\theta)$ is the interaction energy per mole between the field and the dipole. If the dipole moment per mole is $\mathbf{p}_k$ and the electric field is $\mathbf{E}$, then

$$g_k = -|\mathbf{p}||\mathbf{E}|\cos\theta \tag{10.4.2}$$

Other quantities, such as concentration $n_k(\theta)$, entropy density $s(\theta)$, and the "flow" in θ-space can be defined as a function of θ, just as they were defined as functions of x. However, in spherical coordinates, since the volume element is equal to $\sin\theta\,d\theta\,d\phi$, we use the following definitions (Fig. 10.10)

$s(\theta)\sin\theta\,d\theta$ = entropy of molecules with internal coordinate between θ
and $\theta + d\theta$

$n_k(\theta)\sin\theta\,d\theta$ = mole number of molecules with internal coordinate between θ
and $\theta + d\theta$

$J_\theta\sin\theta\,d\theta$ = number of molecules whose orientation is changing from θ to
$\theta + d\theta$ per unit time

For simplicity, we shall consider a unit volume and only one species, dropping the subscript k

With these definitions it is clear that all the formalism developed in section 10.1 for the position x can be directly converted to θ by formally replacing x

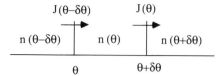

Figure 10.10 Reaction scheme for continuous internal degrees of freedom

with θ. Accordingly, we are led to the equation

$$\frac{d_i s(\theta)}{dt} = -\frac{1}{T}\left(\frac{\partial \tilde{\mu}(\theta)}{\partial \theta}\right) J_N(\theta) > 0 \qquad (10.4.3)$$

which is similar to (10.1.13). In (10.4.3) we can identify the affinity as

$$\tilde{A}(\theta) = -\frac{\partial \tilde{\mu}(\theta)}{\partial \theta} \qquad (10.4.4)$$

for the "reaction" $n(\theta) \rightleftharpoons n(\theta + \delta\theta)$ with corresponding extent of reaction $\xi(\theta)$. The velocity of this reaction $J_N(\theta) = d\xi(\theta)/dt$ (Fig. 10.10) is the number of molecules being transformed from θ to $\theta + d\theta$. With these definitions, the rate of entropy production can also be written as

$$\frac{d_i s(\theta)}{dt} = -\frac{1}{T}\left(\frac{\partial \tilde{\mu}(\theta)}{\partial \theta}\right)\frac{d\xi(\theta)}{dt} > 0 \qquad (10.4.5)$$

For a system with an internal coordinate such as θ, the total rate of entropy change is given by

$$\begin{aligned}
\frac{dS}{dt} &= \frac{1}{T}\frac{dU}{dt} + \frac{p}{T}\frac{dV}{dt} - \frac{1}{T}\int_\theta \frac{\partial \mu(\theta)}{\partial \theta}\frac{d\xi(\theta)}{dt}d\theta \\
&= \frac{1}{T}\frac{dU}{dt} + \frac{p}{T}\frac{dV}{dt} - \frac{1}{T}\int_\theta \frac{\partial \mu(\theta)}{\partial \theta} J_N(\theta)d\theta
\end{aligned} \qquad (10.4.6)$$

For the total entropy production, the Second Law implies that

$$\frac{d_i S}{dt} = -\frac{1}{T}\int_\theta \frac{\partial \mu(\theta)}{\partial \theta} J_N(\theta)d\theta > 0 \qquad (10.4.7)$$

In this formalism we have a more restrictive statement of the Second Law expressed in (10.4.5). When the system reaches equilibrium, since the affinity becomes zero, we have

$$\tilde{A}(\theta) = -\frac{\partial[\mu(\theta) + g\phi(\theta)]}{\partial \theta} = 0 \qquad (10.4.8)$$

Writing the chemical potential more explicitly, we may then conclude that at equilibrium

$$\tilde{\mu}(\theta) = \mu_0(T) + RT\ln[a(\theta)] + g\phi(\theta) = C \qquad (10.4.9)$$

in which C is a constant and $a(\theta)$ is the activity for the molecules having orientation θ with respect to the field $\mathbf{E}$ (Fig. 10.9). Also note that, in the absence of a field, since all orientations are equivalent, μ_0 should be independent of θ. The activity of an ideal mixture can be approximated by the mole fraction. For an internal degree of freedom, each value of θ may be taken as a species and, by analogy, we may define an ideal activity by $a(\theta) = n(\theta)/n_{\text{tot}}$ in which n_{tot} is the total number of dipoles. It is now a matter of elementary calculation to show that, at equilibrium, we have

$$n(\theta) = n_{\text{tot}}F(T)e^{-g\phi(\theta)/RT} = n_{\text{tot}}F(T)e^{|\mathbf{p}||\mathbf{E}|\cos\theta/RT} \tag{10.4.10}$$

in which $F(T)$ is a function of T, expressed in terms of $\mu_0(T)$ and C (exc. 10.8), and in which we have used (10.4.2). Note also that $F(T)$ should be such that $\int_0^\pi n(\theta)\sin\theta\, d\theta = n_{\text{tot}}$.

THE DEBYE EQUATION FOR ELECTRIC DIPOLE RELAXATION

Since the only possible way in which $n(\theta)$ can change is due to the current $J_N(\theta)$, we have a situation analogous to the diffusion in Fig 10.10. As noted earlier, in spherical coordinates we have the following definitions:

$n_k(\theta)\sin\theta\, d\theta$ = mole number of molecules with internal coordinate between θ

and $\theta + d\theta$

$J_\theta \sin\theta\, d\theta$ = number of molecules whose orientation is changing from θ to

$\theta + d\theta$ per unit time

From these definitions it follows that the conservation equation for the dipoles is

$$\frac{\partial n(\theta)\sin\theta}{\partial t} = -\frac{\partial J_N(\theta)\sin\theta}{\partial\theta} \tag{10.4.11}$$

As in the case of diffusion, by looking at the entropy production (10.4.3), we can identify the force corresponding to the flow $J_N(\theta)$ as $-(1/T)(\partial\tilde{\mu}(\theta)/\partial\theta)$. When the system is near equilibrium, there exists a linear relation between the flow and the flux, which can be written as

$$J_N(\theta) = -\frac{L_\theta}{T}\frac{\partial\tilde{\mu}(\theta)}{\partial\theta} \tag{10.4.12}$$

in which L_θ is the linear phenomenological coefficient. In the approximation of an ideal mixture, we have

$$\tilde{\mu}(\theta) = \mu_0(T) + RT\ln(n(\theta)/n_{\text{tot}}) - |\mathbf{p}||\mathbf{E}|\cos\theta \tag{10.4.13}$$

Substituting (10.4.13) into (10.4.12), we obtain

$$J_N(\theta) = -\frac{L_\theta R}{n(\theta)}\frac{\partial n(\theta)}{\partial \theta} + \frac{L_\theta}{T}|\mathbf{p}||\mathbf{E}|\frac{\partial}{\partial \theta}\cos\theta \qquad (10.4.14)$$

In analogy with ordinary diffusion, we may define a rotational diffusion in θ-space, for which expression (10.4.14) corresponds to Fick's law. A rotational diffusion coefficient D_θ may now be identified as

$$D_\theta = \frac{L_\theta R}{n(\theta)} \qquad (10.4.15)$$

With this identification, the flow $J_N(\theta)$ given by (10.4.14) can be written as

$$J_N(\theta) = -D_\theta \frac{\partial n(\theta)}{\partial \theta} - [(D_\theta/RT)|\mathbf{p}||\mathbf{E}|\sin\theta]n(\theta) \qquad (10.4.16)$$

Finally, substituting this expression into (10.4.11) gives

$$\frac{\partial n(\theta)}{\partial t} = \frac{1}{\sin\theta}\frac{\partial}{\partial \theta}\sin\theta\left\{D_\theta\frac{\partial n(\theta)}{\partial \theta} + \left[\left(\frac{D_\theta}{RT}\right)|\mathbf{p}||\mathbf{E}|\sin\theta\right]n(\theta)\right\} \qquad (10.4.17)$$

This is the **Debye equation** for the relaxation of dipoles in an electric field. It has been used for analyzing the relaxation of dipoles in an oscillating electric field.

References

1. Guggenheim, E.A., *Modern Thermodynamics*. 1933, London: Methuen.
2. Prigogine, I. and Mazur, P., *Physica*, **19** (1953) 241.

Data Sources

[A] NBS table of chemical and thermodynamic properties. *J. Phys. Chem. Reference Data*, **11**, Suppl. 2 (1982).
[B] G. W. C. Kaye and Laby, T. H. (eds), *Tables of Physical and Chemical Constants*. 1986, London: Longman.
[C] I. Prigogine and Defay, R., *Chemical thermodynamics*. 1967, London: Longman.
[D] J. Emsley, *The Elements*. 1989, Oxford: Oxford University Press.
[E] L. Pauling, *The Nature of the Chemical Bond*. 1960, Ithaca NY: Cornell University Press.
[F] D. R. Lide (ed.), *CRC Handbook of Chemistry and Physics*, 75th ed. 1994, Ann Arbor MI: CRC Press.
[G] The web site of the National Institute for Standards and Technology, http://webbook.nist.gov.

Examples

Example 10.1   Use the barometric formula to estimate the pressure at an altitude of 3.0 km. The temperature of the atmosphere is not uniform (so it is not in equilibrium). Assume an average temperature $T = 270.0$ K

Solution: The pressure at an altitude h is given by the barometric formula $p(h) = p(0)e^{-gMh/RT}$. Since 78% of the atmosphere consists of N_2, we shall use the molar mass of N_2 for M_k. The pressure at an altitude of 3.0 km will be

$$p(3\,\text{km}) = (1\,\text{atm})\exp\left[-\frac{(9.8\,\text{m s}^{-2})(28.0 \times 10^{-3}\,\text{kg mol}^{-1})3.0 \times 10^{3}\,\text{m}}{(8.314\,\text{J K}^{-1}\,\text{mol}^{-1})(270\,\text{K})}\right]$$

$$= (1\,\text{atm})\exp[-0.366] = 0.69\,\text{atm}$$

Example 10.2   Calculate the membrane potential for the setup shown in Fig. 10.4.

Solution: In this case the expected potential difference across the membrane is

$$V = \phi^{\alpha} - \phi^{\beta} = \frac{RT}{F}\ln\left(\frac{1.0}{0.1}\right) = 0.0257\ln(10)$$

$$= 0.0592\,\text{Volts}$$

Example 10.3   Calculate the cell potential for the cell shown in Fig 10.6. Also calculate the equilibrium constant for the reaction $Zn(s) + 2H^{+} \rightarrow H_2(g) + Zn^{2+}$.

Solution: Considering the two electrode reactions we have

$$2H^{+} + 2e^{-} \longrightarrow H_2(g) \qquad\qquad 0.0$$

$$Zn(s) \longrightarrow Zn^{2+} + 2e^{-} \qquad +0.763$$

The total cell potential is $V_0 = 0 + 0.763V = 0.763\,V$
The equilibrium constant is

$$K = \exp(2FV_0/RT) = \exp\left[\frac{2 \times 9.648 \times 10^{4} \times 0.763}{8.314 \times 298.15}\right] = 6.215 \times 10^{25}$$

Exercises

10.1   Use the chemical potential of an ideal gas in (10.1.9) and obtain the barometric formula (10.1.10). Use the barometric formula to estimate the boiling point of water at an altitude of 2.50 km above sea level. Assume on average that $T = 270$ K.

10.2 A heater coil is run at a voltage of 110 V and draws a current of 2.0 A. If its temperature is equal to 200 °C, what is the rate of entropy production?

10.3 Calculate the equilibrium constants at $T = 25.0\,^\circ$C for the following electrochemical reactions using the standard potentials in Table 10.1:
(i) $Cl_2(g) + 2Li(s) \rightarrow 2Li^+ + 2Cl^-$
(ii) $Cd(s) + Cu^{2+} \rightarrow Cd^{2+} + Cu(s)$
(iii) $2Ag(s) + Cl_2(g) \rightarrow 2Ag^+ + 2Cl^-$
(iv) $2Na(s) + Cl_2(g) \rightarrow 2Na^+ + 2Cl^-$

10.4 If the reaction $Ag(s) + Fe^{3+} + Br^- \rightarrow AgBr(s) + Fe^{2+}$ is not in equilibrium, it can be used to generate an EMF. The "half-cell" reactions that correspond to the oxidation and reduction in this cell are

$$Ag(s) + Br^- \rightarrow AgBr(s) + e^- \quad V_0 = -0.071 \text{ V}$$

$$Fe^{3+} + e^- \rightarrow Fe^{2+} \quad\quad\quad V_0 = 0.771 \text{ V}$$

(a) Calculate V_0 for this reaction.
(b) Determine the EMF for the following activities at $T = 298.15$ K: $a_{Fe^{3+}} = 0.98$, $a_{Br^-} = 0.30$, $a_{Fe^{2+}} = 0.01$.
(c) What will be the EMF when $T = 0.0\,^\circ$C?

10.5 The K^+ concentration inside a nerve cell is much larger than the concentration outside it. Suppose the potential difference across the cell membrane is 90 mV. Assuming the system is in equilibrium, estimate the ratio of the K^+ concentrations inside and outside the cell.

10.6 Verify that

$$n(x, t) = \frac{n(0)}{2\sqrt{\pi Dt}} e^{-x^2/4Dt}$$

is the solution of the diffusion equation (10.3.11). Using Mathematica or Maple, plot this solution over various values of t for one of the gases listed in Table 10.2; assume $n(0) = 1$. This gives you an idea of how far a gas will diffuse in a given time. Obtain a simple expression to estimate the distance a molecule will diffuse in a time t, given its diffusion coefficient D.

10.7 Compute the diffusion current corresponding to the barometric distribution $n(x) = n(0)e^{-gMx/RT}$

10.8 Using (10.4.9) and the ideal activity $a(\theta) = n(\theta)/n_{tot}$ for the dipole orientation, obtain the equilibrium expression (10.4.10). Give an explicit expression for the function $F(T)$ in terms of μ_0 and C.

10.9 The electric dipole moment of water is 6.14×10^{-30} C m. In an electric field of 10.0 Vm^{-1}, find the fraction of molecules oriented with respect to the field in the range $10° < \theta < 20°$ when $T = 298$ K.

11 THERMODYNAMICS OF RADIATION

Introduction

Electromagnetic radiation which interacts with matter also reaches a state of thermal equilibrium with a definite temperature. This state of electromagnetic radiation is called **thermal radiation**, also called *heat radiation* in earlier literature. In fact, today we know that our universe is filled with thermal radiation at a temperature of about 2.8 K.

It has long been observed that heat can pass from one body to another in the form of radiation with no material contact between the two bodies. This was called heat radiation. When it was discovered that motion of charges produced electromagnetic radiation, the idea that heat radiation was a form of electromagnetic radiation was taken up, especially in the works of Gustav Kirchhoff (1824–1887), Ludwig Boltzmann (1844–1906), Josef Stefan (1835–1893) and Wilhelm Wien (1864–1928) and its thermodynamic consequences were investigated [1].

11.1 Energy Density and Intensity of Thermal Radiation

Radiation is associated with energy density u, which is the energy per unit volume, and an intensity I, which is defined as follows: the energy incident on a small area $d\sigma$ due to radiation from a solid angle $d\Omega$ which makes an angle θ with the normal to the surface $d\sigma = I\cos\theta \, d\Omega \, d\sigma$ (Fig. 11.1) [1].

The energy density u and intensity of radiation I could also be defined for a particular frequency as

$u(v)dv$ = energy density of radiation in the frequency range v and $v + dv$.

$I(v)dv$ = Intensity of radiation in the frequency range v and $v + dv$.

The two quantities have a simple relationship [1]:

$$\boxed{u(v) = \frac{4\pi I(v)}{c}} \tag{11.1.1}$$

in which c is the velocity of light. This relation is not particular to electromagnetic radiation; it is generally valid for any energy flow that takes place with a velocity c.

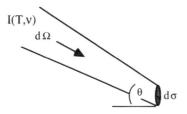

Figure 11.1 Definition of intensity of radiation $I(T, \nu)$. The energy flux incident on the area element $d\sigma$, from a solid angle $d\Omega$ is given by $I(T, \nu)\cos\theta\, d\Omega\, d\sigma$. Here θ is the angle between the normal to $d\sigma$ and the incident radiation

Gustav Kirchhoff (1824–1887) (Courtesy the E. F. Smith Collection, Van Pelt-Dietrich Library, University of Pennsylvania)

As noted by Gustav Kirchhoff (1824–1887), thermal radiation that is in equilibrium simultaneously with several substances should not change with the introduction or removal of a substance. Hence $I(\nu)$ and $u(\nu)$ associated with thermal radiation must be functions only of the temperature T, independent of the substances with which it is in equilibrium.

A body in thermal equilibrium with radiation is continuously emitting and absorbing radiation. The **emissivity**, $e_k(v,T)dv$, of a body k is defined as the intensity of radiation emitted by that body in the frequency range v and $v + dv$ at a temperature T; **absorptivity**, $a_k(v,T)dv$, is defined as the fraction of the incident intensity $I(v)$ that is absorbed in the frequency range v and $v + dv$ at a temperature T. By considering the balance of emitted and absorbed energy (Fig. 11.2), Kirchhoff also deduced that the ratio of emissivity $e_k(v,T)$ and the absorptivity $a_k(v,T)$ of a substance k must equal $I(T,v)$, independent of the substance:

$$\boxed{\frac{e_k(T,v)}{a_k(T,v)} = I(T,v)} \tag{11.1.2}$$

This general relation is called **Kirchhoff's law**. For a perfectly absorbing body, $a(T,v) = 1$. Such a body is called a **blackbody**; *its emissivity is equal to the intensity. $I(T,v)$.*

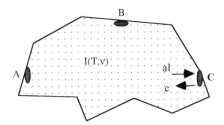

Fig 11.2 Kirchhoff's law states that the ratio $e_k(T,v)/a_k(T,v)$ of emissivity $e_k(T,v)$ to the absorptivity $a_k(T,v)$ is independent of the material k and is equal to the intensity of radiation $I(T,v)$. This is deduced by considering thermal radiation at equilibrium with different substances A, B and C. At equilibrium the total energy absorbed from all angles $\pi I(T,v)a_k(T,v)$ of each elemental area of substance k must equal the energy emitted, $\pi e_k(T,v)$, by the same area. Hence $I(T,v) a_k(T,v) = e_k(T,v)$. But I associated with thermal radiation must be independent of the substances. Hence $e_k(T,v)/a_k(T,v) = I(T,v)$ is independent of the substance k; it is a function only of temperature T and frequency v

Max Planck (1858–1947) (Reproduced, by permission, from the Emilio Segré Visual Archives of the American Institute of Physics)

At the end of the nineteenth century, classical thermodynamics faced the challenge of determining the exact functional form of $u(\nu, T)$ or $I(\nu, T)$. All the deductions based on the principles that were known at that time did not agree with experimental measurements of $u(\nu)$. This fundamental problem remained unsolved until Max Planck (1858–1947) introduced the revolutionary quantum hypothesis. With the quantum hypothesis, according to which matter absorbed and emitted radiation in discrete bundles or "quanta," Planck was able to derive the following expression which agreed well with the observed frequency distribution $u(\nu)$:

$$u(\nu, T) = \frac{8\pi h\nu^3}{c^3} \frac{1}{\left(e^{h\nu/k_{\mathrm{B}}T} - 1\right)}$$

(11.1.3)

Here $h(= 6.626 \times 10^{-34}\,\mathrm{J\,s})$ is Planck's constant and $k_\mathrm{B}(= 1.381 \times 10^{-23}\,\mathrm{J\,K^{-1}})$ is Boltzmann's constant. We shall not present the derivatives of Planck's formula (which requires statistical mechanics); our focus will be on thermodynamic aspects of radiation. Finally, we note that total energy of thermal radiation is

$$u(T) = \int_0^\infty u(\nu, T)d\nu \tag{11.1.4}$$

When the function $u(\nu, T)$ obtained using classical electromagnetic theory was used in this integral, the total energy density, $u(T)$, turned out to be infinite. The Planck formula (11.1.3), however, gives a finite value for $u(T)$.

11.2 The Equation of State

It was clear, even from the classical electromagnetic theory, that a field which interacts with matter and imparts energy and momentum must itself carry energy and momentum. Classical expressions for the energy and momentum associated with the electromagnetic field can be found in texts on electromagnetic theory. To understand the thermodynamic aspects of radiation, we need an equation of state, i.e. an equation that gives the pressure exerted by thermal radiation and its relation to the temperature.

Using classical electrodynamics it can be shown [1] that the pressure exerted by radiation is related to the energy density u by

$$\boxed{p = \frac{u}{3}} \tag{11.2.1}$$

This relation follows from purely mechanical considerations of force exerted by radiation when it is reflected by the walls of a container. Though it was originally derived using classical electrodynamics, equation (11.2.1) can be more easily derived using the physical idea that radiation is a gas of photons (Box 11.1). We shall presently see that when this equation of state is combined with the equations of thermodynamics, we arrive at the following conclusion: the *energy density $u(\nu)$, hence $I(\nu)$, is proportional to the fourth power of the temperature*. The result is credited to Josef Stefan (1835–1893) and Ludwig Boltzmann (1844–1906); it is called the **Stefan–Boltzmann law**. The fact that the energy density $u(T) = \int_0^\infty u(\nu, T)d\nu$ of thermal radiation is only a function of temperature, independent of the volume, implies that in a volume V the total energy is

$$U = Vu(T) \tag{11.2.2}$$

Although thermal radiation is a gas of photons, it differs from an ideal gas. At a fixed temperature T, as the volume of thermal radiation expands, the total energy increases (unlike in an ideal gas, in which it remains constant). As the volume increases, the "heat" that must be supplied to such a system to keep its temperature constant is thermal radiation entering the system. This heat keeps the energy density constant. The change in entropy due to this heat flow is given by

$$d_eS = dQ/T = (dU + p\,dV)/T \qquad (11.2.3)$$

Box 11.1 Photon Gas Pressure: A Heuristic Derivation

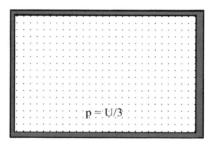

Let $n(v)$ be the number of photons of frequency v. Then the momentum of each photon is given by hv/c. The pressure on the walls is due to the collisions of the photons. Each collision imparts a momentum $2p$ to the wall upon reflection. Since the photons are in random motion, at any instant 1/6 of the photons will be moving in the direction of the wall. Hence the number of photons that will collide with unit area of the wall in one second is $n(v)c/6$. The total momentum imparted to unit area of the wall per second is the pressure. Hence we have

$$p(v) = \left(\frac{n(v)c}{6}\right)\frac{2hv}{c} = \frac{n(v)hv}{3}$$

Now, since the energy density $u(v) = n(v)hv$, we arrive at the result

$$p(v) = \frac{u(v)}{3}$$

A more rigorous derivation, taking all the directions of the photon momentum into consideration, also gives the same result. For photons of all frequencies, we can integrate over the frequency v:

$$p = \int_0^\infty p(v)dv = \int_0^\infty \frac{u(v)}{3}dv = \frac{u}{3}$$

where p is the total pressure due to photons of all frequencies. Note that a similar derivation for the ideal gas gives $p = 2u/3$ in which $u = n(mv^2/2)$, v is the average velocity of the gas molecules.

Once we assign an entropy to the system in this fashion, all the thermodynamic consequences follow. Consider, for example, the Helmholtz equation (5.2.11) (which follows from the fact that entropy is a state function and therefore $\partial^2 S/\partial T\,\partial V = \partial^2 S/\partial V\,\partial T$):

$$\left(\frac{\partial U}{\partial V}\right)_T = T^2\left(\frac{\partial}{\partial T}\left(\frac{p}{T}\right)\right)_V \qquad (11.2.4)$$

Using (11.2.2) and the equation of state $p = u/3$ in this equation, we can obtain

$$4u(T) = T\left(\frac{\partial u}{\partial T}\right) \qquad (11.2.5)$$

Integrating this equation, we arrive at the **Stefan–Boltzmann law**

$$\boxed{u(T) = \beta T^4} \qquad (11.2.6)$$

in which β is a constant. The value of $\beta = 7.56 \times 10^{-16}\,\mathrm{J\,m^{-3}\,K^{-4}}$ is obtained by measuring the intensity of radiation emitted by a blackbody at a temperature T.

Using (11.2.6) we can now write the pressure, $p = u/3$, as a function of temperature:

$$p(T) = \beta T^4/3 \qquad (11.2.7)$$

Equations (11.2.6) and (11.2.7) are the equations of state for thermal radiation. For temperatures of order 10^3 K or less, the radiation pressure is small, but it can be quite large for stellar temperatures. In the interior of stars, where the temperatures can be 10^7 K, if we use (11.2.7) and calculate the pressure due to radiation, we find it is $2.52 \times 10^{12}\,\mathrm{Pa} \approx 2 \times 10^7$ atm!

11.3   Entropy and Adiabatic Processes

For thermal radiation, the change in entropy is entirely due to heat flow:

$$dS = d_e S = \frac{dU + p\,dV}{T}$$

Considering U as a function of V and T, this equation can be written as

$$dS = \frac{1}{T}\left[\left(\frac{\partial U}{\partial V}\right)_T + p\right]dV + \frac{1}{T}\left(\frac{\partial U}{\partial T}\right)_V dT \qquad (11.3.1)$$

Since $U = Vu = V\beta T^4$ and $p = \beta T^4/3$, this equation can be written as

$$dS = \left(\frac{4}{3}\beta T^3\right)dV + \left(4\beta VT^2\right)dT \qquad (11.3.2)$$

In this equation we can identify the derivatives of S with respect to T and V:

$$\left(\frac{\partial S}{\partial V}\right)_T = \frac{4}{3}\beta T^3 \qquad \left(\frac{\partial S}{\partial T}\right)_V = 4\beta VT^2 \qquad (11.3.4)$$

By integrating these two equations and setting $S = 0$ at $T = 0$ and $V = 0$, it is easy to see (exc. 11.3) that

$$\boxed{S = \frac{4}{3}\beta VT^3} \qquad (11.3.5)$$

The above expression for entropy and the equations of state (11.2.6) and (11.2.7) are basic; all other thermodynamic quantities for thermal radiation can be obtained from them. Unlike other thermodynamic systems we have studied so far, the temperature T is sufficient to specify all the thermodynamic quantities for thermal radiation; the energy density $u(T)$, the entropy density $s(T) = S(T)/V$ and all other thermodynamic quantities are entirely determined by T. There is no term involving a chemical potential in the expressions for S or U. If we consider the particle nature of thermal radiation, i.e. a gas of photons, the *chemical potential must be assumed to equal zero* — as we shall see in section 11.5.

In an adiabatic process the entropy remains constant. From the expression for entropy (11.3.5) the relation between volume and temperature in an adiabatic process immediately follows:

$$\boxed{VT^3 = \text{constant}} \qquad (11.3.6)$$

The radiation filling the universe is currently at about 2.8 K. The effect of the expansion of the universe on the radiation that fills it can be approximated by an adiabatic process. (During the evolution of the universe its total entropy is not a constant. Irreversible processes generate entropy, but the increase in entropy of radiation due to these irreversible processes is small.) Using (11.3.6) and the current value of T, one can compute the temperature when the volume was only a small fraction of the present volume. Thus thermodynamics gives us the relation between the volume of the universe and the temperature of the thermal radiation that fills it.

11.4   Wien's Theorem

At the end of the nineteenth century, one of the most outstanding problems was the frequency dependence of the energy density $u(v, T)$. Wilhelm Wien (1864–1928) made an important contribution in his attempt to obtain $u(v, T)$. Wien developed a method with which he could analyze what may be called the *microscopic consequences* of the laws of thermodynamics. He began by considering an adiabatic compression of thermal radiation. Such a compression keeps the system in thermal equilibrium but changes the temperature so that $VT^3 = $ constant (11.3.6). On a microscopic level, he analyzed the shift of each frequency v to a new frequency v' due to its interaction with the compressing piston. Since this frequency shift corresponds to a change in temperature such that $VT^3 = $ constant, he could obtain a relation that described how $u(v, T)$ changed with v and T [1]. This led Wien to the conclusion that $u(v, T)$ must have the following functional form:

$$\boxed{u(v, T) = v^3 f(v/T)} \tag{11.4.1}$$

i.e. $u(v, T)$ is a function of the *ratio* (v/T) multiplied by v^3. This conclusion follows from the laws of thermodynamics. We shall refer to (11.4.1) as **Wien's theorem**. Note that (11.4.1) is in agreement with Planck's formula (11.1.3)

Experimentally it was found that, for a given T, as a function of v, $u(v, T)$ has a maximum. Let v_{max} be the value of v at which $u(v, T)$ reaches its maximum value. Then because $u(v, T)$ is a function of the ratio $(v/T)^*$, it follows that value of v at which $u(v/T)$ reaches its maximum depends only on the value of the ratio (v/T). So at the maximum, (v_{max}/T) will have the same value for all temperatures. In other words,

$$\frac{T}{v_{max}} = \text{constant} \tag{11.4.2}$$

or since $v_{max} = c/\lambda_{max}$, we see that, $T\lambda_{max} = $ constant. This constant can be calculated using Planck's formula:

$$\boxed{T\lambda_{max} = 2.8979 \times 10^{-3} \, \text{m K}} \tag{11.4.3}$$

The two equations, (11.4.2) and (11.4.3), are often referred to as **Wien's displacement law**.

Wien's method is general and it can be applied, for example, to an ideal gas. Here the objective would be to obtain the energy density u as a function of the velocity v and the temperature. It can be shown [2] that $u(v, T) = v^4 f(v^2/T)$, which shows us that thermodynamics implies the velocity distribution is a

---

* For a given T, $u(v, T) = T^3 (v/T)^3 f(v/T)$.

function of (v^2/T). Thus thermodynamics does also tell us about microscopic statistical aspects of systems.

All classical attempts to obtain the form of $u(v, T)$ for thermal radiation gave results that neither agreed with experiments nor gave finite values for $u(v, T)$ when all frequencies v (0 to ∞) were included. It is now well known that to solve this problem Planck introduced his quantum hypothesis in 1901.

11.5 Chemical Potential for Thermal Radiation

The equations of state for thermal radiation are

$$p = \frac{u}{3} \qquad u = \beta T^4 \qquad (11.5.1)$$

where u is the energy density and p is the pressure.

If all the material particles in a volume are removed, what was classically thought to be a vacuum is not empty but filled with thermal radiation at the temperature of the walls of the container. There is no distinction between heat and thermal radiation in the following sense. If we consider a volume filled with thermal radiation in contact with a heat reservoir (Fig. 11.3), then if the volume is enlarged, the system temperature T, hence the energy density u, is maintained constant by the flow of heat into the system from the reservoir. The heat that flows into the system is thermal radiation.

From the particle viewpoint, thermal radiation consists of photons which we shall refer to as *thermal photons*. Unlike in an ideal gas, the total number of thermal photons is not conserved during isothermal changes of volume. The change in the total energy, $U = uV$, due to the flow of thermal photons from or to the heat reservoir must be interpreted as a flow of heat. Thus, for thermal

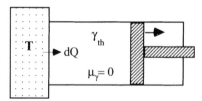

Figure 11.3 Heat radiation in contact with a heat reservoir: the energy entering or leaving the system is thermal radiation; although the number of photons is changing, $dU = dQ - p\,dV$

radiation $ds = des = dQ/T$ and,

$$dU = dQ - p\,dV = T\,dS - pd\,V \qquad (11.5.2)$$

This equation remains valid even though the number of photons in the system is changing. Comparing this equation with the equation introduced by Gibbs, $dU = T\,dS - p\,dV + \mu\,dN$, we conclude that the chemical potential $\mu = 0$. *The state in which $\mu = 0$ is a state in which the partial pressure or the particle density is a function only of the temperature.* Indeed in the expression for the chemical potential, $\mu_k = \mu_k^0(T) + RT \ln(p_k/p_0)$, if we set $\mu_k = 0$, we see that the partial pressure p_k is only a function of T.

TWO-LEVEL ATOM IN EQUILIBRIUM WITH RADIATION

With the above observations that the chemical potential of thermal radiation is zero, the interaction of a two-level atom with blackbody radiation (which Einstein used to obtain the ratio of the rates of spontaneous and stimulated radiation) can be analyzed in a somewhat different light. If A and A^* are the two states of the atom, and γ_{th} is a thermal photon, then the spontaneous and stimulated emission of radiation can be written as

$$A^* \rightleftharpoons A + \gamma_{th} \qquad (11.5.3)$$
$$A^* + \gamma_{th} \rightleftharpoons A + 2\gamma_{th} \qquad (11.5.4)$$

From the viewpoint of chemical equilibrium, the above two reactions are the same. The condition for chemical equilibrium is

$$\mu_{A^*} = \mu_A + \mu_\gamma \qquad (11.5.5.)$$

Since $\mu_\gamma = 0$, we have $\mu_{A^*} = \mu_A$. As we have seen in Chapter 9, if we use the expression $\mu_k = \mu_k(T) + RT \ln(p_k/p_0)$ for the chemical potential, and note that the concentration is proportional to the partial pressure, the law of mass action takes the form

$$\frac{[A]}{[A^*]} = K(T) \qquad (11.5.6)$$

On the other hand, looking at reactions (11.5.3) and (11.5.4) as elementary chemical reactions, we may write

$$\frac{[A][\gamma_{th}]}{[A^*]} = K'(T) \qquad (11.5.7)$$

But because $[\gamma_{th}]$ is a function of temperature only, it can be absorbed into definition of the equilibrium constant; we can define $K(T) \equiv K'(T)/[\gamma_{th}]$ and recover equation (11.5.6), which follows from thermodynamics.

Similarly we may consider any exothermic reaction

$$A + B \rightleftharpoons 2C + \text{heat} \qquad (11.5.8)$$

from the viewpoint of thermal photons, and write it as

$$A + B \rightleftharpoons 2C + \gamma_{th} \qquad (11.5.9)$$

The condition for equilibrium can now be written as

$$\mu_A + \mu_B = 2\mu_C + \mu_\gamma \qquad (11.5.10)$$

Since $\mu_\gamma = 0$ we recover the condition for chemical equilibrium derived in Chapter 9. For this reaction too, one can obtain $K'(T)$ similar to (11.5.7).

11.6   Matter, Radiation and Zero Chemical Potential

When we consider interconversion of particles and radiation, as in the case of particle–antiparticle pair creation and annihilation, the chemical potential of thermal photons becomes more significant (Fig. 11.4). Consider thermal photons in equilibrium with electron–positron pairs:

$$2\gamma \rightleftharpoons e^+ + e^- \qquad (11.6.1)$$

At thermal equilibrium we have

$$\mu_{e^+} + \mu_{e^-} = 2\mu_\gamma \qquad (11.6.2)$$

For reasons of symmetry we may assert that $\mu_{e^+} = \mu_{e^-}$. Since $\mu_\gamma = 0$ we must conclude that for particle–antiparticle pairs that can be created by thermal photons $\mu_{e^+} = \mu_{e^-} = 0$.

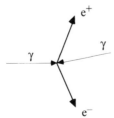

Figure 11.4   Creation of particle–antiparticle pairs by thermal photons

It is interesting to consider the state of matter for which $\mu = 0$. For simplicity, let us consider $\mu = 0$ in an ideal gas mixture for which

$$\mu_k = \frac{U_k - TS_k + p_kV}{N_k}$$

$$= \frac{N_k\left(\frac{3}{2}RT + W_k\right) - N_kRT\left[\frac{3}{2}\ln T + \ln\left(\frac{V}{N_k}\right) + s_0\right] + N_kRT}{N_k} = 0$$

(11.6.3)

in which we use the internal energy $U_k = N_k\left(\frac{3}{2}RT + W_k\right)$ and the entropy $S_k = N_kR\left[\frac{3}{2}\ln T + \ln(V/N_k) + s_0\right]$ for a component of an ideal gas mixture, along with the ideal gas equation $p_kV = N_kRT$. As we have noted in Chapters 2 and 3, the theory of relativity gives us the energy constant $W_k = M_kc^2$, in which M_k is the molar rest mass of the particles of an ideal gas, and quantum theory gives us the entropy constant s_0. Using equation (11.6.3) we can write the particle density (N_k/V) as

$$\frac{N_k}{V} = Q_k(T)e^{(\mu_k - M_kc^2)/RT}$$

(11.6.4)

in which $Q_k(T)$ is a function of temperature only (and which is closely related to the "partition function" in the statistical mechanics of an ideal gas). If the process of pair production is in thermal equilibrium $\mu = 0$, and the thermal particle density is given by

$$\left(\frac{N_k}{V}\right)_{th} = Q_k(T)e^{-M_kc^2/RT}$$

(11.6.5)

The corresponding partial pressure is given by

$$p_{k,th} = RTQ_k(T)e^{-M_kc^2/RT}$$

(11.6.6)

The physical meaning of these equations can be understood as follows. Just as photons of energy $h\nu$ are excitations of the electromagnetic field, in modern field theory particles of energy $E = \sqrt{m^2c^4 + p^2c^2}$ are also excitations of a quantum field. In the nonrelativistic approximation $E \approx mc^2 + p^2/2m$. According to the Boltzmann principle, the probability $P(E)$ of an excitation of energy E is given by

$$P(E) \propto \rho(E)e^{-E/kT} = \rho(E)e^{-[mc^2 + (p^2/2m)]/k_BT}$$

(11.6.7)

where $\rho(E)$ is the density of states of energy E. If we approximate the statistics of these excitations by classical Boltzmann statistics, the density of particles of mass m can be obtained by integrating (11.6.7) over all momenta p. We then obtain an expression of the form (11.6.5) in which the molar mass $M_k = N_A m_k$. Thus equations (11.6.5) and (11.6.6) give the density and partial pressure due to particles that appear spontaneously at temperature T as thermal excitations of quantum fields. In this state in which $\mu = 0$ there is no distinction between thermal radiation and matter; just as for thermal photons, the particle density is entirely determined by the temperature

At ordinary temperatures, this thermal particle density is extremely small. But quantum field theory has now revealed the thermodynamic importance of the state $\mu = 0$. It is a state of thermal equilibrium that matter could reach; indeed matter was in such a state during the early part of the universe. Had matter stayed in thermal equilibrium with radiation, at the current temperature of the universe the density of protons and electrons, given by (11.6.5) or its modifications, would be virtually zero. The existence of particles at their present temperatures has to be viewed as a nonequilibrium state. As a result of the particular way in which the universe has evolved, matter was not able to convert to radiation and stay in thermal equilibrium with it.

From (11.6.4) we see that assigning a nonzero value for the chemical potential is a way of fixing the particle density at a given temperature. Since we have an understanding of the **absolute zero of chemical potential**, we can write the chemical potential as

$$\mu_k = RT \ln\left(\frac{p_k}{p_{k,th}}\right) \tag{11.6.8}$$

in which $p_{k,\text{th}}$ is the thermal pressure defined above. In principle, one may adopt this scale of chemical potential.

References

1. Planck, M., *Theory of Heat Radiation.* (*History of Modern Physics*, Vol. 11), 1988, Washington, DC, Am. Inst. of Physics.
2. Kondepudi, D. K., *Foundations of Physics*, **17** (1987) 713–722.
3. Reichl, L. E., A Modern Course in Statistical Physics, 2nd ed. 1998, New York: Wiley.

Example

Example 11.1   Using the equation of state, calculate the energy density and pressure of thermal radiation at 6000 K (approximately the temperature of the radiation from the sun). Also calculate the pressure at $T = 10^7$ K
Solution: The energy density is given by the Stefan–Boltzmann law $u = \beta T^4/3$, in which $\beta = 7.56 \times 10^{-16}\,\mathrm{J\,m^{-3}\,K^{-4}}$ (11.2.6). Hence the energy

The pressure due to thermal radiation is given by $p = u/3 = (0.98/3)\,\mathrm{J\,m^{-3}} = 0.33\mathrm{Pa} \approx 3 \times 10^{-6}$ atm. At $T = 10^7$ K the energy density and pressure are

$$u = 7.56 \times 10^{-16}\,\mathrm{J\,m^{-3}\,K^{-4}}(10^7\,\mathrm{K})^4 = 7.56 \times 10^{12}\,\mathrm{J\,m^{-3}}$$
$$p = u/3 = 2.52 \times 10^{12}\,\mathrm{Pa} = 2.5 \times 10^7\,\mathrm{atm}$$

Exercises

11.1 Obtain (11.2.5) by using (11.2.1) and (11.2.2) in the Helmholtz equation (11.2.4).

11.2 Using Planck's formula (11.1.3) for $u(\nu, T)$ in (11.1.4), obtain the Stefan–Boltzmann law (11.2.6) and an expression for the Stefan–Boltzmann constant β.

11.3 Show that (11.3.5) follows from (11.3.4).

11.4 At an early stage of its evolution, the universe was filled with thermal radiation at a very high temperature. As the universe expanded adiabatically, the temperature of the radiation decreased. Using the current value of $T = 2.8$ K, obtain the ratio of the present volume to the volume of the universe when $T = 10^{10}$ K.

11.5 Plot the Planck energy distribution $u(\nu)$ for $T = 6000$ K; $\lambda_{\max} = 483$ nm for the solar thermal radiation. What will $\lambda_{\max}$ be if the sun's surface temperature is 10 000 K.

11.6 Express Planck's formula (11.1.3) as a function of the wavelength λ instead of the frequency ν.

11.7 Assuming the average surface temperature of the Earth is about 288 K, estimate the amount of energy radiated as thermal radiation. Since the total energy of the Earth is nearly a constant, the solar radiation absorbed equals energy radiated. Assuming the temperature of the solar radiation is 6000 K, estimate the total entropy production on the surface of the Earth due to absorption and emission of solar radiation.

PART III

FLUCTUATIONS AND STABILITY

12 THE GIBBS STABILITY THEORY

12.1 Classical Stability Theory

The random motion of molecules causes all thermodynamic quantities such as temperature, concentration and partial molar volume to fluctuate. In addition, due to its interaction with the exterior, the state of a system is subject to constant perturbations. The state of equilibrium must remain stable in the face of all fluctuations and perturbations. In this chapter we shall develop a theory of stability for isolated systems in which the total energy U, volume V and mole numbers N_k are constant. The stability of the equilibrium state leads us to conclude that certain physical quantities, such as heat capacities, have a definite sign. This will be an introduction to the theory of stability as was developed by Gibbs. Chapter 13 contains some elementary applications of this stability theory. In Chapter 14, we shall present a more general theory of stability and fluctuations based on the entropy production associated with a fluctuation. The more general theory is applicable to a wide range of systems, including nonequilibrium systems.

For an isolated system the entropy reaches its maximum value at equilibrium; thus, any fluctuation can only reduce the entropy. In response to a fluctuation, entropy-producing irreversible processes spontaneously drive the system back to equilibrium. Hence, *the state of equilibrium is stable to any perturbation that results in a decrease in entropy.* Conversely, if fluctuations can grow, the system is not in equilibrium. The fluctuations in temperature, volume, etc., are quantified by their magnitudes such as δT and δV. The entropy of the system is a function of these parameters. In general, the entropy can be expanded as a power series in these parameters, so we have

$$S = S_{\text{eq}} + \delta S + \frac{1}{2}\delta^2 S + \dots \qquad (12.1.1)$$

In such an expansion, the term δS represents the *first-order* terms containing δT, δV, etc., the term $\delta^2 S$ represents the *second order* terms containing $(\delta T)^2$, $(\delta V)^2$, etc., and so on. This notation will be made explicit in the examples that follow. Also, as we shall see below, since the entropy is a maximum, the first-order term δS vanishes. The change in entropy is due to the second- and higher-order terms, the leading contribution coming from the second-order term $\delta^2 S$.

We shall look at the stability conditions associated with fluctuations in different quantities such as temperature, volume and mole numbers in an isolated system in which U, V and N_k are constant.

12.2 Thermal Stability

For the fluctuations in temperature, we shall consider a simple situation without loss of generality. Let assume that the fluctuation occurs in a small part of the system (Fig. 12.1). Due to a fluctuation there is a flow of energy δU from one part to the other, resulting in a small temperature fluctuations δT in the smaller part. The subscripts 1 and 2 identify the two parts of the system. The total entropy of the system is

$$S = S_1 + S_2 \qquad (12.2.1)$$

Here entropy S_1 is a function of U_1, V_1, etc., and S_2 is a function of U_2, V_2, etc. If we express S as a Taylor series about its equilibrium value, S_{eq}, we can express the change in entropy ΔS from its equilibrium value as

$$S - S_{eq} = \Delta S = \left(\frac{\partial S_1}{\partial U_1}\right)\delta U_1 + \left(\frac{\partial S_2}{\partial U_2}\right)\delta U_2 + \left(\frac{\partial^2 S_1}{\partial U_1^2}\right)\frac{(\delta U_1)^2}{2}$$
$$+ \left(\frac{\partial^2 S_2}{\partial U_2^2}\right)\frac{(\delta U_2)^2}{2} + \cdots \qquad (12.2.2)$$

where all the derivatives are evaluated at the equilibrium state.

Since the total energy of the system remains constant, $\delta U_1 = -\delta U_2 = \delta U$. Also, recall that $(\partial S/\partial U)_{V,N} = 1/T$. Hence (12.2.2) can be written as

$$\Delta S = \left(\frac{1}{T_1} - \frac{1}{T_2}\right)\delta U + \left[\frac{\partial}{\partial U_1}\frac{1}{T_1} + \frac{\partial}{\partial U_2}\frac{1}{T_2}\right]\frac{(\delta U)^2}{2} + \cdots \qquad (12.2.3)$$

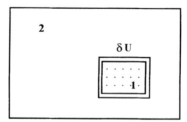

Figure 12.1 Thermal fluctuations in the equilibrium state. We consider a fluctuation that results in a flow of energy δU from one part to the other, causing the temperatures to change by a small amount δT

We can now identify the first and second variations of entropy, δS and $\delta^2 S$, and write them explicitly in terms of the perturbation δU:

$$\delta S = \left(\frac{1}{T_1} - \frac{1}{T_2} \right) \delta U \qquad (12.2.4)$$

$$\frac{1}{2} \delta^2 S = \left[\frac{\partial}{\partial U_1} \frac{1}{T_1} + \frac{\partial}{\partial U_2} \frac{1}{T_2} \right] \frac{(\delta U)^2}{2} \qquad (12.2.5)$$

At equilibrium, since all thermodynamic forces must vanish, the entire system should be at the same temperature. Hence $T_1 = T_2$, and the first variation of entropy $\delta S = 0$. (If it is taken as a postulate that entropy is a maximum at equilibrium, then the first variation should vanish. One then concludes that $T_1 = T_2$.) The changes in entropy due to fluctuations in the equilibrium state are due to the second variation $\delta^2 S$ (the smaller higher-order terms in the Taylor series are neglected). At equilibrium, since S is a maximum, fluctuations can only decrease S, i.e. $\delta^2 S < 0$, and spontaneous, entropy-increasing irreversible processes drive the system back to the state of equilibrium. Now let us write (12.2.5) explicitly in terms of the physical properties of the system and see what the condition for stability implies. First we note that, at constant V,

$$\frac{\partial}{\partial U} \frac{1}{T} = -\frac{1}{T^2} \frac{\partial T}{\partial U} = -\frac{1}{T^2} \frac{1}{C_V N} \qquad (12.2.6)$$

in which C_V is the molar heat capacity. Also, if the change in the temperature of the smaller system is δT then we have $\delta U_1 = C_{V_1}(\delta T)$ where $C_{V_1} = C_V N_1$ is the heat capacity of the smaller part. $C_{V_2} = C_V N_2$ is the heat capacity of the larger part. Using (12.2.6) for the two parts in (12.2.5), writing $\delta U = C_{V_1}(\delta T)$, and noting that all the derivatives are evaluated at equilibrium, so that $T_1 = T_2 = T$, we obtain

$$\frac{1}{2} \delta^2 S = -\frac{C_{V_1}(\delta T)^2}{2T^2} \left(1 + \frac{C_{V_1}}{C_{V_2}} \right) \qquad (12.2.7)$$

If system 1 is small compared to system 2, $C_{V_1} \ll C_{V_2}$ so that the second term in the parentheses can be ignored. As the number of moles N_1 in the heat capacity is arbitrary we shall use C_V in its place with the understanding that it is the heat capacity, not the molar heat capacity. For stability of the equilibrium state, we thus have

$$\boxed{ \frac{1}{2} \delta^2 S = -\frac{C_V(\delta T)^2}{2T^2} < 0 } \qquad (12.2.8)$$

This condition requires that the heat capacity $C_V > 0$. *Thus, the state of equilibrium is stable to thermal fluctuations because the heat capacity at constant volume is positive.* Conversely, if the heat capacity is negative, the system is not in stable equilibrium.

12.3   Mechanical Stability

We now turn to stability of the system with respect to fluctuations in the volume of a subsystem with N remaining constant, i.e. fluctuations in the molar volume. As in the previous case consider the system divided into two parts (Fig. 12.2) but this time assume there is a small change in volume δV_1 of system 1 and a small change δV_2 of system 2. Since the total volume of the system remains fixed, $\delta V_1 = -\delta V_2 = \delta V$. To compute the change in entropy associated with such a fluctuation, we can write an equation similar to (12.2.3), with V taking the place of U. Since $(\partial S/\partial V)_{U,N} = p/T$, a calculation similar to that in section 12.2 (exc. 12.2) leads to

$$\delta S = \left(\frac{p_1}{T_1} - \frac{p_2}{T_2}\right)\delta V \qquad (12.3.1)$$

$$\frac{1}{2}\delta^2 S = \left[\frac{\partial}{\partial V_1}\frac{p_1}{T_1} + \frac{\partial}{\partial V_2}\frac{p_2}{T_2}\right]\frac{(\delta V)^2}{2} \qquad (12.3.2)$$

Because the derivatives are evaluated at equilibrium $p_1/T_1 = p_2/T_2 = p/T$. The first variation δS vanishes (as it must if S is a maximum at equilibrium). To understand the physical meaning of the conditions for stability $\delta^2 S < 0$, the second variation can be written in terms of the isothermal compressibility. The isothermal compressibility κ_T is defined by $\kappa_T = -(1/V)(\partial V/\partial p)$. During the fluctuation in V we assume that T remains unchanged. With these observations it

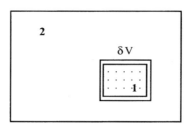

Figure 12.2 Fluctuation in volume of a system for fixed N and V

is easy to see that (12.3.2) can be written as

$$\delta^2 S = -\frac{1}{T\kappa_T} \frac{(\delta V)^2}{V_1} \left[1 + \frac{V_1}{V_2} \right] \qquad (12.3.3)$$

As before, if one part is much larger than another, $V_2 \gg V_1$, this expression can be simplified and the condition for stability can be written as:

$$\boxed{\delta^2 S = -\frac{1}{T\kappa_T} \frac{(\delta V)^2}{V} < 0} \qquad (12.3.4)$$

where we have used V for the arbitrary volume V_1. This is valid when $\kappa_T > 0$. *Thus the state of equilibrium is stable to volume or mechanical fluctuations because the isothermal compressibility is positive. If $\kappa_T < 0$, the system is in an unstable nonequilibrium state.*

12.4   Stability and Fluctuations in Mole Number

Fluctuations in the mole numbers of the various components of a system occur due to chemical reactions and due to transport, such as diffusion. We shall consider each case separately.

CHEMICAL STABILITY

These fluctuations can be identified as the fluctuations in the extent of reaction ξ, about its equilibrium value. Considering a fluctuation $\delta\xi$, the change in entropy is

$$S - S_{eq} = \Delta S = \delta S + \frac{1}{2}\delta^2 S = \left(\frac{\partial S}{\partial \xi} \right)_{U,V} \delta\xi + \frac{1}{2}\left(\frac{\partial^2 S}{\partial \xi^2} \right)_{U,V} (\delta\xi)^2 \quad (12.4.1)$$

We have seen in Chapter 4 that $(\partial S/\partial \xi)_{U,V} = A/T$. Hence equation (12.4.1) can be written as

$$\Delta S = \delta S + \frac{1}{2}\delta^2 S = \left(\frac{A}{T} \right)_{eq} \delta\xi + \frac{1}{2T}\left(\frac{\partial A}{\partial \xi} \right)_{eq} (\delta\xi)^2 \qquad (12.4.2)$$

(T is constant.) In this equation the identification of the first and second variation of entropy is obvious. At equilibrium the affinity A vanishes, so that once again $\delta S = 0$. For the stability of the equilibrium state, we then require the second variation $\delta^2 S$ to be negative:

$$\frac{1}{2}\delta^2 S = \frac{1}{2T}\left(\frac{\partial A}{\partial \xi}\right)_{eq}(\delta \xi)^2 < 0 \qquad (12.4.3)$$

Since $T > 0$, the condition for stability of the equilibrium state is

$$\boxed{\left(\frac{\partial A}{\partial \xi}\right)_{eq} < 0}^{*} \qquad (12.4.4)$$

When many chemical reactions take place simultaneously, condition (12.4.3) can be generalized to the following statement [1, 2]:

$$\frac{1}{2}\delta^2 S = \sum_{i,j}\frac{1}{2T}\left(\frac{\partial A_i}{\partial \xi_j}\right)_{eq}\delta\xi_i\,\delta\xi_j < 0 \qquad (12.4.5)$$

STABILITY TO FLUCTUATIONS DUE TO DIFFUSION

The fluctuations in mole numbers considered so far were only due to chemical reactions. The fluctuation in mole number can also occur due to exchange of matter between a part of a system and the rest (Fig. 12.3). As in the case of exchange of energy, we consider the total change in entropy of the two parts of the system:

$$S = S_1 + S_2 \qquad (12.4.6)$$

$$S - S_{eq} = \Delta S = \sum_{k}\left[\left(\frac{\partial S_1}{\partial N_{1k}}\right)\delta N_{1k} + \left(\frac{\partial S_2}{\partial N_{2k}}\right)\delta N_{2k}\right]$$
$$+ \sum_{i,j}\left[\left(\frac{\partial^2 S_1}{\partial N_{1i}\partial N_{1j}}\right)\frac{\delta N_{1i}\delta N_{1j}}{2} + \left(\frac{\partial^2 S_2}{\partial N_{2i}\partial N_{2j}}\right)\frac{\delta N_{2i}\delta N_{2j}}{2}\right] + \cdots$$

$$(12.4.7)$$

Now we note that $\delta N_{1k} = -\delta N_{2k} = \delta N_k$ and $(\partial S/\partial N_k) = -\mu_k/T$. Equation (12.4.7) can then be written so that the first and second variation of the entropy can be identified:

$$\Delta S = \delta S + \frac{\delta^2 S}{2} = \sum_{k}\left(\frac{\mu_{2k}}{T} - \frac{\mu_{1k}}{T}\right)\delta N_k - \sum_{i,j}\left(\frac{\partial}{\partial N_j}\frac{\mu_{1i}}{T} + \frac{\partial}{\partial N_j}\frac{\mu_{2i}}{T}\right)\frac{\delta N_i\,\delta N_j}{2}$$

$$(12.4.8)$$

---

* This condition can be used to derive the Le Chatelier–Braun principle discussed in section 9.3.

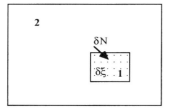

Figure 12.3 Fluctuations in mole number can occur due to chemical reactions and exchange of molecules between the two systems. The state of equilibrium is stable if the entropy change associated with fluctuations is negative

As before, if the derivatives are evaluated at the state of equilibrium, the chemical potentials of the two parts must be equal. Hence the first term vanishes. Furthermore, if system 1 is small compared to system 2, the change in the chemical potential (which depends on the concentrations) with respect to N_k of system 2 will be small compared to the corresponding change in system 1. That is,

$$\left(\frac{\partial}{\partial N_j} \frac{\mu_{1i}}{T}\right) \gg \left(\frac{\partial}{\partial N_j} \frac{\mu_{2i}}{T}\right)$$

if system 1 is much smaller than system 2. We then have

$$\delta^2 S = -\sum_{i,j} \left(\frac{\partial}{\partial N_j} \frac{\mu_{1i}}{T}\right) \delta N_i \, \delta N_j < 0 \qquad (12.4.9)$$

as the condition for the stability of an equilibrium state when fluctuations in mole number are considered.

In fact, this condition is general and can be applied to fluctuations due to chemical reactions as well. By assuming the fluctuations $\delta N_k = \nu_k \delta\xi$, in which ν_k is the stoichiometric coefficient, we can obtain (exc. 12.4) condition (12.4.5). *Thus a system that is stable to diffusion is also stable to chemical reactions.* This is called the **Duhem–Jougeut theorem** [3, 4] A more detailed discussion of this theorem and many other aspects of stability theory can be found in the literature [2].

In summary, the general condition for the stability of the equilibrium state to thermal, volume and mole number fluctuations can be expressed by combining (12.2.8), (12.3.4) and (12.4.9)

$$\delta^2 S = -\frac{C_V(\delta T)^2}{T^2} - \frac{1}{T\kappa_T} \frac{(\delta V)^2}{V} - \sum_{i,j} \left(\frac{\partial}{\partial N_j} \frac{\mu_i}{T}\right) \delta N_i \, \delta N_j < 0 \qquad (12.4.10)$$

Here C_V is the heat capacity of the system with arbitrary volume V and chemical potential μ_j. Though we have derived the above results by assuming S to be a function of U, V and N_k, and a system in which U, V and N are constant, the results derived have a more general validity in that they are also valid for other situations in which p and/or T are maintained constant. The corresponding results are expressed in terms of the enthalpy H, Helmholtz free energy F and Gibbs free energy G. In fact, a more general theory of stability that is valid for a wide range of conditions can be developed using the entropy production d_iS as the basis. This more general approach will be presented in Chapter 14. The Gibbs stability theory is valid only for well-defined boundary conditions such as $T = \text{const}$. In contrast, the approach of Chapter 14 is independent of such conditions; it depends on irreversible processes inside the system.

References

1. Glansdorff, P. and Prigogine, I., *Thermodynamics of Structure Stability and Fluctuations*. 1971, New York: Wiley.
2. Prigogine, I. and Defay, *Chemical Thermodynamics*. 1954, London: Longman.
3. Jouguet, E., Notes de mécanique chimique. *J. Ecole Polytech. (Paris) Ser. 2*, **21** (1921) 61.
4. Duhem, P., *Traité élémentaire de Mécanique Chimique, 4 Vols.* 1899, Paris: Gauthiers-Villars.

Exercises

12.1   For an ideal gas at equilibrium at $T = 300$ K and $p = 1.0$ atm, calculate the change in entropy due to a fluctuation of $\delta T = 1.0 \times 10^{-3}$ K in a volume $V = 1.0 \times 10^{-6}$ mL.

12.2   Obtain expressions (12.3.1) and (12.3.2) for the first and second-order entropy changes due to fluctuations of volume at constant N.

12.3   Explain the physical meaning of condition (12.4.4) for stability with respect to a chemical reaction.

12.4   In expression (12.4.9) assume that the change in mole number is due to a chemical reaction and obtain expression (12.4.3) and generalize it to (12.4.5).

13 CRITICAL PHENOMENA AND CONFIGURATIONAL HEAT CAPACITY

Introduction

In this chapter we shall consider applications of stability theory to critical phenomena of liquid–vapor transitions and separation of binary mixtures. When the applied pressure and temperature are altered, systems can become unstable, causing their physical state to transform into another distinct state. For example, when the temperature of a two-component liquid mixture (such as hexane and nitrobenzene) changes, the mixture may become unstable to changes in the composition; the mixture then separates into two phases, each rich in one of the components. In Chapters 18 and 19 we shall see that in far-from-equilibrium systems, loss of stability can lead to a wide variety of complex nonequilibrium states. We shall also look at how a system that can undergo internal transformations responds to a rapid change in temperature. This leads us to the concept of **configurational heat capacity**.

13.1 Stability and Critical Phenomena

In Chapter 7 we briefly looked at the critical behavior of a pure substance. If its temperature is above the critical temperature T_c, there is no distinction between the gas and the liquid states, regardless of the pressure. Below the critical temperature, at low pressures the substance is in the form of a gas, but as the pressure is increased liquid begins to form. We can understand this transformation in terms of stability.

As shown in Fig. 13.1 by the arrows, using an appropriate path it is possible to go from a gaseous state to a liquid state in a continuous fashion. This was noted by James Thomson, who also suggested that the isotherms below the critical point were also continuous, as shown in Fig. 13.2 by the curve IAJKLBM. This suggestion was pursued by van der Waals, whose equation, introduced in Chapter 1, indeed gives the Thompson curve. However, the region JKL in Fig. 13.2 cannot be physically realized because it is an unstable region, i.e. it is not mechanically stable. Section 12.3 showed that the condition for mechanical stability is ensured when compressibility

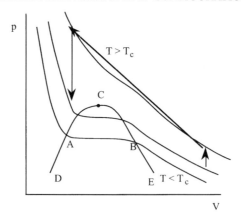

Figure 13.1 The critical behavior of a pure substance. Below the critical temperature, at a fixed temperature, a decrease in volume results in a transition to a liquid state in the region AB in which the two phases coexist. The envelope of the segments AB for the family of isotherms has the shape ECD. Above critical temperature T_c there is no such transition. The gas becomes more and more dense, there being no distinction between gas and liquid phases. Following the path shown by the arrows, it possible to go from a gas to a liquid state without going through a transition

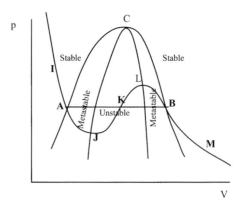

Figure 13. 2 Stable, metastable and unstable regions for a liquid–vapor transition. In the region JKL we have $(\partial p/\partial V)_T > 0$, so the system is unstable

$\kappa_T \equiv (-1/V)(\partial V/\partial p) > 0$. In Fig. 13.2 it implies the system is stable only if

$$\boxed{\left(\frac{\partial p}{\partial V}\right)_T < 0} \tag{13.1.1}$$

This condition is satisfied for segments IA and BM, and for all the isotherms above the critical temperature; these are stable regions. But along segment JKL we see that $(\partial p/\partial V)_T > 0$, which means these states are unstable; if the volume of the system is kept fixed, small fluctuations in pressure, depending on the initial state, will either cause the vapor to condense or the liquid to evaporate. The system will collapse to a point along segment AB, where liquid and vapor coexist. As shown in section 7.4, the amount of the substance in the two phases is given by the "lever rule."

   In region BL of Fig. 13.2 the system is a supersaturated vapor and may begin to condense if nucleation can occur. This is a **metastable state**. Similarly, in region AJ we have a superheated liquid that will vaporize if there is nucleation of the vapor phase. The stable, metastable and unstable regions are indicated in Fig. 13.2. Finally, at the critical point C, both the first and second derivatives of p with respect to V equal zero. Here the stability is determined by the higher-order derivatives. For stable mechanical equilibrium at the critical point, we have

$$\left(\frac{\partial p}{\partial V}\right)_{T_c} = 0 \qquad \left(\frac{\partial^2 p}{\partial V^2}\right)_{T_c} = 0 \qquad \left(\frac{\partial^3 p}{\partial V^3}\right)_{T_c} < 0 \tag{13.1.2}$$

which is an inflection point. The inequality $(\partial^3 p/\partial V^3) < 0$ is obtained by considering terms of higher order than $\delta^2 S$.

13.2   Stability and Critical Phenomena in Binary Solutions

In solutions, depending on the temperature, the various components can segregate into separate phases. For simplicity, we shall only consider binary mixtures. The phenomenon is similar to the critical phenomenon in a liquid–vapor transition in that across one range of temperature the system is in one homogeneous phase (solution), but across an another range of temperature the system becomes unstable and the two components separate into two phases. The **critical temperature** that separates these two ranges depends on the composition of the mixture. This can happen in three ways, as illustrated by the following examples.

   At atmospheric pressure the liquids *n*-hexane and nitrobenzene are miscible in all proportions when the temperature is above 19 °C. Below 19 °C the mixture separates into two distinct phases, one rich in nitrobenzene and the other in *n*-

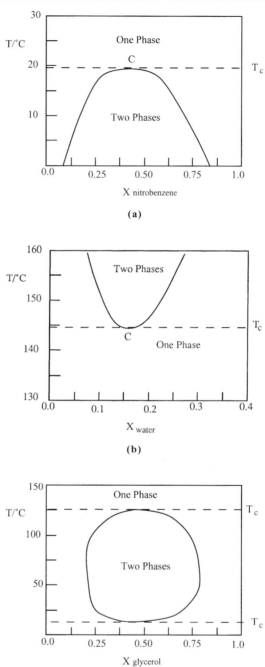

Figure 13.3 Three types of phase diagram showing critical phenomena in binary solutions: (a) A mixture of hexane and nitrobenzene, (b) a mixture of diethylamine and water, (c) a mixture of *m*-toluidine and glycerol

hexane. The corresponding phase diagram is shown in Fig. 13.3(a). At about $10\,^{\circ}$C one of the phases has a mole fraction of nitrobenzene equal to 0.18 and the other phase has a mole fraction equal to 0.70. As the temperature is increased, at $T = T_c$ the two liquid layers become identical in composition, indicated by the point C. The point C is called the **critical solution point** or **consolute point** and its location depends on the applied pressure. In this example, above the critical temperature the two liquids are miscible in all proportions. This is the case of **upper critical temperature**. But this is not always the case, as shown in Fig. 13.3(b) and (c). The critical temperature can be such that below T_c the two components become miscible in all proportions; an example is diethylamine and water. Such a mixture is said to have a **lower critical solution temperature**. Binary systems can have both upper and lower critical solution temperatures, as shown in Fig. 13.3(c); an example is a mixture of *m*-toluidine and glycerol.

Let us now look at the phase separation in binary mixtures from the viewpoint of stability. The separation of phases occurs when the system becomes unstable with respect to diffusion of the two components, i.e. if the separation of the two components produces an increase in entropy; then the fluctuations in the mole number due to diffusion in a given volume grow, resulting in the separation of the two components. As we have seen in section 12.4, the condition for stability against diffusion of the components is

$$\delta^2 S = -\sum_{i,k} \frac{\partial}{\partial N_k}\left(\frac{\mu_i}{T}\right)\delta N_k\, \delta N_i < 0 \qquad (13.2.1)$$

At a fixed T, for binary mixtures, this can be written in the explicit form

$$\mu_{11}(\delta N_1)^2 + \mu_{22}(\delta N_2)^2 + \mu_{21}(\delta N_1)(\delta N_2) + \mu_{12}(\delta N_1)(\delta N_2) > 0 \quad (13.2.2)$$

in which

$$\mu_{11} = \frac{\partial \mu_1}{\partial N_1} \qquad \mu_{22} = \frac{\partial \mu_2}{\partial N_2} \qquad \mu_{21} = \frac{\partial \mu_2}{\partial N_1} \qquad \mu_{12} = \frac{\partial \mu_1}{\partial N_2} \qquad (13.2.3)$$

Condition (13.2.2) is mathematically identical to the statement that the matrix with elements μ_{ij}, is *positive definite*. Also, because

$$\mu_{21} = \frac{\partial \mu_2}{\partial N_1} = \frac{\partial}{\partial N_1}\frac{\partial G}{\partial N_2} = \frac{\partial}{\partial N_2}\frac{\partial G}{\partial N_1} = \mu_{12} \qquad (13.2.4)$$

this matrix is symmetric. Thus the stability of the system is assured if the symmetric matrix

$$\begin{bmatrix} \mu_{11} & \mu_{12} \\ \mu_{21} & \mu_{22} \end{bmatrix} \qquad (13.2.5)$$

is positive definite. The necessary and sufficient conditions for the positivity of (13.2.5) are:

$$\mu_{11} > 0, \quad \mu_{22} > 0 \quad \text{and} \quad (\mu_{11}\mu_{22} - \mu_{21}\mu_{12} > 0 \tag{13.2.6}$$

If these conditions are not satisfied, then condition (13.2.2) will be violated and the system will become unstable. Note that (13.2.4) and (13.2.6) imply that $\mu_{12} = \mu_{21} < 0$, to assure stability for all positive values of μ_{11} and μ_{22}.

If we have an explicit expression for the chemical potential, then conditions (13.2.6) can be related to the activity coefficients of the system. This can be done for a class of solutions called **strictly regular solutions**, which were studied by Hildebrandt and by Fowler and Guggenheim. The two components of these solutions strongly interact and their chemical potentials are of the form

$$\mu_1(T,p,x_1,x_2) = \mu_1^0(T,p) + RT \ln x_1 + \alpha x_2^2 \tag{13.2.7}$$

$$\mu_2(T,p,x_1,x_2) = \mu_2^0(T,p) + RT \ln x_2 + \alpha x_1^2 \tag{13.2.8}$$

in which

$$x_1 = \frac{N_1}{N_1 + N_2} \quad \text{and} \quad x_2 = \frac{N_2}{N_1 + N_2} \tag{13.2.9}$$

are the mole fractions. The factor α is related to the difference in interaction energy between two similar molecules (two molecules of component 1 or two of component 2) and two dissimilar molecules (one molecule of component 1 and one of component 2). For solutions that are nearly perfect, α is nearly zero. From these expressions it follows that activity coefficients are given by $RT \ln \gamma_1 = \alpha x_2^2$ and $RT \ln \gamma_2 = \alpha x_1^2$. We can now apply the stability conditions (13.2.6) to this system. On explicit evaluation of the derivative, the condition $\mu_{11} = \partial \mu_1 / \partial N_1 > 0$ can be written as (exc. 13.5):

$$\frac{RT}{2\alpha} - x_1(1 - x_1) > 0 \tag{13.2.10}$$

For nearly perfect solutions, since $\alpha \to 0$, this inequality is always satisfied.

For a given composition specified by x_1, if $(R/2\alpha)$ is positive, for sufficiently large T this condition will be satisfied. However, for smaller T it can be violated. The maximum value of $x_1(1 - x_1)$ is 0.25. Thus for $(RT/2\alpha)$ less than 0.25 there must be a range of x_1 for which the inequality (13.2.10) is not valid. When this happens, the system becomes unstable and separates into two phases. In this case we have an upper critical solution temperature. From (13.2.10) it follows that the relation between the mole fraction and the temperature below which the

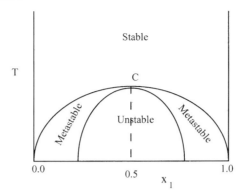

Figure 13.4 The phase diagram for strictly regular solutions

system becomes unstable is

$$\frac{RT_c}{2\alpha} - x_1(1 - x_1) = 0 \tag{13.2.11}$$

This gives us the plot of T_c as a function of x_1 shown in Fig. 13.4. It is easy to see that the maximum of T_c occurs at $x_1 = 0.5$. Thus the critical temperature and the critical mole fraction are

$$(x_1)_c = 0.5 \quad \text{and} \quad T_c = \frac{\alpha}{2R} \tag{13.2.12}$$

The equation $T = (2\alpha/R)x_1(1 - x_1)$ gives the boundary between the metastable region and the unstable region. The boundary between the stable region and the metastable region is the coexistence curve. The coexistence curve of the two phases can be obtained by writing the chemical potentials μ_1 and μ_2 in both phases and equating them. This is left as an exercise.

13.3 Configurational Heat Capacity

The thermodynamic state of a chemically reacting system may be specified, in addition to p and T, by the extent of reaction ξ. For such a system the heat capacity must also involve the changes in ξ due to the change in temperature. For example, we may consider a compound that may exist in two isomeric forms. Then the extent of reaction for the transformation is the variable ξ. The heat absorbed by such a system produces changes not only in p and T but also in ξ, causing it to relax to a new state of equilibrium with respect to the transformation. If the system is in equilibrium with respect to the extent of reaction ξ, the corresponding affinity $A = 0$. Now since the heat exchanged

$dQ = dU + p\,dV = dH - V\,dp$, we can write

$$dQ = [h_{T,\xi} - V]dp + C_{p,\xi}\,dT + h_{T,p}\,d\xi \qquad (13.3.1)$$

in which

$$h_{T,\xi} = \left(\frac{\partial H}{\partial p}\right)_{T,\xi} \qquad C_{p,\xi} = \left(\frac{\partial H}{\partial T}\right)_{p,\xi} \qquad h_{T,p} = \left(\frac{\partial H}{\partial \xi}\right)_{T,p} \qquad (13.3.2)$$

At constant pressure, we can write the heat capacity $C'_p = C_p N$ as:

$$C'_p = \left(\frac{dQ}{dT}\right)_p = C_{p,\xi} + h_{T,p}\left(\frac{d\xi}{dT}\right)_p \qquad (13.3.3)$$

Now, for an equilibrium transformation it can be shown (exc. 13.6) that

$$\left(\frac{\partial \xi}{\partial T}\right)_{p,A=0} = -\frac{h_{T,p}}{T\left(\dfrac{\partial A}{\partial \xi}\right)_{T,p}} \qquad (13.3.4)$$

Substituting (13.3.4) into (13.3.3) we obtain the following result for a system that remains in equilibrium while it is receiving heat:

$$C'_{p,A=0} = C_{p,\xi} - T\left(\frac{\partial A}{\partial \xi}\right)_{T,p}\left(\frac{\partial \xi}{\partial T}\right)^2_{p,A=0} \qquad (13.3.5)$$

But we have seen in section 12.4 that the condition for stability of a system with respect to chemical reactions is that $(\partial A/\partial \xi) < 0$. Hence the second term on the right-hand side of (13.3.5) is positive. The term $C_{p,\xi}$ is the heat capacity at constant composition. There may be situations, however, in which the relaxation of the transformation represented by ξ is very slow. In this case we measure the heat capacity at constant composition. This leads us to a general conclusion:

"The heat capacity at a constant composition is always less than heat capacity of a system that remains in equilibrium with respect to ξ as it receives heat."

The term $h_{T,p}(d\xi/dT)$ is called the **configurational heat capacity**, as it refers to the heat capacity due to the relaxation of the system to the equilibrium configuration. The configurational heat capacity can be observed in systems such as glycerin when it is a supercooled liquid. The molecules can vibrate but not rotate freely as they do in the liquid state. This restricted motion is called *libration*. As the temperature is increased, a greater fraction of the molecules begin to rotate. For this system the variable ξ is the extent of reaction for the libration–rotation transformation. For glycerin there exists a state called the *vitreous state* in which the libration–rotation equilibrium is reached rather slowly. If this system is heated rapidly, the equilibrium is not maintained and the measured heat capacity will be $C_{p,\xi}$, which will be lower than the heat capacity measured through slow heating during which the system remains in equilibrium.

Further Reading

- Hildebrandt, J. M., Prausnitz J. M. and Scott R. L., *Regular and Related Solutions.* 1970, New York: Van Nostrand Reinhold.
- Van Ness, H. C. and Abbott M. M., *Classical Thermodynamics of Nonelectrolyte Solutions.* 1982, New York: McGraw-Hill.
- Prigogine, I. and Defay R., *Chemical Thermodynamics*, 4th ed. 1967, London: Longman.

Exercises

13.1 Using the Gibbs–Duhem equation at constant p and T, $\sum N_k \, d\mu_k = 0$ and $(\partial \mu_k / \partial N_i) = (\partial \mu_i / \partial N_k)_{p,T}$, and using $d\mu_k = \sum_i (\partial \mu_k / \partial N_i)_{p,T} \, dN_i$, show that

$$\sum_i \left(\frac{\partial \mu_k}{\partial N_i} \right)_{p,T} N_i = 0$$

This means that the determinant of the matrix with elements $\mu_{ki} = (\partial \mu_k / \partial N_i)$ is equal to zero. Consequently, one of the eigenvalues of the matrix (13.2.5) is zero.

13.2 Show that if the 2×2 matrix (13.2.5) has a negative eigenvalue, inequality (13.2.2) can be violated.

13.3 Show that if the matrix (13.2.5) has positive eigenvalues, then $\mu_{11} > 0$ and $\mu_{22} > 0$.

13.4 In a strictly binary solution, assuming the two phases are symmetric, i.e. the dominant mole fraction in either phase is the same, obtain the coexistence curve by equating the chemical potentials in the two phases.

13.5 Using (13.2.7) and (13.2.9), show that the condition $\mu_{11} = \partial \mu_1 / \partial N_1 > 0$ leads to equation (13.2.10).

13.6 For an equilibrium transformation show that

$$\left(\frac{\partial \xi}{\partial T}\right)_{p,A=0} = -\frac{h_{T,p}}{T\left(\dfrac{\partial A}{\partial \xi}\right)_{T,p}}$$

by using $A(\xi, p, T) = 0$ along the transformation path.

14 STABILITY AND FLUCTUATIONS BASED ON ENTROPY PRODUCTION

14.1 Stability and Entropy Production

In chapter 12 we considered fluctuations in an isolated system in which U, V and N_k are constant and we obtained conditions for the stability of the equilibrium state. These conditions, in fact, have a more general validity in that they remain valid when other types of boundary condition are imposed on the system. For example, instead of constant U and V, we may consider systems maintained at constant T and V, constant p and S or constant T and p. The main reason for the general validity of the stability conditions is that all these conditions are a direct consequence of the fact that for all spontaneous processes $d_i S > 0$. As we have seen in Chapter 5, when each of these three pairs of variables is held constant, one of the thermodynamic potentials F, H or G is minimized. In each case we have shown that

$$dF = -Td_i S \leq 0 \qquad (T, V = \text{constant}) \qquad (14.1.1)$$

$$dG = -Td_i S \leq 0 \qquad (T, p = \text{constant}) \qquad (14.1.2)$$

$$dH = -Td_i S \leq 0 \qquad (S, p = \text{constant}) \qquad (14.1.3)$$

Through these relations, the change of the thermodynamic potentials ΔF, ΔG or ΔH due to a fluctuation can be related to the entropy production $\Delta_i S$. The system is stable to all fluctuations that result in $\Delta_i S < 0$, because they do not correspond to the spontaneous evolution of a system due to irreversible processes. From the above relations it is clear how one could also characterize stability of the equilibrium state by stating that the system is stable to fluctuations for which $\Delta F > 0, \Delta G > 0$ or $\Delta H > 0$. For fluctuations in the equilibrium state, these conditions can be written more explicitly in terms of the second-order variations $\delta^2 F > 0, \delta^2 G > 0$ and $\delta^2 H > 0$, which in turn can be expressed using the second-order derivatives of these potentials. The conditions for stability obtained in this way are identical to those obtained in Chapter 12.

A theory of stability that is based on the positivity of entropy production in spontaneous processes is more general than the classical Gibbs–Duhem theory of stability [1, 2], which is limited to the constraints expressed in (14.1.1) to (14.1.3) and the associated thermodynamic potentials. In addition, stability theory based on entropy production can also be used to obtain conditions for the stability of nonequilibrium states.

In our more general approach, the main task is to obtain an expression for the entropy production $\Delta_i S$ associated with a fluctuation. A system is stable to fluctuations if the associated $\Delta_i S < 0$. In Chapter 3 we have seen that the general form of entropy production due to irreversible processes takes the quadratic form

$$\frac{d_i S}{dt} = \sum_k F_k \frac{dX_k}{dt} = \sum_k F_k J_k \qquad (14.1.4)$$

in which the F_k are the "thermodynamic forces" and where we have represented dX_k/dt as the "flow" or "current" J_k. Thermodynamic forces arise when there is nonuniformity of temperature, pressure or chemical potential. If we denote the equilibrium state by E, and the state to which a fluctuation has driven the system by F, then

$$\Delta_i S = \int_E^F d_i S = \int_E^F \sum_k F_k \, dX_k \qquad (14.1.5)$$

In this section we shall present some simple cases for the calculation of $\Delta_i S$ and defer the more general theory to later chapters in which we consider the stability of nonequilibrium states.

CHEMICAL STABILITY

Let us look at entropy production associated with a fluctuation in a chemical reaction specified by change $\delta\xi$ in the extent of reaction (Fig. 14.1). In Chapter 4

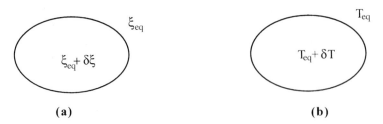

Figure 14.1 (a) A local fluctuation in the extent of reaction; the entropy change associated with such a fluctuation can be calculated using (14.1.6). (b) A local fluctuation in temperature; the associated entropy change can be calculated using (14.1.10)

we have seen that entropy production due to a chemical reaction is

$$d_i S = \frac{A}{T} d\xi \qquad (14.1.6)$$

At equilibrium the affinity $A_{eq} = 0$. For a small change $\alpha = (\xi - \xi_{eq})$ of the extent of reaction from the equilibrium state, we may approximate A by

$$A = A_{eq} + \left(\frac{\partial A}{\partial \xi}\right)_{eq} \alpha = \left(\frac{\partial A}{\partial \xi}\right)_{eq} \alpha \qquad (14.1.7)$$

The entropy production $\Delta_i S$ for the fluctuation is then given by

$$\Delta_i S = \int_0^{\delta\xi} d_i S = \int_0^{\delta\xi} \frac{A}{T} d\xi = \frac{1}{T} \int_0^{\delta\xi} \left(\frac{\partial A}{\partial \xi}\right)_{eq} \alpha \, d\alpha = \left(\frac{\partial A}{\partial \xi}\right)_{eq} \frac{(\delta\xi)^2}{2T} \qquad (14.1.8)$$

where we have used $d\xi = d\alpha$. The **stability condition** $\Delta_i S < 0$ now takes the form

$$\boxed{\Delta_i S = \left(\frac{\partial A}{\partial \xi}\right)_{eq} \frac{(\delta\xi)^2}{2T} < 0} \qquad (14.1.9a)$$

which is identical to (12.4.3). If we consider r chemical reactions, we have

$$\boxed{\Delta_i S = \sum_{i,j}^r \frac{1}{2T} \left(\frac{\partial A_i}{\partial \xi_j}\right)_{eq} \delta\xi_i \, \delta\xi_j < 0} \qquad (14.1.9b)$$

Note how this condition was arrived at by assuming only that for spontaneous processes $\Delta_i S > 0$. It does not depend on the boundary conditions imposed on the system.

THERMAL STABILITY

As a second example, let us consider stability to thermal fluctuations. Let the temperature of a local region of interest be $T_{eq} + \alpha$, where T_{eq} is the equilibrium temperature and α is a small deviation. As we have seen in Chapter 3, the entropy production due to heat flow is

$$\frac{d_i S}{dt} = \left(\frac{1}{T_{eq} + \alpha} - \frac{1}{T_{eq}}\right) \frac{dQ}{dt}$$

$$= -\frac{\alpha}{T_{eq}^2} \frac{dQ}{dt} \qquad (14.1.10)$$

For small changes in temperature, we can write $dQ = C_V d\alpha$ where C_V is the heat capacity* at constant volume. Then for a temperature change of δT, we have

$$\Delta_i S = \int_0^{\delta T} d_i S = \int_0^{\delta T} -\frac{C_V}{T_{eq}^2} \alpha \, d\alpha = -\frac{C_V}{T_{eq}^2} \frac{(\delta T)^2}{2} \qquad (14.1.11)$$

The condition for stability can now be written as

$$\boxed{\Delta_i S = -\frac{C_V}{T_{eq}^2} \frac{(\delta T)^2}{2} < 0} \qquad (14.1.12)$$

which is identical to condition (12.2.8). As before, we conclude this condition is satisfied only if $C_V > 0$.

In this way, a general thermodynamic theory of stability can be formulated using the entropy production as the basis. In general, it can be shown that

$$\Delta_i S = \int_E^F d_i S = \int_E^F \sum_k F_k \, dX_k$$

$$= -\frac{C_v(\delta T)^2}{2T^2} - \frac{1}{T\kappa_T} \frac{(\delta V)^2}{2V} - \sum_{i,j} \left(\frac{\partial}{\partial N_j} \frac{\mu_i}{T} \right) \frac{\delta N_i \, \delta N_j}{2} < 0 \qquad (14.1.13)$$

Finally, we note that this entropy term is second order in the perturbations, δT, δV and δN_k, consistent with the theory of Chapter 12.

In expression (14.1.13) the independent variables are T, V and N. The following more general expression through which the entropy change due to any other set of independent variables can be derived from (14.1.13):

$$\Delta_i S = \frac{\delta^2 S}{2} = \frac{-1}{2T} \left[\delta T \, \delta S - \delta p \, \delta V + \sum_i \delta \mu_i \, \delta N_i \right] < 0 \qquad (14.1.14)$$

In (14.1.13) the first term $C_V \delta T/T = \delta Q/T = \delta S$; similarly in the second term $\delta V/\kappa_T V = -\delta p$; and in the third term $\sum_j (\partial \mu_i/\partial N_j) dN_j = \delta \mu_i$. Using these relations it is easy to see that (14.1.13) can be written in the form (14.1.14)

That the entropy production should be second order in the perturbation follows from the fact that the forces F_k and fluxes J_k vanish at equilibrium. If $\delta J_k = (dX_k/dt)$ and δF_k are forces and fluxes associated with the fluctuation

---

* For notational simplicity we use C_V for the heat capacity though we used it to denote molar heat capacity in earlier chapters.

close to equilibrium, entropy production takes the form

$$\frac{d\Delta_i S}{dt} = \frac{d_i S}{dt} = \sum_k \delta F_k \, \delta J_k = \sum_k F_k J_k > 0 \qquad (14.1.15)$$

From this expression it is clear that the leading contribution to the entropy change due to a fluctuation in the equilibrium state is of second order and we may make this explicit by using $\delta^2 S/2$ in place of $\Delta_i S$. In terms of $\delta^2 S/2$, equations (14.1.14) and (14.1.15) can be written as

$$\boxed{\delta^2 S < 0} \qquad \boxed{\frac{1}{2}\frac{d\delta^2 S}{dt} = \sum_k \delta F_k \, \delta J_k > 0} \qquad (14.1.16)$$

in which the second equation represents the Second Law. These two equations express the essence of stability of the equilibrium state: the fluctuations decrease the entropy whereas the irreversible processes restore the system to its initial state. These equations are specific cases of a more general theory of stability formulated by Lyapunov which we will discuss in Chapters 17 and 18.

14.2 Thermodynamic Theory of Fluctuations

THE PROBABILITY DISTRIBUTION

In the previous sections we have discussed stability of a thermodynamic state in the face of fluctuations. But the theory that we presented does not give us the probability for a fluctuation of a given magnitude. To be sure, our experience tells us that fluctuations in thermodynamic quantities are extremely small in macroscopic systems except near critical points; still we would like to have a theory that relates these fluctuations to thermodynamic quantities and gives us the conditions under which they become important.

In an effort to understand the relation between microscopic behavior of matter, which was in the realm of mechanics, and macroscopic laws of thermodynamics, Ludwig Boltzmann (1844–1906) introduced his famous relation that related entropy and probability (Box 3.1):

$$\boxed{S = k_B \ln W} \qquad (14.2.1)$$

in which $k_B = 1.38 \times 10^{-23} \mathrm{J\,K^{-1}}$ is the Boltzmann constant and W is the number of microscopic states corresponding to the macroscopic thermodynamic state. The variable W is called **thermodynamic probability** (as suggested by Max Planck) because, unlike the usual probability, it is a number larger than one—in fact it is a very large number. Thus, Boltzmann introduced the idea of probability into thermodynamics—a controversial idea whose true meaning could only be understood through the modern theories of unstable dynamical systems [3].

Albert Einstein (1879–1955) proposed a formula for the probability of a fluctuation in thermodynamic quantities by using Boltzmann's idea in reverse; whereas Boltzmann used "microscopic" probability to derive thermodynamic entropy, Einstein used thermodynamic entropy to obtain the probability of a fluctuation through the following relation:

$$\boxed{P(\Delta S) = Z e^{\Delta S/k_B}}\tag{14.2.2}$$

in which ΔS is the entropy change associated with the fluctuation from the state of equilibrium, and Z is a normalization constant that ensures the sum of all probabilities equals one. Though relations (14.2.1) and (14.2.2) may be mathematically close, it is important to note that conceptually one is the opposite of the other. In (14.2.1) the probability of a state is the fundamental quantity and entropy is derived from it; in (14.2.2) entropy as defined in thermodynamics is the fundamental quantity and the probability of a fluctuation is derived from it. Thermodynamic entropy also gives us the probability of fluctuations.

To obtain the probability of a fluctuation, we must obtain the entropy change associated with it (Fig. 14.2). The basic problem then is to obtain the ΔS in terms of the fluctuations δT, δp, etc. But this has already been done in the previous sections. Expression (14.1.13) gives us the entropy associated with a fluctuation:

$$\Delta S = -\frac{C_V(\delta T)^2}{2T^2} - \frac{1}{2T\kappa_T}\frac{(\delta V)^2}{V} - \sum_{i,j}\left(\frac{\partial}{\partial N_j}\frac{\mu_i}{T}\right)\frac{\delta N_i \, \delta N_j}{2}\tag{14.2.3}$$

This expression can be made more explicit if the derivative of the chemical potential is expressed in terms of the mole numbers. For ideal gases this can easily be done because the chemical potential of a component k is

$$\mu_k = \mu_{k0}(T) + RT \ln (p_k/p_0)$$
$$= \mu_{k0}(T) + RT \ln (N_k RT/Vp_0)\tag{14.2.4}$$

in which p_0 is the standard pressure (usually 1 bar). Substituting this expression into (14.2.3), we obtain

$$\boxed{\Delta S = -\frac{C_V(\delta T)^2}{2T^2} - \frac{1}{T\kappa_T}\frac{(\delta V)^2}{2V} - \sum_{i}\frac{R(\delta N_i)^2}{2N_i}}\tag{14.2.5}$$

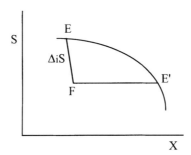

Figure 14.2 The entropy change ΔS associated with a fluctuation. Entropy S is shown as a function of a thermodynamic variable X. The reference equilibrium state is denoted by E. The fluctuation which results in a decrease in entropy drives the system to the point F. We compute the entropy change ΔS associated with the fluctuation by computing the entropy produced, $\Delta_i S$, as the system relaxes back to the equilibrium state. In classical formalisms that do not use $d_i S$, the entropy change is calculated by first determining an equilibrium state E′ that has the same entropy as the state F and then considering a reversible path along the equilibrium trajectory E′E

Here C_V is the heat capacity of the ideal gas mixture. In this expression the N_i are expressed in moles. Through multiplying them by the Avogadro number N_A, they can be converted to *numbers of molecules* $\tilde{N}_i$ (note $k_B N_A = R$). The same expression can be derived for an ideal system for which $\mu_k = \mu_{k0}(T) + RT \ln x_k$, in which x_k is the mole fraction (exc. 14.2). Now using the Einstein formula (14.2.2), the probability of a fluctuation in T, V and $\tilde{N}_i$ can be written as

$$P(\delta T, \delta V, \delta \tilde{N}_i) = Z \exp(\delta S / k_B)$$

$$= Z \exp\left[-\frac{C_V (\delta T)^2}{2k_B T^2} - \frac{1}{2k_B T \kappa_T} \frac{(\delta V)^2}{V} - \sum_i \frac{(\delta \tilde{N}_i)^2}{2\tilde{N}_i} \right]$$

$$(14.2.6)$$

in which we have replaced R by $k_B N_A$. In this expression the normalization factor Z is defined as

$$\frac{1}{Z} = \int \int \int P(x, y, z) dx\, dy\, dz \qquad (14.2.7)$$

The probability distribution is a Gaussian in each of the independent variables δT, δV, $\delta \tilde{N}_k$. The integral of a Gaussian is given by

$$\int_{-\infty}^{\infty} e^{-x^2/a}\, dx = \sqrt{\pi a} \qquad (14.2.8)$$

With this formula, the probability can be explicitly written and the root mean square value of the fluctuations can be obtained (exc. 14.3). A more general form of the probability distribution can be obtained from (14.1.14) in which the change of entropy due to fluctuation is expressed in terms of products of pairs of variables:

$$P = Z \exp\left[\frac{\delta^2 S}{2 k_B}\right] = Z \exp\left[\frac{-1}{2 k_B T}\left(\delta T\, \delta S - \delta p\, \delta V + \sum_k \delta \mu_k\, \delta N_k\right)\right] \qquad (14.2.9)$$

in which Z is the normalization constant. For any set of independent variables Y_k, equation (14.2.9) can be used to obtain the probability distribution for the fluctuation of these variables by expressing δT, δS, etc., in terms of the fluctuations in Y_k. For notational convenience, let us denote the deviations of the independent variables Y_k from their equilibrium values by α_k. Then, in general, $\delta^2 S$ will be a quadratic function in α_k:

$$\frac{\delta^2 S}{2} = -\frac{1}{2} \sum_{i,j} g_{ij} \alpha_i \alpha_j \qquad (14.2.10)$$

in which the g_{ij} are appropriate coefficients; the negative sign is introduced only to emphasize the fact that $\delta^2 S$ is a negative quantity. For a single α, the probability $P(\alpha) = \sqrt{(g/2\pi k_B)} \exp(-g\alpha^2/2 k_B)$. In the more general case, the corresponding probability distribution can be written explicitly as

$$P(\alpha_1, \alpha_2, \ldots, \alpha_m) = \sqrt{\frac{\det [\mathbf{g}]}{(2\pi k_B)^m}} \exp\left[-\frac{1}{2 k_B} \sum_{i,j=1}^{m} g_{ij} \alpha_i \alpha_j\right] \qquad (14.2.11)$$

in which $\det[\mathbf{g}]$ is the determinant of the matrix g_{ij}. In the rest of this section we shall derive some important general results that we will use in Chapter 16 for deriving a set of fundamental relations in nonequilibrium thermodynamics, the Onsager reciprocal relations.

AVERAGE VALUES AND CORRELATIONS

In general, given a probability distribution for a set of variables α_k, one can compute average values and correlations between pairs of variables. We shall use the notation $\langle f \rangle$ to denote the average value of any function $f(\alpha_1, \alpha_2, \ldots, \alpha_m)$ of the variables α_k; it is computed using the integral

$$\langle f \rangle = \int f(\alpha_1, \alpha_2, \ldots, \alpha_m) P(\alpha_1, \alpha_2, \ldots, \alpha_m) d\alpha_1 \, d\alpha_2 \ldots d\alpha_m \qquad (14.2.12)$$

The correlation between two variables f and g is defined as

$$\langle fg \rangle = \int f(\alpha_1, \alpha_2, \ldots, \alpha_m) g(\alpha_1, \alpha_2, \ldots, \alpha_m) P(\alpha_1, \alpha_2, \ldots, \alpha_m) d\alpha_1 \, d\alpha_2 \ldots d\alpha_m$$

$$(14.2.13)$$

We shall soon give a general expression for the correlation $\langle \alpha_i \alpha_j \rangle$ between any two variables α_i and α_j, but first we shall calculate other correlation functions that will lead us to the result.

We have seen earlier that the entropy production (14.1.15) associated with a small fluctuation from equilibrium can be written as

$$\frac{d\Delta_i S}{dt} = \frac{1}{2}\frac{d\delta^2 S}{dt} = \sum_k F_k J_k \qquad (14.2.14)$$

in which the F_k are thermodynamic forces that drive the flows $J_k = dX_k/dt$, both of which vanish at equilibrium. Now if we compute the time derivative of the general quadratic expression (14.2.10) since $g_{ij} = g_{ji}$, we obtain

$$\frac{1}{2}\frac{d\delta^2 S}{dt} = -\sum_{i,j} g_{ij}\alpha_i\frac{d\alpha_j}{dt} \qquad (14.2.15)$$

If we identify the derivative $(d\alpha_j/dt)$ as a "thermodynamic flow" (similar to δJ_j close to equilibrium), comparison of (14.2.14) and (14.2.15) shows that the term

$$F_j \equiv -\sum_i g_{ij}\alpha_i \qquad (14.2.16)$$

will be the corresponding thermodynamic force. Furthermore, due to the Gaussian form of the probability distribution (14.2.11), we also have the relation

$$F_i = k_B \frac{\partial \ln P}{\partial \alpha_i} \tag{14.2.17}$$

We will first show that

$$\boxed{\langle F_i \alpha_j \rangle = -k_B \delta_{ij}} \tag{14.2.18}$$

in which δ_{ij} is the "Kronecker delta": $\delta_{ij} = 0$ if $i \neq j$ and $\delta_{ij} = 1$ if $i = j$. This shows that each fluctuation is correlated to its corresponding force but not to other forces. By definition

$$\langle F_i \alpha_j \rangle = \int F_i \alpha_j P \, d\alpha_1 \, d\alpha_2 \ldots d\alpha_m$$

Using (14.2.17) this integral can be written as

$$\langle F_i \alpha_j \rangle = \int k_B \left(\frac{\partial \ln P}{\partial \alpha_i} \right) \alpha_j P \, d\alpha_1 \, d\alpha_2 \ldots d\alpha_m$$

$$= \int k_B \left(\frac{\partial P}{\partial \alpha_i} \right) \alpha_j \, d\alpha_1 \, d\alpha_2 \ldots d\alpha_m$$

which on integration by parts gives

$$\langle F_i \alpha_j \rangle = k_B P \alpha_j \Big|_{-\infty}^{+\infty} - k_B \int \left(\frac{\partial \alpha_j}{\partial \alpha_i} \right) P \, d\alpha_1 \, d\alpha_2 \ldots d\alpha_m$$

The first term vanishes because $\lim_{\alpha \to \pm\infty} \alpha_j P(\alpha_j) = 0$. The second term vanishes if $i \neq j$ and equals $-k_B$ if $i = j$. Thus we arrive at the result (14.2.18).

Also useful is the following general result. By substituting (14.2.16) into (14.2.18) we arrive at

$$\langle F_i \alpha_j \rangle = \left\langle -\sum_k g_{ik} \alpha_k \alpha_j \right\rangle = -\sum_k g_{ik} \langle \alpha_k \alpha_j \rangle = -k_B \delta_{ij}$$

which simplifies to

$$\sum_k g_{ik} \langle \alpha_k \alpha_j \rangle = k_B \delta_{ij} \tag{14.2.19}$$

This implies that the matrix $\langle \alpha_k \alpha_j \rangle / k_B$ is the inverse of the matrix g_{ik}:

$$\boxed{\langle \alpha_i \alpha_j \rangle = k_B (g^{-1})_{ij}} \tag{14.2.20}$$

One particularly interesting result is the average value of the entropy fluctuations associated with the m independent variables α_i:

$$\langle \Delta_i S \rangle = \left\langle -\frac{1}{2} \sum_{i,j=1}^{m} g_{ij} \alpha_i \alpha_j \right\rangle = -\frac{1}{2} \sum_{i,j=1}^{m} g_{ij} \langle \alpha_i \alpha_j \rangle = -\frac{k_B}{2} \sum_{i,j=1}^{m} g_{ij} (g^{-1})_{ji}$$

$$= -\frac{k_B}{2} \sum_{i=1}^{m} \delta_{ii} = -\frac{m k_B}{2} \tag{14.2.21}$$

Thus, we see that the average value of entropy fluctuations due to m independent variables is given by the simple expression

$$\boxed{\langle \Delta_i S \rangle = -\frac{m k_B}{2}} \tag{14.2.22}$$

Each independent process that contributes to entropy is associated with a fluctuation $-k_B/2$ at equilibrium. This result is analogous to the equipartition theorem in statistical mechanics, which states that each degrees of freedom carries with it an average energy of $k_B T/2$.

As a simple example, let us consider entropy fluctuations due to r chemical reactions. As was shown in (14.1.9), we have

$$\Delta_i S_{\text{chem}} = \sum_{i,j}^{r} \frac{1}{2T} \left(\frac{\partial A_i}{\partial \xi_j} \right)_{\text{eq}} \delta \xi_i \, \delta \xi_j = -\frac{1}{2} \sum_{i,j}^{r} g_{ij} \, \delta \xi_i \, \delta \xi_j \tag{14.2.23}$$

in which we have made the identification $g_{ij} = -(1/T)(\partial A_i/\partial \xi_j)_{\text{eq}}$. From the general result (14.2.22) we see that the average value of the entropy fluctuations due to r chemical reaction is

$$\boxed{\langle \Delta_i S_{\text{chem}} \rangle = -r \frac{k_B}{2}} \tag{14.2.24}$$

This shows how fluctuations in ξ_i decrease entropy. In Chapter 16 we shall use (14.2.16) and (14.2.20) to derive the Onsager reciprocal relations.

References

1. Gibbs, J. W., *The Scientific Papers of J. Willard Gibbs, Vol. 1: Thermodynamics*, A. N. Editor (ed.). 1961, New York: Dover.
2. Callen, H. B., *Thermodynamics'* 2nd ed. 1985, New York: John Wiley.
3. Petrosky, T., and Prigogine, I., *Chaos, Solitons and Fractals*, **7** (1996), 441–497.

Exercises

14.1 By considering the change $\delta^2 F$, obtain the condition for stability with respect to thermal fluctuations when N_k and V are constant.

14.2 Obtain the expression

$$\Delta_i S = -\frac{C_V (\delta T)^2}{2T^2} - \frac{1}{T\kappa_T}\frac{(\delta V)^2}{2V} - \sum_i \frac{R(\delta N_i)^2}{2N_i}$$

for an ideal system, where $\mu_k = \mu_{k0}(T) + RT \ln x_k$.

14.3 (a) Evaluate the normalization constant Z for (14.2.6).
(b) Obtain the probability $P(\delta T)$ for the fluctuations of the one variable δT.
(c) Obtain average values for the square of the fluctuations by evaluating $\int_{-\infty}^{\infty} (\delta T)^2 P(\delta T) d(\delta T)$.

14.4 Obtain (14.2.17) from (14.2.11).

14.5 Consider an ideal gas at a temperature T and $p = 1$ atm. Assume this ideal gas has two components A and B in equilibrium with respect to interconversion, $A \rightleftharpoons B$. In a small volume δV, calculate the number of molecules that should convert from A to B to change the entropy by k_B in the considered volume. Equation (14.2.24) then gives the expected fluctuations.

PART IV

LINEAR NONEQUILIBRIUM THERMODYNAMICS

15 NONEQUILIBRIUM THERMODYNAMICS: THE FOUNDATIONS

15.1 Local Equilibrium

As emphasized earlier, we live in a world that is not in thermodynamic equilibrium. The 2.8 K thermal radiation that fills the universe is not in thermal equilibrium with the matter in the galaxies. On a smaller scale, the earth, its atmosphere, biosphere and the oceans are all in a nonequilibrium state due to the constant influx of energy from the sun. In the laboratory, most of the time we encounter phenomena exhibited by systems not in thermodynamic equilibrium, while equilibrium systems are the exception.

Yet, thermodynamics that describes equilibrium states is of great importance and extremely useful. This is because almost all systems are *locally* in thermodynamic equilibrium. For almost every macroscopic system we can meaningfully assign a temperature, and other thermodynamic variables to every "elemental volume" ΔV. In most situations we may assume that *equilibrium thermodynamic relations are valid for the thermodynamic variables assigned to an elemental volume*. This is the concept of **local equilibrium**. In the following paragraphs we shall make this concept of local equilibrium precise. When this is done, we have a theory in which all intensive thermodynamic variables T, p, μ, become functions of position $\mathbf{x}$ and time t:

$$T = T(\mathbf{x}, t) \qquad p = p(\mathbf{x}, t) \qquad \mu = \mu(\mathbf{x}, t)$$

The extensive variables are replaced by densities s, u and n_k:

$$s(\mathbf{x}, t) = \text{entropy per unit volume}$$
$$u(\mathbf{x}, t) = \text{energy per unit volume} \tag{15.1.1}$$
$$n_k(\mathbf{x}, t) = \text{mole number per unit volume of reactant } k$$

(In some formulations the extensive quantities are replaced by entropy, energy and volume per unit mass.) The Gibbs relation $dU = TdS - pdV + \sum_k \mu_k dN_k$ is assumed to be valid for small volume elements. With $U = Vu$ and $S = sV$ it follows that relations such as

$$\left(\frac{\partial u}{\partial s}\right)_{n_k} = T \qquad Tds = du - \sum_k \mu_k \, dn_k \tag{15.1.2}$$

for the densities are valid (exc. 15.1) at every position $\mathbf{x}$ and time t. In these equations the volume does not appear because s, u and n_k are densities. The entire system is viewed as a collection of systems characterized by different values of T, μ, etc., and interacting with each other.

Let us look at the physical conditions which make local equilibrium a valid assumption. First we must look at the concept of temperature. From statistical mechanics it can be seen that temperature is well defined when the velocity distribution is "Maxwellian." According to the **Maxwell distribution of velocities**, the probability $P(\mathbf{v})$ that a molecule has a velocity $\mathbf{v}$ is given by

$$\boxed{P(\mathbf{v})d^3\mathbf{v} = \left(\frac{\beta}{\pi}\right)^{3/2} e^{-\beta \mathbf{v}^2} d^3\mathbf{v}} \tag{15.1.3}$$

$$\beta = \frac{m}{2k_{\mathrm{B}}T} \tag{15.1.4}$$

The temperature is identified through relation (15.1.4), in which m is the mass of the molecule and k_{B} is the Boltzmann constant. In practice, only under very extreme conditions do we find significant deviations from the Maxwell distribution. Any initial distribution of velocities quickly becomes Maxwellian due to molecular collisions. Computer simulations of molecular dynamics have revealed that the Maxwell distribution is reached in less than 10 times the average time between collisions, which in a gas at a pressure of 1 atm is about 10^{-8} s [1]. Consequently, physical processes that perturb the system significantly from the Maxwell distribution have to be very rapid. A detailed statistical mechanical analysis of the assumption of local equilibrium can be found in [2].

Chemical reactions are of particular interest to us. In almost all reactions only a very small fraction of molecular collisions produce a chemical reaction. Collisions between molecules that produce a chemical reaction are called **reactive collisions**. For a gas at a pressure of 1 atm the collision frequency is 10^{31} collisions per litre per second. If nearly every collision produced a chemical reaction, the resulting rate would be of the order of $10^8 \mathrm{~mol\,L^{-1}\,s^{-1}}$! Reaction rates that approach such a large value are extremely rare. Most of the reaction rates we encounter indicate that reactive collision rates are several orders of magnitude smaller than overall collision rates. Between reactive collisions the system quickly relaxes to equilibrium, redistributing the change in energy due to the chemical reaction. In other words, any perturbation of the Maxwell distribution due to a chemical reaction quickly relaxes back to the Maxwellian with a slightly different temperature. Hence, on the timescale of chemical reactions, temperature is locally well defined. (Small corrections to the rate laws due to small deviations from the Maxwell distribution in highly exothermic reactions can be theoretically obtained [3–6]. These results have been found to agree well with the results of molecular dynamics simulations done on modern computers [7].)

Next, let us look at the sense in which thermodynamic variables, such as entropy and energy, may be considered functions of position. As we have seen in Chapters 12 and 14, every thermodynamic quantity undergoes fluctuations. For a small elemental volume ΔV we can meaningfully associate a value for a thermodynamic quantity Y only when the size of the fluctuations, e.g. the root mean square (rms) value, δY is very small compared to Y. Clearly, if ΔV is too small, this condition will not be satisfied. From (14.2.6) it follows that if $\tilde{N}$ is the number of particles in the considered volume, then the rms value of the fluctuations $\delta \tilde{N} = \tilde{N}^{1/2}$. As an example, let us consider an ideal gas for which $N = \tilde{N}/N_\mathrm{A} = (p/RT)\Delta V$. For a given ΔV it is easy to compute the relative value of the fluctuation $\delta \tilde{N}/\tilde{N} = 1/\tilde{N}^{1/2}$. To understand how small ΔV can be, we consider a gas at a pressure $p = 1$ atm and $T = 298$ K, and compute the fluctuations in the number of particles $\tilde{N}$ in a volume $\Delta V = (1 \, \mu\mathrm{m})^3 = 10^{-15}$ L. We find that $\delta \tilde{N}/\tilde{N} \approx 4 \times 10^{-7}$. For liquids and solids the same value of $\delta \tilde{N}/\tilde{N}$ will correspond to an even smaller volume. Hence it is meaningful to assign a mole number density to a volume with a characteristic size of a micrometer. The same is generally true for other thermodynamic variables. If we are to assign a number density to a volume ΔV, then the number density in this volume should be nearly uniform. This means that the variation of number density with position on the scale of a micrometer should be very nearly zero, a condition satisfied by most macroscopic systems. This shows that a theory based on local equilibrium is applicable to a wide range of macroscopic systems.

EXTENDED THERMODYNAMICS

In the above approach, an implicit assumption is that the thermodynamic quantities do not depend on the gradients in the system, i.e. it is postulated that entropy s is a function of the temperature T and the mole numbers density n_k, but not their gradients. However, flows represent a level of organization. This implies that the local entropy in a nonequilibrium system may be expected to be smaller than the equilibrium entropy. In the recently developed formalism of **extended thermodynamics**, gradients are included in the basic formalism and there appears a small correction to the local entropy due to the flows. We shall not be discussing this more advanced formalism. For a detailed exposition of extended thermodynamics, we refer the reader to some recent books [8–11]. Extended thermodynamics finds application in systems where there are large gradients, such as in shock waves. For almost all systems that we encounter, thermodynamics based on local equilibrium has excellent validity.

15.2   Local Entropy Production

As we noted in the previous section, the Second Law of thermodynamics must be a local law. If we divide a system into r parts, then not only is

$$d_i S = d_i S^1 + d_i S^2 + \ldots + d_i S^r \geq 0 \qquad (15.2.1)$$

in which $d_i S^k$ is the entropy production in the kth part, but also

$$d_i S^k \geq 0 \qquad (15.2.2)$$

for every k. Clearly, this statement that the entropy production due to irreversible processes is positive in every part is stronger than the classical statement of the Second Law that the entropy of an isolated system can only increase or remain unchanged.[*] And note that the Second Law as stated by (15.2.2) does not require the system to be isolated. *It is valid for all systems, regardless of the boundary conditions.*

The local increase of entropy in continuous systems can be defined by using the entropy density $s(\mathbf{x}, t)$. As was the case for the total entropy, $ds = d_i s + d_e s$, with $d_i s \geq 0$. We define local entropy production as

$$\boxed{\sigma(\mathbf{x}, t) \equiv \frac{d_i s}{dt} \geq 0} \qquad (15.2.3)$$

$$\frac{d_i S}{dt} = \int_V \sigma(\mathbf{x}, t) dV \qquad (15.2.4)$$

Nonequilibrium thermodynamics is founded on the explicit expression for σ in terms of the irreversible processes that we can identify and study experimentally. Before we begin deriving this expression, however, we shall write the explicit local forms of balance equations for energy and concentrations.

---

[*] One general point to note about the First Law and the Second Law is that both laws must be *local laws*. In fact, to be compatible with the principle of relativity, and to be valid regardless of the observer's state of motion, these laws *must* be local. Nonlocal laws of energy conservation or of entropy production are inadmissible because the notion of simultaneity is relative. Consider two parts of a system spatially separated by some nonzero distance. If changes in energy δu_1 and δu_2 occur in these two parts *simultaneously* in one frame of reference so that $\delta u_1 + \delta u_2 = 0$, the energy is conserved. However, in another frame of reference that is in motion with respect to the first, the two changes in energy *will not occur simultaneously*. Thus, during the time between one change of u and the other, the law of conservation of energy will be violated. Similarly, the entropy changes in a system, δS_1 and δS_2 at two spatially separated parts of a system must be independently positive. It is inadmissible to have the simultaneous decrease of one and increase of the other so that their sum is positive.

Box 15.1 Differential Form of the Balance Equation

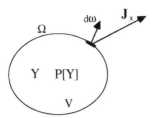

Consider a quantity Y whose density is denoted by y. The change in the amount of Y in a volume V is the sum of the net flow of Y into the volume V and the production of Y within that volume. If $\mathbf{J}_Y$ is the current density (flow through unit area that is perpendicular to $\mathbf{J}_Y$ per unit time), then *the change in Y due to the flow* $= \int_\Omega \mathbf{J}_Y \cdot d\boldsymbol{\omega}$ in which $d\boldsymbol{\omega}$ is the vector representing an area element, as illustrated. The magnitude of $d\boldsymbol{\omega}$ equals the area of the element; its direction is perpendicular to the area *pointing outwards*. If $P[Y]$ is the amount of Y produced per unit volume per unit time, we have *the change in Y due to production* $= \int_V P[Y]dV$. Then the balance equation for the change in Y in the considered volume can be written as

$$\int_V \left(\frac{\partial y}{\partial t}\right) dV = \int_V P[Y] dV - \int_\Omega \mathbf{J}_Y \cdot d\boldsymbol{\omega}$$

The negative sign in the second term is because $d\boldsymbol{\omega}$ points outwards.

According to Gauss' theorem, for any vector field $\mathbf{J}$:

$$\int_\Omega \mathbf{J} \cdot d\boldsymbol{\omega} = \int_V (\nabla \cdot \mathbf{J}) dV$$

Applying this theorem to the surface integral of $\mathbf{J}_Y$ in the balance equation, we see that

$$\int_V \left(\frac{\partial y}{dt}\right) dV = \int_V P[Y] dV - \int_V (\nabla \cdot \mathbf{J}_Y) dV$$

Since this relation should be valid for an any volume, we can equate the integrands. This gives us the differential form of the balance equation for Y:

$$\left(\frac{\partial y}{\partial t}\right) + (\nabla \cdot \mathbf{J}_Y) = P[Y]$$

15.3 Balance Equation for Concentration

The mole-number balance equation can easily be obtained using the general balance equation described in Box 15.1. The changes in n_k (mole number per unit volume) are due to transport of particles, through processes such as diffusion and convection, $d_e n_k$, and due to chemical reactions, $d_i n_k$; the total change $dn_k = d_e n_k + d_i n_k$. Denoting by $\mathbf{v_k}(x, t)$ the velocity of the kth component at location x at time t, the balance equation can be written as

$$\frac{\partial n_k}{\partial t} = \frac{\partial_e n_k}{\partial t} + \frac{\partial_i n_k}{\partial t} = -\nabla \cdot (n_k \mathbf{v}_k) + P[n_k] \qquad (15.3.1)$$

in which $P[n_k]$ is the production (which could be positive or negative) of the kth component per unit volume per unit time due to chemical reactions. As we have seen in Chapter 9, if ν_k is the stoichiometric coefficient of the reactant k in a particular reaction, the number of moles of k produced per unit time per unit volume is given by $\nu_k (1/V)(d\xi/dt)$ in which ξ is the extent of reaction. If there are several reactions, we can identify them by a subscript j. The velocity of the jth reaction is

$$v_j = \frac{1}{V} \frac{d\xi_j}{dt} \qquad (15.3.2)$$

The velocities of reaction v_j are specified by empirical laws, as discussed in Chapter 9. The production of component k can now be expressed in terms of the reaction velocities v_j and the corresponding stoichiometric coefficients ν_{jk}:

$$P[n_k] \equiv \sum_j \nu_{jk} v_j \qquad (15.3.3)$$

The equation for particle-number balance can now be written as

$$\frac{\partial n_k}{\partial t} = \frac{\partial_e n_k}{\partial t} + \frac{\partial_i n_k}{\partial t} = -\nabla \cdot n_k \mathbf{v}_k + \sum_j \nu_{jk} v_j \qquad (15.3.4)$$

Convective flow is the motion of the center of mass whereas the flow with respect to the center of mass accounts for transport, such as diffusion, that is apart from convection. The center of mass velocity $\mathbf{v}$ is given by

$$\boxed{\mathbf{v} \equiv \frac{\sum_k M_k n_k \mathbf{v}_k}{\sum_k M_k n_k}} \qquad (15.3.5)$$

in which M_k is the molar mass of component k. The nonconvective flow or **diffusion flow** $\mathbf{J}_k$ of component k is then defined as*

$$\mathbf{J}_k = n_k(\mathbf{v}_k - \mathbf{v}) \qquad (15.3.6)$$

---

* In the thermodynamics of superfluids it is more convenient to keep the motions of the components separate. Also, diffusion flow with respect to mean volume velocity, defined by replacing M_k in (15.3.5) with specific volume, is also used.

The convectional and diffusional parts of the flow can be made explicit by using (15.3.6) in (15.3.4):

$$\frac{\partial n_k}{\partial t} = -\nabla \cdot \mathbf{J}_k - \nabla \cdot (n_k \mathbf{v}) + \sum_j v_{jk} v_j \qquad (15.3.7)$$

By virtue of their definitions, the nonconvective flows $\mathbf{J}_k$ must obey the relation

$$\boxed{\sum_k M_k \mathbf{J}_k = 0} \qquad (15.3.8)$$

i.e. these flows should not result in center of mass motion. Thus the $\mathbf{J}_k$ are not all independent, a point that we shall return to in Chapter 16 while considering coupling between different flows. Also, based on the definition of $d_e n_k$ and $d_i n_k$, we can make the identification

$$\frac{\partial_e n_k}{\partial t} = -\nabla \cdot \mathbf{J}_k - \nabla \cdot (n_k \mathbf{v}) \quad \text{and} \quad \frac{\partial_i n_k}{\partial t} = \sum_j v_{jk} v_j \qquad (15.3.9)$$

In the absence of convection, the flow is entirely $\mathbf{J}_k$. We then have

$$\boxed{\frac{\partial n_k}{\partial t} = -\nabla \cdot \mathbf{J}_k + \sum_j v_{jk} v_j} \qquad (15.3.10)$$

In the presence of an external field, such as a static electric field, $\mathbf{J}_k$ may have a part that depends on the field. When no field is present, $\mathbf{J}_k$ is entirely due to diffusion; in Chapters 18 and 19 we shall study such *diffusion–reaction systems* under far-from-equilibrium conditions in some detail.

15.4   Energy Conservation in Open Systems

In Chapter 2 we saw the foundations of the concept of energy and its conservation. We have also noted how this conservation law must be local. We can express the local form of the law in a differential form. The total energy density e is a sum of the kinetic and the internal energies:

$$\boxed{e = \frac{1}{2} \sum_k (M_k n_k) \mathbf{v}_k^2 + u} \qquad (15.4.1)$$

in which $(M_k n_k)$ is the mass per unit volume and $\mathbf{v}_k$ is the velocity of component k. Equation (15.4.1) may be considered as the definition of the internal energy u, i.e. energy not associated with bulk motion. Using the center of mass velocity $\mathbf{v}$

defined by (15.3.5), also called the "barycentric velocity," equation (15.4.1) can be written as

$$e = \frac{\rho}{2}\mathbf{v}^2 + \frac{1}{2}\sum_k (M_k n_k)(\mathbf{v}_k - \mathbf{v})^2 + u \qquad (15.4.2)$$

in which the density $\rho = \sum_k M_k n_k$. The term $\frac{1}{2}\sum_k (M_k n_k)(\mathbf{v}_k - \mathbf{v})^2$ is sometimes referred to as the **kinetic energy of diffusion** [12]. Thus the total energy density is the sum of the kinetic energy associated with convection and diffusion, and the internal energy. In some formalisms, the sum of the last two terms is defined as the internal energy [12], in which case the internal energy includes the kinetic energy of diffusion.

When an external field is present, the energy of interaction $\sum_k \tau_k n_k \psi$, in which τ_k is the "coupling constant" and ψ is the potential, should also be considered. This energy can be introduced either as an additional term in (15.4.1) [12] or assumed to be included in the definition of u. Following our formalism in Chapters 2 and 10, we shall assume that the term $\sum_k \tau_k n_k \psi$ is included in the definition of the internal energy u.

Since energy is conserved, there is no source term in the balance equation. Therefore the formal differential form for the conservation of energy is

$$\boxed{\frac{\partial e}{\partial t} + \nabla \cdot \mathbf{J}_e = 0} \qquad (15.4.3)$$

in which $\mathbf{J}_e$ is the energy current density. In order to make this expression more explicit in terms of the processes in the system, we begin by looking at the change in u. Being a function of T and n_k, the change in the energy density $u(T, n_k)$ is

$$du = \left(\frac{\partial u}{\partial T}\right)_{n_k} dT + \sum_k \left(\frac{\partial u}{\partial n_k}\right)_T dn_k$$
$$= c_V\, dT + \sum_k u_k\, dn_k \qquad (15.4.4)$$

in which $u_k \equiv (\partial u / \partial n_k)_T$ is the partial molar energy of the kth component and c_V is the constant-volume heat capacity per unit volume. For the time variation of the internal energy density we can write

$$\frac{\partial u}{\partial t} = c_V \frac{\partial T}{\partial t} + \sum_k u_k \frac{\partial n_k}{dt} \qquad (15.4.5)$$

Using the mole-number balance equation (15.3.10), we can rewrite this equation as

$$\frac{\partial u}{\partial t} = c_V \frac{\partial T}{\partial t} + \sum_{kj} u_k \nu_{jk} v_j - \sum_k u_k \nabla \cdot \mathbf{J}_k \tag{15.4.6}$$

The quantity $\sum_k u_k \nu_{jk} = \sum_k (\partial u/\partial n_k)_T \nu_{jk}$ is the change in the internal energy per unit volume, at constant T, due to a chemical reaction. It is the heat of reaction of the jth reaction at constant volume and temperature; we shall denote it by $(r_{V,T})_j$. For exothermic reactions $(r_{V,T})_j$ is negative. Furthermore, to relate (15.4.6) to the conservation equation (15.4.3), we can make use of the identity $u_k \nabla \cdot \mathbf{J}_k = \nabla \cdot (u_k \mathbf{J}_k) - \mathbf{J}_k \cdot (\nabla u_k)$ and rewrite (15.4.6) as

$$\frac{\partial u}{\partial t} = c_V \frac{\partial T}{\partial t} + \sum_j (r_{V,T})_j v_j + \sum_k \mathbf{J}_k \cdot (\nabla u_k) - \sum_k \nabla \cdot (u_k \mathbf{J}_k) \tag{15.4.7}$$

Using (15.4.2) and (15.4.7) the energy conservation equation (15.4.3) can be more explicitly written as

$$\begin{aligned} \frac{\partial e}{\partial t} &= c_V \frac{\partial T}{\partial t} + \sum_j (r_{V,T})_j v_j + \sum_k \mathbf{J}_k \cdot (\nabla u_k) - \sum_k \nabla \cdot (u_k \mathbf{J}_k) \\ &\quad + \frac{\partial}{\partial t}(\mathrm{KE}) = -\nabla \cdot \mathbf{J}_e \end{aligned} \tag{15.4.8}$$

in which (KE) is the kinetic energy associated with convection and diffusion:

$$(\mathrm{KE}) \equiv \left(\frac{\rho}{2} \mathbf{v}^2 + \frac{1}{2} \sum_k M_k n_k (\mathbf{v}_k - \mathbf{v})^2 \right) \tag{15.4.9}$$

The energy flow $\mathbf{J}_e$ can now be identified by *defining* a **heat flow $\mathbf{J}_q$**:

$$\boxed{-\nabla \cdot \mathbf{J}_q \equiv c_V \frac{\partial T}{\partial t} + \sum_j (r_{V,T})_j v_j + \sum_k \mathbf{J}_k \cdot (\nabla u_k) + \frac{\partial}{\partial t}(\mathrm{KE})} \tag{15.4.10}$$

Finally, substituting (15.4.10) into (15.4.8), we can identify the energy flow as

$$\mathbf{J}_e = \mathbf{J}_q + \sum_k u_k \mathbf{J}_k \tag{15.4.11}$$

The definition of heat flow (15.4.10) leads to a physically appropriate interpretation of the processes that change the internal energy and the temperature. Using (15.4.7) in (15.4.10) we obtain

$$-\nabla \cdot \mathbf{J}_q = \frac{\partial u}{\partial t} + \nabla \cdot \left(\sum_k u_k \mathbf{J}_k \right) + \frac{\partial}{\partial t}(\mathrm{KE})$$

which can be rewritten as

$$\boxed{\frac{\partial u}{\partial t} + \nabla \cdot \mathbf{J}_u = -\frac{\partial}{\partial t}(\mathrm{KE})} \qquad (15.4.12)$$

where

$$\boxed{\mathbf{J}_u = \sum_k (u_k \mathbf{J}_k + \mathbf{J}_q)} \qquad (15.4.13)$$

This is the balance equation for the internal energy. It shows that the internal energy as defined above may be associated with a flow $\mathbf{J}_u = (\sum_k u_k \mathbf{J}_k + \mathbf{J}_q)$ and a source term on the right-hand side of (15.4.12), which is the rate at which the kinetic energy is dissipated. Equation (15.4.12) shows that the changes in u are a result of heat flow $\mathbf{J}_q$, matter flow $u_k \mathbf{J}_k$, and due to dissipation of kinetic energy of the bulk motion. The dissipation of kinetic energy can be related to viscous forces in the fluid.

The definition of heat flow (15.4.10) also gives an equation for the change in temperature

$$c_V \frac{\partial T}{\partial t} + \nabla \cdot \mathbf{J}_q = \sigma_{\text{heat}} \qquad (15.4.14)$$

$$\sigma_{\text{heat}} = -\sum_j (r_{V,T})_j \, v_j - \sum_{k,i} \mathbf{J}_k \cdot (\nabla u_k) - \frac{\partial}{\partial t}(\mathrm{KE}) \qquad (15.4.15)$$

Equation (15.4.14) is an extension of the Fourier equation for heat transport with the addition of a heat source σ_{heat}. It is useful to note that the term $\nabla u_k = \sum_i (\partial u_k / \partial n_i) \nabla n_i + (\partial u_k / \partial T) \nabla T$. For ideal systems, in the absence of temperature gradients, since the partial molar energy u_k is independent of n_k, this term will vanish; it is the heat generated or absorbed due to molecular interaction when the number density of a nonideal system changes. In the following chapters we shall not consider systems with convection. In addition, we will only consider situations in which the kinetic energy of diffusion remains small, so the term $\partial(\mathrm{KE})/\partial t = 0$.

Definition (15.4.10) of $\mathbf{J}_q$ is one of the many equivalent ways of defining the heat flow. Depending on the particular physical conditions and the experimental quantities that are measured, different definitions of $\mathbf{J}_q$ are used. A more extensive discussion of this aspect may be found in the literature [12]. The various definitions of $\mathbf{J}_q$, of course, give the same physical results.

When an external field is present, as noted earlier (Chapter 10), the energy of interaction $\sum_k \tau_k n_k \psi$, in which τ_k is the "coupling constant" and ψ is the potential, should also be included in u so that

$$u(T, n_k) = u^0(T, n_k) + \sum_k n_k \tau_k \psi \qquad (15.4.16)$$

where $u^0(T, n_k)$ is the energy density in the absence of the field. For an electric field $\tau_k = Fz_k$, where F is the Faraday constant and z_k is the ion number; ψ is the electrical potential ϕ. For a gravitational field $\tau_k = M_k$, the molar mass; ψ is the gravitational potential. For the time derivative of u, in place of (15.4.5) we now have

$$\frac{\partial u}{\partial t} = c_V \frac{\partial T}{\partial t} + \sum_k (u_k^0 + \tau_k \psi) \frac{\partial n_k}{dt} \qquad (15.4.17)$$

in which $u_k^0 = (\partial u^0 / \partial n_k)_T$. Equation (15.4.17) differs from (15.4.5) only in that the term u_k is replaced by $(u_k^0 + \tau_k \psi)$. This means that the corresponding expressions for $\mathbf{J}_q$ and $\mathbf{J}_e$ can be obtained by simply replacing u_k with $(u_k^0 + \tau_k \psi)$. Thus we arrive at the conservation equation

$$\boxed{\frac{\partial e}{\partial t} + \nabla \cdot \mathbf{J}_e^\psi = 0} \qquad (15.4.18)$$

In which

$$\boxed{\mathbf{J}_e^\psi = \mathbf{J}_q + \sum_k (u_k^0 + \tau_k \psi) \mathbf{J}_k} \qquad (15.4.19)$$

In this case the heat current is defined by

$$-\nabla \cdot \mathbf{J}_q \equiv c_V \frac{\partial T}{\partial t} + \sum_j (r_{V,T})_j v_j + \sum_k \mathbf{J}_k \cdot (\nabla u_k) + \frac{\partial}{\partial t}(\text{KE}) + \sum_k \tau_k \mathbf{J}_k \cdot \nabla \psi \qquad (15.4.20)$$

Comparing (15.4.20) with (15.4.10) we see the following. In the last term, $\nabla \psi$ is the negative of the field strength. In the case of an electric field, the last term becomes $-\mathbf{I} \cdot \mathbf{E}$ in which $\mathbf{E} = -\nabla \psi$ is the electric field and $\mathbf{I} = \sum_k \tau_k \mathbf{J}_k$ is the total current density; $\mathbf{I} \cdot \mathbf{E}$ is the Ohmic heat produced by an electric current. For the balance equation of u, in place of (15.4.12) we have

$$\boxed{\frac{\partial u}{\partial t} + \nabla \cdot \mathbf{J}_u = -\frac{\partial}{\partial t}(\text{KE}) + \mathbf{I} \cdot \mathbf{E}} \qquad (15.4.21)$$

in which $\mathbf{J}_u = \sum_k u_k^0 \mathbf{J}_k + \mathbf{J}_q$. Similarly (15.4.14) is modified such that the source of heat will now contain an additional term due to the Ohmic heat

$$c_V \frac{\partial T}{\partial t} + \nabla \cdot \mathbf{J}_q = \sigma_{\text{heat}} \tag{15.4.22}$$

$$\sigma_{\text{heat}} = -\sum_j (r_{V,T})_j v_j - \sum_k \mathbf{J}_k \cdot (\nabla u_k) - \frac{\partial}{\partial t}(\text{KE}) + \mathbf{I} \cdot \mathbf{E} \tag{15.4.23}$$

In this text we will only consider systems in mechanical equilibrium in which the kinetic energy of diffusion is small.

15.5 The Entropy Balance Equation

The balance equation for entropy can be derived using the conservation of energy and the balance equation for the concentrations. This gives us an explicit expression for entropy production σ—which can be related to irreversible processes such as heat conduction, diffusion and chemical reactions—and the entropy current $\mathbf{J}_S$. The formal entropy balance equation is

$$\boxed{\frac{\partial s}{\partial t} + \nabla \cdot \mathbf{J}_S = \sigma} \tag{15.5.1}$$

To obtain the explicit forms of $\mathbf{J}_S$ and σ, we proceed as follows. For simplicity, we shall consider a system with no dissipation of kinetic energy due to convection or diffusion, and no external field. From the Gibbs relation $Tds = du - \sum \mu_k \, dn_k$ it follows that

$$\frac{\partial s}{\partial t} = \frac{1}{T}\frac{\partial u}{\partial t} - \sum_k \frac{\mu_k}{T}\frac{\partial n_k}{\partial t} \tag{15.5.2}$$

Now using the mole-number balance equation (15.3.10) and the balance equation (15.4.12) for the internal energy with $\partial(\text{KE})/\partial t = 0$, expression (15.5.2) can be written as

$$\frac{\partial s}{\partial t} = -\frac{1}{T}\nabla \cdot \mathbf{J}_u + \sum_k \frac{\mu_k}{T}\nabla \cdot \mathbf{J}_k - \sum_{k,j} \frac{\mu_k}{T} v_{jk} v_j \tag{15.5.3}$$

This equation can be simplified and written in the form (15.5.1) by making the following observations. First, the affinity of reaction j is

$$A_j = -\sum_k \mu_k v_{jk} \tag{15.5.4}$$

Second, if g is a scalar function and $\mathbf{J}$ is a vector, then

$$\nabla \cdot (g\mathbf{J}) = \mathbf{J} \cdot (\nabla g) + g(\nabla \cdot \mathbf{J}) \tag{15.5.5}$$

Through an elementary calculation using (15.5.4) and (15.5.5), we can rewrite (15.5.3) to arrive at the following equation for the entropy balance:

$$\boxed{\frac{\partial s}{\partial t} + \nabla \cdot \left(\frac{\mathbf{J}_u}{T} - \sum_k \frac{\mu_k \mathbf{J}_k}{T} \right) = \mathbf{J}_u \cdot \nabla \frac{1}{T} - \sum_k \mathbf{J}_k \cdot \nabla \frac{\mu_k}{T} + \sum_j \frac{A_j v_j}{T}} \tag{15.5.6}$$

By comparing this equation with (15.5.1), we can make the identifications

$$\boxed{\mathbf{J}_s = \left(\frac{\mathbf{J}_u}{T} - \sum_k \frac{\mu_k \mathbf{J}_k}{T} \right)} \tag{15.5.7}$$

and

$$\boxed{\sigma = \mathbf{J}_u \cdot \nabla \frac{1}{T} - \sum_k \mathbf{J}_k \cdot \nabla \frac{\mu_k}{T} + \sum_j \frac{A_j v_j}{T} \geq 0} \tag{15.5.8}$$

where we have emphasized the Second Law, $\sigma \geq 0$.

If we identify the heat current by $\mathbf{J}_u = \mathbf{J}_q + \sum_k u_k \mathbf{J}_k$, then using the relation $u_k = \mu_k + Ts_k$ (exc. 15.2) where $s_k = (\partial s / \partial n_k)_T$ is the partial molar entropy of component k, the entropy current $\mathbf{J}_s$ can be written as

$$\mathbf{J}_S = \left(\frac{\mathbf{J}_q}{T} + \sum_k \frac{u_k - \mu_k}{T} \mathbf{J}_k \right) = \left(\frac{\mathbf{J}_q}{T} + \sum_k s_k \mathbf{J}_k \right) \tag{15.5.9}$$

As was the case for the energy current, the expression for the entropy current consists of two parts, one due to heat flow and the other due to matter flow.

If an external field with potential ψ is included, from the Gibbs equation $Tds = du - \sum_k \mu_k dn_k - \sum_k \tau_k \psi dn_k$, it follows that

$$T\frac{\partial s}{\partial t} = \frac{\partial u}{\partial t} - \sum_k (\mu_k + \tau_k \psi) \frac{\partial n_k}{\partial t} \tag{15.5.10}$$

Comparing (15.5.2) with (15.5.10), we see that the only difference is that the chemical potential μ_k is replaced by the term $(\mu_k + \tau_k \psi)$. Correspondingly, the

Table 15.1   Table of thermodynamic forces and flows

	Force F_α	Flow (Current) J_α
Heat conduction	$\nabla \dfrac{1}{T}$	Energy flow $\mathbf{J}_u$
Diffusion	$-\nabla \dfrac{\mu_k}{T}$	Diffusion current $\mathbf{J}_k$
Electrical conduction	$\dfrac{-\nabla \phi}{T} = \dfrac{\mathbf{E}}{T}$	Ion current densities $\mathbf{I}_k$
Chemical reactions	$\dfrac{A_j}{T}$	Velocity of reaction $v_j = \dfrac{1}{V}\dfrac{d\xi_j}{dt}$

entropy current (15.5.7) and the entropy production (15.5.8) now become

$$\mathbf{J}_S = \left(\frac{\mathbf{J}_k^\psi}{T} - \sum_k \frac{(\tau_k \psi + \mu_k)}{T} \mathbf{J}_k \right) \tag{15.5.11}$$

$$\boxed{\sigma = \mathbf{J}_u \cdot \nabla \frac{1}{T} - \sum_k \mathbf{J}_k \cdot \nabla \left(\frac{\mu_k}{T} \right) + \frac{\mathbf{I} \cdot (-\nabla \psi)}{T} + \sum_j \frac{A_j v_j}{T}} \tag{15.5.12}$$

in which $\mathbf{J}_u = \mathbf{J}_q + \sum_k u_k^0 \mathbf{J}_k$, in which u_k^0 is the partial molar energy in the absence of the field, and $\mathbf{I} = \sum_k \tau_k \mathbf{J}_k$. For a static electric field $\mathbf{E}$, we have $-\nabla \psi = \mathbf{E}$ and $\mathbf{I} = \sum_k \mathbf{I}_k$ is the total current density.

Expression (15.5.12) for the entropy production is fundamental to nonequilibrium thermodynamics. It shows that entropy production σ has the bilinear form

$$\boxed{\sigma = \sum_\alpha F_\alpha J_\alpha} \tag{15.5.13}$$

of forces F_α and currents or flows J_α. It is through this expression that we identify the thermodynamic forces and the flows they drive. For example, the force $\nabla(1/T)$ drives the flow $\mathbf{J}_u$; the chemical affinities A_j drive the chemical reactions with velocities v_j. These forces and the corresponding flows are identified in Table 15.1. A transformation that leaves σ invariant and alternative forms of writing σ are discussed in Appendix 15.1.

Appendix 15.1. Entropy Production

TRANSFORMATION THAT LEAVES σ INVARIANT

The entropy production remains invariant under certain transformations. One theorem [13] is that *under mechanical equilibrium, σ is invariant under the* *transformation*

$$\mathbf{J}_k \rightarrow \mathbf{J}'_k = \mathbf{J}_k + \mathbf{V} n_k \qquad \text{(A15.1.1)}$$

in which the $\mathbf{J}_k$ are the matter currents, n_k is the concentration of component k and $\mathbf{V}$ is an arbitrary velocity. This statement implies that a uniform "drift velocity" imposed on all the components of the system leaves the entropy production unchanged.

To prove this theorem, we first obtain a relation that the chemical potentials must satisfy in a system at mechanical equilibrium. If $n_k \mathbf{f}_k$ is the force acting on component k, then for mechanical equilibrium we have

$$\sum_k n_k \mathbf{f}_k - \nabla p = 0 \qquad \text{(A15.1.2)}$$

This condition can be written in terms of the chemical potential using the Gibbs–Duhem equation:

$$s\,dT - dp + \sum_k n_k\,d\mu_k = 0 \qquad \text{(A15.1.3)}$$

Since

$$dp = (\nabla p) \cdot d\mathbf{r} \quad \text{and} \quad d\mu_k = (\nabla \mu_k) \cdot d\mathbf{r} \qquad \text{(A15.1.4)}$$

under isothermal conditions $(dT = 0)$, substituting (A15.1.4) into (A15.1.3) we obtain the relation

$$\nabla p = \sum_k n_k \nabla \mu_k \qquad \text{(A15.1.5)}$$

Using this expression, condition (A15.1.2) for mechanical equilibrium can now be written in terms of the chemical potential as

$$\sum_k (n_k \mathbf{f}_k - n_k \nabla \mu_k) = 0 \qquad \text{(A15.1.6)}$$

With this result, the invariance of entropy production σ under the transformation (A15.1.1) can be shown as follows. In the presence of an external force $\mathbf{f}_k$ per mole, acting on the component k, under isothermal conditions, the entropy production per unit volume (15.5.12) can be written by identifying $\mathbf{f}_k = -\tau_k \nabla \psi$. It takes the simple form

$$\sigma = \sum_k \frac{\mathbf{J}_k}{T} \cdot (\mathbf{f}_k - \nabla \mu_k) \qquad \text{(A15.1.7)}$$

The transformation (A15.1.1) implies that $\mathbf{J}_k = \mathbf{J}_k' - \mathbf{V}n_k$. If we substitute this expression into (A15.1.7), the entropy production becomes

$$\sigma = \sum_k \frac{\mathbf{J}_k'}{T} \cdot (\mathbf{f}_k - \nabla\mu_k) - \mathbf{V} \cdot \sum_k (n_k\mathbf{f}_k - n_k\nabla\mu_k) \qquad (A15.1.8)$$

Due to the condition for mechanical equilibrium (A15.1.6), the second summation on the right-hand side is zero. Thus we have the invariance theorem, according to which

$$\sigma = \sum_k \frac{\mathbf{J}_k}{T} \cdot (\mathbf{f}_k - \nabla\mu_k) = \sum_k \frac{\mathbf{J}_k'}{T} \cdot (\mathbf{f}_k - \nabla\mu_k) \qquad (A15.1.9)$$

for a transformation $\mathbf{J}_k \rightarrow \mathbf{J}_k' = \mathbf{J}_k + \mathbf{V}n_k$ in which $\mathbf{V}$ is an arbitrary velocity.

ALTERNATIVE FORMS FOR ENTROPY PRODUCTION

Different definitions of the heat current $\mathbf{J}_q$ give somewhat different expressions for σ. An elementary example is when $\mathbf{J}_q$ is defined to be $\mathbf{J}_u$; then the flow associated with $\nabla(1/T)$ will be the heat current. Another form of σ arises when the force associated with the matter flow $\mathbf{J}_k$ is written as $-\nabla\mu_k$ instead of $-\nabla(\mu_k/T)$. By separating the gradient of μ_k from the gradient of $(1/T)$, it is straightforward to show that (15.5.12) can be rewritten as

$$\sigma = \mathbf{J}_u' \cdot \nabla\frac{1}{T} - \sum_k \frac{\mathbf{J}_k \cdot \nabla\mu_k}{T} + \sum_k \frac{\mathbf{I}_k \cdot (-\nabla\psi)}{T} + \sum_j \frac{A_j v_j}{T} \qquad (A15.1.10a)$$

in which

$$\mathbf{J}_u' = \mathbf{J}_u - \sum_k \mu_k\mathbf{J}_k = \mathbf{J}_q + \sum_k (u_k^0 - \mu_k)\mathbf{J}_k = \mathbf{J}_q + \sum_k Ts_k\mathbf{J}_k \qquad (A15.1.10b)$$

When concentration gradients appear at constant T, it is useful to write this expression in terms of the gradient of μ at constant T. This can be done by noting that

$$\frac{\partial\mu_k}{\partial x} = \left(\frac{\partial\mu_k}{\partial T}\right)_{n_k} \frac{\partial T}{\partial x} + \sum_k \left(\frac{\partial\mu_k}{\partial n_k}\right)_T \frac{\partial n_k}{\partial x}.$$

Since this is also true for the y and the z derivatives, it follows that

$$\nabla\mu_k = \frac{\partial\mu_k}{\partial T}\nabla T + (\nabla\mu)_T \qquad (A15.1.11)$$

where $(\nabla\mu_k)_T = \sum_k (\partial\mu_k/\partial n_k)_T \nabla n_k$. As a consequence of (A15.1.11), we have

$$\nabla\frac{\mu_k}{T} = \left[\mu_k - T\left(\frac{\partial\mu_k}{\partial T}\right)\right]\nabla\frac{1}{T} + \frac{(\nabla\mu)_T}{T}$$

$$= u_k^0 \nabla\frac{1}{T} + \frac{(\nabla\mu)_T}{T} \qquad (A15.1.12)$$

where we have used the relation $u_k^0 \equiv (\partial u^0/\partial n_k)_T = \mu_k + Ts_k = \mu_k -T(\partial\mu/\partial T)_{n_k}$ (exc. 15.2). Substitution of (A15.1.12) into (15.5.12) gives

$$\sigma = \mathbf{J}_q \cdot \nabla\frac{1}{T} - \sum_k \frac{\mathbf{J}_k \cdot (\nabla\mu_k)_T}{T} + \sum_k \frac{\mathbf{I}_k \cdot (-\nabla\psi)}{T} + \sum_j \frac{A_j v_j}{T} \quad (A15.1.13)$$

(we can establish the relation between the heat currents in this text and those used in the classic text of de Groot and Mazur [12]: $\mathbf{J}_u = \mathbf{J}_q^{\mathrm{DM}}$ and $\mathbf{J}_q = \mathbf{J}_q'^{\mathrm{DM}}$ in which superscript DM indicates the quantity used by de Groot and Mazur.)

References

1. Alder, B. J. and Wainright, T., in *Transport Processes in Statistical Mechanics*, 1958. New York: Interscience.
2. Prigogine, I., *Physica*, **15** (1949), 272–284.
3. Prigogine, I. and Xhrouet, E., *Physica*, **XV** (1949), 913.
4. Prigogine, I. and Mahieu, M., *Physica*, **XVI** (1950), 51.
5. Present, R. D., *J. Chem. Phys.*, **31** (1959), 747.
6. Ross, J. and Mazur, P., *J. Chem. Phys.*, **35** (1961), 19.
7. Baras, F. and Malek-Mansour, M., *Physica A*, **188** (1992), 253–276.
8. Jou, D., Casas-Vázquez, J. and Lebon, G., *Extended Irreversible Thermodynamics*, 1996, New York, Berlin: Springer-Verlag.
9. Recent bibliography on Extended Thermodynamics may be found in Jou, D., Casas-Vázquez, J. and Lebon, G., *J. of Non-Equilibrium Thermodynamics*, **23** (1998) 277.
10. Müller, I. and Ruggeri, T., *Extended Thermodynamics*. 1993, New York: Springer-Verlag.
11. Salamon, P. and Sieniutycz, S. (eds), *Extended Thermodynamic Systems*, 1992, New York: Taylor & Francis.
12. de Groot, S. R. and Mazur, P., *Non-Equilibrium Thermodynamics*, 1969, Amsterdam: North Holland.
13. Prigogine, I., *Etude Thermodynamique des Processus Irreversibles*, 1947, Liège: Desoer.

Exercises

15.1 Assume that the Gibbs relation $dU = TdS - pdV + \sum_k \mu_k dN_k$ valid for a small volume element V. Show the validity of the relation $Tds = du - \sum_k \mu_k dn_k$ in which $s = (S/V)$, $u = (U/V)$ and $n_k = (N_k/V)$

15.2 (a) Using the Helmholtz energy density f and an appropriate Maxwell relation, show that

$$u_k \equiv \left(\frac{\partial u}{\partial n_k}\right)_T = \mu_k + T s_k = \mu_k - T\left(\frac{\partial \mu_k}{\partial T}\right)_{n_k},$$

in which

$$s_k = \left(\frac{\partial s}{\partial n_k}\right)_T$$

(b) We have seen that in the presence of a field $u = u^0 + \sum_k \tau_k n_k \psi$. Show that $f_k = (\mu_k + \tau_k \psi)$ and

$$u_k^0 \equiv \left(\frac{\partial u^0}{\partial n_k}\right)_T = \mu_k + T s_k = \mu_k - T\left(\frac{\partial \mu_k}{\partial T}\right)_{n_k}$$

15.3 Using the law of conservation of energy (15.4.3) and the concentration balance equation (15.3.10), show that the current as defined by (15.4.11) satisfies the energy conservation equation (15.4.8)

15.4 From (15.4.16) and (15.4.17) obtain (15.4.18) and (15.4.19).

15.5 Obtain (A15.1.10a) and (A15.1.10b) from (15.5.12).

16 NONEQUILIBRIUM THERMODYNAMICS: THE LINEAR REGIME

16.1 Linear Phenomenological Laws

When a system is close to equilibrium, a general theory based on linear relations between forces and flows could be formulated. In the previous chapter we have seen that the entropy production per unit volume, σ, can be written as

$$\sigma = \sum_k F_k J_k \qquad (16.1.1)$$

in which F_k are forces, such as the gradient of $(1/T)$, and J_k are flows, such as the heat flow. The forces drive the flows; a nonvanishing gradient of $(1/T)$ causes the flow of heat. At equilibrium, all the forces and the corresponding flows vanish, i.e. the flows J_k are functions of forces F_k such that they vanish when $F_k = 0$. For a small deviation in the forces from their equilibrium value of zero, the flows can be expected to be linear functions of the forces. (In other words, the flows are assumed to be analytic functions of the forces, as is the case with most physical variables.) Accordingly, the following relation between the flows and the forces is assumed:

$$\boxed{J_k = \sum_j L_{kj} F_j} \qquad (16.1.2)$$

Here the L_{kj} are constants called **phenomenological coefficients**. Note how (16.1.2) implies that not only can a force such as the gradient of $(1/T)$ cause the flow of heat but it can also drive other flows, such as a flow of matter or an electrical current. The thermoelectric effect is one such cross effect, in which a thermal gradient drives not only a heat flow but also an electrical current and vice versa (Fig. 16.1). Another example is cross diffusion, in which a gradient in the concentration of one compound drives a diffusion current of another. Such cross effects were known long before thermodynamics of irreversible processes was formulated. Each cross effect was studied on an individual basis, but without a unifying formalism. For example, the thermoelectric phenomenon was investigated in the 1850s and William Thomson (Lord Kelvin) [2] gave theoretical explanations for the observed Seebeck and Peltier effects (Fig. 16.1). (Kelvin's reasoning was later found to be incorrect.) Similarly, other cross

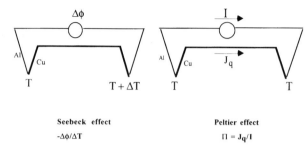

Figure 16.1 The thermoelectric effect is a "cross effect" between thermodynamic forces and flows. (a) In the Seebeck effect, two dissimilar metal wires are joined and the junctions are maintained at different temperatures. As a result an EMF is generated. The EMF generated is generally of the order of 10^{-5} V per Kelvin of temperature difference and it may vary from sample to sample. (b) In the Peltier effect, the two junctions are maintained at the same temperature and a current is passed through the system. The current flow drives a heat flow J_q from one junction to the other. The Peltier heat current is generally of the order of 10^{-5} J s^{-1} per amp. [1]

effects were observed and studied in the nineteenth century. Neglecting the cross effects, some of the well-established phenomenological laws are as follows:

$$\text{Fourier's law of heat conduction:} \quad \mathbf{J}_q = -\kappa \nabla T(x) \qquad (16.1.3)$$

$$\text{Fick's law of diffusion:} \quad \mathbf{J}_k = -D_k \nabla n_k(x) \qquad (16.1.4)$$

$$\text{Ohm's law of electrical conduction:} \quad I = \frac{V}{R} \qquad (16.1.5a)$$

$$\text{Alternative form of Ohms' law:} \quad \mathbf{I} = \frac{\mathbf{E}}{\rho} \qquad (16.1.5b)$$

In these equations, κ is the heat conductivity, D_k is the diffusion coefficient of compound k and n_k is the concentration of compound k. Ohm's law is usually stated as (16.1.5a) in which I is the electrical current, R is the resistance and V is the voltage. It can also be stated in terms of the electric current density $\mathbf{I}$, the electric field $\mathbf{E}$ and the resistivity ρ (resistance per unit length per unit area of cross section). Other quantities in the above equations are as defined in Table 15.1

As a specific example of the general relation (16.1.2), let us consider the thermoelectric phenomenon mentioned above (Fig. 16.1). The equations that describe thermoelectric cross-coupling are*

$$\mathbf{J}_q = L_{qq} \nabla \left(\frac{1}{T} \right) + L_{qe} \frac{\mathbf{E}}{T} \tag{16.1.6}$$

$$\mathbf{I}_e = L_{ee} \frac{\mathbf{E}}{T} + L_{eq} \nabla \frac{1}{T} \tag{16.1.7}$$

in which L_{qq}, L_{qe}, etc., correspond to the L_{ij}'s in (16.1.2). Experimentally these coefficients can be measured for various conductors. We shall discuss this and other examples in detail in later sections of this chapter.

Phenomenological laws and the cross effects between the flows were independently studied and, until the formalism presented here was developed in the 1930s, there was no *unified* theory of all the cross effects. Relating the entropy production to the phenomenological laws is the first step in developing a unified theory. For conditions under which the linear phenomenological laws (16.1.2) are valid, entropy production (16.1.1) takes the quadratic form

$$\boxed{\sigma = \sum_{jk} L_{jk} F_j F_k > 0} \tag{16.1.8}$$

Here F_k can be positive or negative. A matrix L_{jk} that satisfies the condition (16.1.8) is said to be positive definite. The properties of **positive definite** matrices are well characterized. For example, a two-dimensional matrix L_{ij} is positive definite only when the following conditions are satisfied (exc. 16.1):

$$L_{11} > 0 \qquad L_{22} > 0 \qquad (L_{12} + L_{21})^2 < 4 L_{11} L_{22} \tag{16.1.9}$$

In general, the diagonal elements of a positive definite matrix must be positive. In addition, a necessary and sufficient condition for a matrix L_{ij} to be positive definite is that its determinant and all the determinants of lower dimension obtained by deleting one or more rows and columns must be positive. Thus, according to the Second Law, the "proper coefficients" L_{kk} should be positive; the "cross coefficients," L_{ik} $(i \neq k)$, can have either sign. Furthermore, as we shall see in the next section, the elements L_{jk} also obey the *Onsager reciprocal relations* $L_{jk} = L_{kj}$. The positivity of entropy production and the Onsager relations form the foundation for linear nonequilibrium thermodynamics.

16.2   Onsager Reciprocal Relations and The Symmetry Principle

That reciprocal relations, $L_{ij} = L_{ji}$, were associated with cross effects was noticed by William Thomson (Lord Kelvin) and others even during the last century. The early explanations of the reciprocal relations were based on thermodynamic reasoning that was not on a firm footing. For this reason,

---

* Note that $\mathbf{J}_u = \mathbf{J}_q$ when $\sum u_k \mathbf{J}_k = 0$.

Lars Onsager (1903–1976) (Reproduced, by permission, from the Emilio Segré Visual Archives of the American Institute of Physics)

William Thomson and others regarded the reciprocal relations only as conjectures. A well-founded theoretical explanation for these relations was developed by Lars Onsager (1903–1976) in 1931 [3]. Onsager's theory is based on the *principle of detailed balance or microscopic reversibility* that is valid for systems at equilibrium.

The principle of detailed balance or microscopic reversibility is formulated using the general thermodynamic theory of equilibrium fluctuations that we discussed in section 14.2. A summary of the main results of this section is as follows.

- The entropy $\Delta_i S$ associated with fluctuations α_i can be written as

$$\Delta_i S = -\frac{1}{2} \sum_{k,j} g_{kj} \alpha_j \alpha_k = \frac{1}{2} \sum_k F_k \alpha_k \qquad (16.2.1)$$

in which

$$F_k = \frac{\partial \Delta_i S}{\partial \alpha_k} = -\sum_j g_{kj} \alpha_j \qquad (16.2.2)$$

is the conjugate thermodynamic force for the thermodynamic flow $d\alpha_k/dt$.

- According to the Einstein formula (14.2.2), the entropy associated with the fluctuations gives the following probability distribution for the fluctuations:

$$P(\alpha_1, \alpha_2, \ldots, \alpha_m) = Z \, \mathrm{Exp}(\Delta_i S/k_B) = Z \, \mathrm{Exp}\left[-\frac{1}{2k_B} \sum_{i,j} g_{ij} \alpha_i \alpha_j \right] \qquad (16.2.3)$$

in which k_B is the Boltzmann constant and Z is the normalization constant.

- As was shown in section 14.2, using the probability distribution (16.2.3), the following expressions for correlations between F_i and α_j can be obtained:

$$\boxed{\langle F_i \alpha_j \rangle = -k_B \delta_{ij}} \qquad (16.2.4)$$

$$\boxed{\langle \alpha_i \alpha_j \rangle = k_B (g^{-1})_{ij}} \qquad (16.2.5)$$

in which $(g^{-1})_{ij}$ is the inverse of the matrix g_{ij}.

These are the basic results of the theory of fluctuations needed to derive the reciprocal relations $L_{ik} = L_{ki}$.

THE ONSAGER RECIPROCAL RELATIONS

Onsager's theory begins with the assumption that, where linear phenomenological laws are valid, a deviation α_k, decays according to the linear law

$$J_k = \frac{d\alpha_k}{dt} = \sum_j L_{kj} F_j \qquad (16.2.6a)$$

which, by virtue of (16.2.2), can also be written as

$$J_k = \frac{d\alpha_k}{dt} = -\sum_{j,i} L_{kj} g_{ji} \alpha_i = \sum_i M_{ki} \alpha_i \qquad (16.2.6b)$$

in which the matrix M_{ki} is the product of the matrices L_{kj} and g_{ji}. The equivalence of (16.2.6a) and (16.2.6b) shows that phenomenological equations for the flows that are usually written in the form (16.2.6b) can be transformed into (16.2.6a) in which the flows are linear functions of the forces F_k.

As we shall see, according to the principle of detailed balance, the effect of α_i on the flow $(d\alpha_k/dt)$ is the same as the effect of α_k on the flow $(d\alpha_i/dt)$. This condition can be expressed in terms of the correlation $\langle \alpha_i d\alpha_k/dt \rangle$ between α_i and $(d\alpha_k/dt)$ as

$$\boxed{\left\langle \alpha_i \frac{d\alpha_k}{dt} \right\rangle = \left\langle \alpha_k \frac{d\alpha_i}{dt} \right\rangle} \qquad (16.2.7)$$

In a way, this correlation isolates that part of the flow $(d\alpha_k/dt)$ that depends on the variable α_i. Once the validity of (16.2.7) is accepted, the reciprocal relations directly follow from (16.2.6a). Multiplying (16.2.6a) by α_i and taking the average, we obtain

$$\left\langle \alpha_i \frac{d\alpha_k}{dt} \right\rangle = \sum_j L_{kj} \langle \alpha_i F_j \rangle = -k_{\mathrm{B}} \sum_j L_{kj} \delta_{ij} = -k_{\mathrm{B}} L_{ki} \qquad (16.2.8)$$

where we have used $\langle F_i \alpha_j \rangle = -k_{\mathrm{B}} \delta_{ij}$. Similarly

$$\left\langle \alpha_k \frac{d\alpha_i}{dt} \right\rangle = \sum_j L_{ij} \langle \alpha_k F_j \rangle = -k_{\mathrm{B}} \sum_j L_{ij} \delta_{kj} = -k_{\mathrm{B}} L_{ik} \qquad (16.2.9)$$

If the equality (16.2.7) is valid, we immediately obtain the **Onsager reciprocal relations**

$$\boxed{L_{ki} = L_{ik}}^* \qquad (16.2.10)$$

We are then naturally led to ask why (16.2.7) is valid. Onsager argued that this equality is valid because of **microscopic reversibility**, which, according to Onsager, is

"the assertion that transitions between two (classes of) configurations A and B should take place equally often in the directions A $\rightarrow$ B and B $\rightarrow$ A in a given time τ." [3 p. 418]

This statement is the same as the **principle of detailed balance** that was discussed in Chapter 9. According to this principle, if α_i has a value $\alpha_i(t)$ at time t, and if at time $t + \tau$ a correlated variable α_k has a value $\alpha_k(t + \tau)$, then the time-reversed transition should occur equally often. This means

$$\boxed{\langle \alpha_i(t)\alpha_k(t + \tau) \rangle = \langle \alpha_k(t)\alpha_i(t + \tau) \rangle} \qquad (16.2.11)$$

Note that (16.2.11) remains unchanged if τ is replaced by $-\tau$.

---

* In the presence of a magnetic field **B**, L_{ij} may be functions of **B**. In this case the reciprocal relations take the form $L_{ki}(\mathbf{B}) = L_{ik}(-\mathbf{B})$

From this equality, relation (16.2.7) can be obtained by noting that

$$\frac{d\alpha_k}{dt} \approx \frac{\alpha_k(t+\tau) - \alpha_k(t)}{\tau}$$

so that

$$\left\langle \alpha_i \frac{d\alpha_k}{dt} \right\rangle = \left\langle \alpha_i(t) \left\{ \frac{\alpha_k(t+\tau) - \alpha_k(t)}{\tau} \right\} \right\rangle = \frac{1}{\tau} \langle \alpha_i(t)\alpha_k(t+\tau) - \alpha_i(t)\alpha_k(t) \rangle$$

(16.2.12)

$$\left\langle \alpha_k \frac{d\alpha_i}{dt} \right\rangle = \left\langle \alpha_k(t) \left\{ \frac{\alpha_i(t+\tau) - \alpha_i(t)}{\tau} \right\} \right\rangle = \frac{1}{\tau} \langle \alpha_k(t)\alpha_i(t+\tau) - \alpha_k(t)\alpha_i(t) \rangle$$

(16.2.13)

If we now use the relation $\langle \alpha_i(t)\alpha_k(t+\tau) \rangle = \langle \alpha_k(t)\alpha_i(t+\tau) \rangle$, and use the fact that $\langle \alpha_i(t)\alpha_k(t) \rangle = \langle \alpha_k(t)\alpha_i(t) \rangle$ in (16.2.12) and (16.2.13), equality (16.2.7) follows.

Thus we see that the principle of detailed balance or microscopic reversibility, expressed as $\langle \alpha_i(t)\alpha_k(t+\tau) \rangle = \langle \alpha_k(t)\alpha_i(t+\tau) \rangle$, leads to the reciprocal relations $L_{ij} = L_{ji}$.

THE SYMMETRY PRINCIPLE

Though forces and flows are coupled in general, the possible coupling is restricted by a general symmetry principle. This principle, which states that *macroscopic causes always have fewer or equal symmetries than the effects they produce*, was originally stated by Pierre Curie [4] but not in the context of thermodynamics. It was introduced by Prigogine [5] into nonequilibrium thermodynamics because it enables us to eliminate the possibility of coupling between certain forces and flows on the basis of symmetry. We shall refer to this principle as the **symmetry principle**; in some texts it is also called the **Curie principle**. For example, a scalar thermodynamic force such as chemical affinity, which has the high symmetry of isotropy, cannot drive a heat current which has lower symmetry because of its directionality. As an explicit example, let us consider a system in which there is heat transport and chemical reaction. The entropy production is ($\mathbf{J}_u = \mathbf{J}_q$)

$$\sigma = \mathbf{J}_q \cdot \nabla \frac{1}{T} + \frac{A}{T} v$$

(16.2.14)

The general linear phenomenological laws that follow from this are

$$\mathbf{J}_q = L_{qq} \nabla \frac{1}{T} + L_{qc} \frac{A}{T}$$

(16.2.15)

$$v = L_{cc} \frac{A}{T} + L_{cq} \nabla \frac{1}{T}$$

(16.2.16)

According to the symmetry principle, the scalar process of chemical reaction, due to its higher symmetry of isotropy and homogeneity, cannot generate a heat current which has a direction — and hence is anisotropic. Another way of stating this principle is that a scalar cause cannot produce a vectorial effect. Therefore $L_{qc} = 0$. As a consequence of the reciprocal relations, we have $L_{qc} = L_{cq} = 0$. In general, irreversible processes of different tensorial character (scalars, vectors and higher-order tensors) do not couple to each other.

Because of the symmetry principle, the entropy production due to scalar, vectorial and tensorial processes should each be positive. In the above case, we must have

$$\mathbf{J}_q \cdot \nabla \frac{1}{T} \geq 0 \qquad \frac{A}{T} v \geq 0 \qquad (16.2.17)$$

(Also, the entropy production due to chemical reactions in each phase should be separately positive.) Thus, the symmetry principle provides constraints for the coupling of, and the entropy production due to, irreversible processes.

In the following sections we shall present several cross effects in detail to illustrate the experimental implications of Onsager's reciprocal relations.

16.3   Thermoelectric Phenomena

As a first illustration of the theory presented in the last two sections, let us consider thermoelectric effects which involve the flow of heat $\mathbf{J}_q$ and electric current $\mathbf{I}_e$ in conducting wires (in which the subscript indicates that the flow corresponds to the flow of electrons). The entropy production per unit volume due to these two irreversible processes and the linear phenomenological laws associated with it are

$$\sigma = \mathbf{J}_q \cdot \nabla \left(\frac{1}{T} \right) + \frac{\mathbf{I}_e \cdot \mathbf{E}}{T} \qquad (16.3.1)$$

$$\boxed{\mathbf{J}_q = L_{qq} \nabla \left(\frac{1}{T} \right) + L_{qe} \frac{\mathbf{E}}{T}} \qquad (16.3.2)$$

$$\boxed{\mathbf{I}_e = L_{ee} \frac{\mathbf{E}}{T} + L_{eq} \nabla \frac{1}{T}} \qquad (16.3.3)$$

where $\mathbf{E}$ is the electric field. For a one-dimensional system, such as a conducting wire, the vectorial aspect of $\mathbf{J}_q$ and $\mathbf{I}_e$ is unimportant and both may be treated as scalars. To relate the coefficients L_{qq} and L_{ee} with the heat conductivity κ and resistance R, we can write equations (16.3.2) and (16.3.3) in a one-dimensional system as

$$J_q = -\frac{1}{T^2} L_{qq} \frac{\partial}{\partial x} T + L_{qe} \frac{E}{T} \qquad (16.3.4)$$

$$I_e = L_{ee} \frac{E}{T} - \frac{1}{T^2} L_{eq} \frac{\partial}{\partial x} T \qquad (16.3.5)$$

Fourier's law (16.1.3) of heat conduction is valid when the electric field $E = 0$. Comparing the heat conduction term $J_q = -(1/T^2)L_{qq}\partial T/\partial x$ to Fourier's law (16.1.3), leads to the identification

$$\kappa = \frac{L_{qq}}{T^2} \qquad (16.3.6)$$

We can now specify more precisely what is meant by the **near-equilibrium linear regime**. It means that L_{qq}, L_{ee}, etc., may be treated as constants. Since $T(x)$ is a function of position, such an assumption is strictly not valid. It is valid only in the approximation that the change in T from one end of the system to another is small compared to the average T, i.e. if the average temperature is T_{avg}, then $|T(x) - T_{avg}|/ T_{avg} \ll 1$ for all x. Hence we may approximate $T^2 \approx T^2_{avg}$ and use κT^2_{avg} in place of κT^2.

To find the relation between L_{ee} and the resistance R, we note that $V = -\Delta\phi = \int_0^l E \, dx$ in which l is the length of the system. The current I_e is independent of x. At constant temperature ($\partial T/\partial x = 0$), the current is entirely due to electrical potential difference. Integrating (16.3.5) over the length of the system, we obtain

$$\int_0^l I_e dx = \frac{L_{ee}}{T} \int_0^l E \, dx \quad \text{or} \quad I_e l = \frac{L_{ee}}{T} V \qquad (16.3.7)$$

Comparing this equation with Ohm's law (16.1.5a), we make the identification

$$L_{ee} = \frac{T}{(R/l)} = \frac{T}{r} \qquad (16.3.8)$$

in which *r is the resistance per unit length*. As noted in (16.1.5b), Ohm's can also be stated in general as

$$\mathbf{I} = \frac{\mathbf{E}}{\rho} \qquad (16.3.9)$$

in which ρ *is the specific resistance*, $\mathbf{I}$ is the current density and $\mathbf{E}$ is the electric field. Comparing (16.3.5) with (16.3.9) we have the general relation

$$L_{ee} = \frac{T}{\rho} \qquad (16.3.10)$$

When we consider a one-dimensional system, ρ is replaced by r, resistance per unit length.

THE SEEBECK EFFECT

The cross coefficients L_{qe} and L_{eq} can also be related to experimentally measured quantities. In the Seebeck effect (Box 16.1), a temperature difference between two junctions of dissimilar metals produces an EMF. This EMF is measured at zero current. For this system, equations (16.3.4) and (16.3.5) may be used. Setting $I_e = 0$ in (16.3.5) we obtain

$$0 = L_{ee}ET - L_{eq}\frac{\partial}{\partial x}T \qquad (16.3.11)$$

This equation may now be integrated to obtain a relation between the temperature difference ΔT and the EMF generated due to this temperature difference, $\Delta\phi = -\int E dx$. In doing this integration, we shall assume that the total variation ΔT is small and make the approximation $\int TE\,dx \approx T\int E\,dx = -T\Delta\phi$. This gives us the relation

$$L_{eq} = -L_{ee}T\left(\frac{\Delta\phi}{\Delta T}\right)_{I=0} \qquad (16.3.12)$$

Experimentally the ratio $-(\Delta\phi/\Delta T)_{I=0}$, called **thermoelectric power**, is measured. Some typical values of thermoelectric power are shown in Table 16.1; its sign may be positive or negative. Using (16.3.12) the coefficient L_{eq} can be related to the measured quantities.

THE PELTIER EFFECT

In the Peltier effect, the two junctions are maintained at a constant temperature while a current I is passed through the system (Box 16.1). This causes a flow of heat from one junction to another. The two junctions are maintained at the same temperature only by removing heat from one of the junctions and thus maintaining a steady heat flow J_q. Under these conditions, the ratio

$$\Pi = \left(\frac{J_q}{I_e}\right) \qquad (16.3.13)$$

which can be measured, is called the **Peltier heat**. Some typical values of (Π/T) are shown in Table 16.1. The phenomenological coefficient L_{qe} can be related to the Peltier heat as follows. Since there is no temperature difference between the two junctions, $\partial T/\partial x = 0$, and equations (16.3.4) and (16.3.5) become

$$J_q = L_{qe} \frac{E}{T} \qquad (16.3.14)$$

$$I_e = L_{ee} \frac{E}{T} \qquad (16.3.15)$$

Dividing one equation by the other and using (16.3.8) and (16.3.13), we obtain

$$L_{qe} = \Pi L_{ee} = \Pi \frac{T}{(R/l)} = \Pi \frac{T}{r} \qquad (16.3.16)$$

In this manner, all the phenomenological coefficients L_{qe} and L_{eq} can be related to the experimental parameters of the cross effects.

Box 16.1 Onsager Reciprocal Relations in Thermoelectric Phenomena

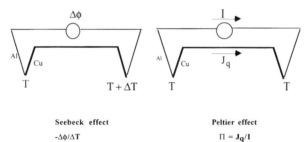

Seebeck effect

$-\Delta\phi/\Delta T$

Peltier effect

$\Pi = J_q/I$

Table 16.1 Some experimental data confirming Onsager reciprocal relations[*]

Thermocouple	$T(°C)$	$\Pi/T(\mu V\,K^{-1})$	$-\Delta\phi/\Delta T(\mu V\,K^{-1})$	L_{qe}/L_{eq}
Cu–Al	15.8	2.4	3.1	0.77
Cu–Ni	0	18.6	20.0	0.930
Cu–Ni	14	20.2	20.7	0.976
Cu–Fe	0	−10.16	−10.15	1.000
Cu–Bi	20	−71	−66	1.08
Fe–Ni	16	33.1	31.2	1.06
Fe–Hg	18.4	16.72	16.66	1.004

[*] More extensive data can be found in [1].

Having identified all the linear phenomenological coefficient in terms of the experimentally measured quantities, we can now turn to the reciprocal relations, according to which one must find

$$L_{qe} = L_{eq} \qquad (16.3.17)$$

Upon using (16.3.12) for L_{eq} and (16.3.16) for L_{qe}, we find

$$-L_{ee}T\left(\frac{\Delta\phi}{\Delta T}\right) = \Pi L_{ee} \quad \text{or} \quad \boxed{-\left(\frac{\Delta\phi}{\Delta T}\right) = \frac{\Pi}{T}} \tag{16.3.18}$$

Experimental data verifying this prediction for some pairs of conductors are shown in Table 16.1.

16.4 Diffusion

In this section we will apply the theory of linear nonequilibrium thermo-dynamics to the process of diffusion. When several species are simultaneously diffusing, it is found that the flow of one species influences the flow of another, i.e. there are cross effects between diffusing species. The entropy production per unit volume associated with simultaneous diffusion of several species is

$$\sigma = -\sum_k \mathbf{J}_k \cdot \nabla\left(\frac{\mu_k}{T}\right) \tag{16.4.1}$$

in which $\mathbf{J}_k$ is the matter current and μ_k is the chemical potential of species k. Under isothermal conditions, the associated linear laws are

$$\mathbf{J}_i = -\sum_k \frac{L_{ik}}{T}\nabla\mu_k \tag{16.4.2}$$

Our first task is to relate the linear coefficients L_{ik} to the experimentally measured diffusion coefficients D_{ij}. For simultaneous diffusion of several species (under isothermal conditions), a "generalized Fick's law" may be written as:

$$\mathbf{J}_i = -\sum_k D_{ik}\nabla n_k(\mathbf{x}) \tag{16.4.3}$$

in which $n_k(\mathbf{x})$ is the concentration of the component k at position $\mathbf{x}$. As an example, diffusion coefficients D_{ij} in a molten silicate solution of CaO–Al$_2$O$_3$–SiO$_2$ at various temperatures [6, 7] are shown in Table 16.2. (Diffusion coefficients for some gases and liquids are given in Chapter 10.) Let us consider a system with two components. The Gibbs–Duhem relation tells us that the chemical potentials, hence the forces $-\nabla(\mu_k/T)$, are not all independent. For a two-component system, when T and p are constant, we have

$$n_1\,d\mu_1 + n_2\,d\mu_2 = 0 \tag{16.4.4}$$

Since $d\mu_k = d\mathbf{r}\cdot\nabla\mu_k$ for an arbitrary $d\mathbf{r}$, equation (16.4.4) leads to the following

Table 16.2 Some Fick's law diffusion coefficients showing cross effects in a molten silicate solution*

$T(K)$	$D_{11}(m^2\,s^{-1})$	$D_{12}(m^2s^{-1})$	$D_{21}(m^2\,s^{-1})$	$D_{22}(m^2s^{-1})$
1723	$(6.8 \pm 0.3) \times 10^{-11}$	$(-2.0 \pm 0.5) \times 10^{-11}$	$(-3.3 \pm 0.5) \times 10^{-11}$	$(4.1 \pm 0.7) \times 10^{-11}$
1773	$(1.0 \pm 0.1) \times 10^{-10}$	$(-2.8 \pm 0.8) \times 10^{-11}$	$(-4.2 \pm 0.8) \times 10^{-11}$	$(7.3 \pm 0.4) \times 10^{-11}$
1823	$(1.8 \pm 0.2) \times 10^{-10}$	$(-4.6 \pm 0.6) \times 10^{-11}$	$(-6.4 \pm 0.5) \times 10^{-11}$	$(1.5 \pm 0.1) \times 10^{-10}$

* The composition of the silicate is 40% CaO, 20% Al_2O_3 and 40% SiO by weight [6, 7].

relation between the gradients of the chemical potentials:

$$n_1 \nabla \mu_1 + n_2 \nabla \mu_2 = 0 \tag{16.4.5}$$

This shows that the thermodynamic forces are not all independent. Nor are all the flows $\mathbf{J}_k$ independent. In most physical situations, the relation between the flows is more conveniently expressed as the condition for "no volume flow"[1]. For a two-component system, this is expressed as

$$\mathbf{J}_1 v_1 + \mathbf{J}_2 v_2 = 0 \tag{16.4.6}$$

in which the v_k are partial molar volumes. For notational simplicity we use v_k for partial molar volume instead of $V_{m,k}$. This equation is the statement that the diffusion flows do not result in any change in volume (Fig. 16.2).

As a consequence of (16.4.5) the entropy production due to diffusion under isothermal conditions can be written (exc. 16.4) as*

$$\sigma = -\frac{1}{T} \left(\mathbf{J}_1 - \frac{n_1}{n_2} \mathbf{J}_2 \right) \cdot \nabla \mu_1 \tag{16.4.7}$$

Now, using condition (16.4.6) for no volume flow, the expression for entropy production can be written as

$$\sigma = -\frac{1}{T} \left(1 + \frac{v_1 n_1}{v_2 n_2} \right) \mathbf{J}_1 \cdot \nabla \mu_1 \tag{16.4.8}$$

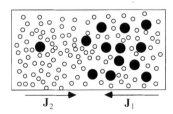

Figure 16.2 Diffusion in a two-component system. In most physical situations the flow of the components due to diffusion does not produce a change in volume

* As shown in Appendix 15.1, even if there is a constant volume flow σ remains unchanged.

The linear phenomenological law that relates the flux $\mathbf{J}_1$ to the conjugate force in (16.4.8) is

$$\mathbf{J}_1 = -L_{11} \frac{1}{T} \left(1 + \frac{v_1 n_1}{v_2 n_2} \right) \nabla \mu_1 \qquad (16.4.9)$$

We can relate this equation to Fick's law, usually written $\mathbf{J}_1 = -D_1 \nabla n_1$, by noting that $\nabla \mu_1 = (\partial \mu_1 / \partial n_1) \nabla n_1$. We then have

$$\mathbf{J}_1 = -L_{11} \frac{1}{T} \left(1 + \frac{v_1 n_1}{v_2 n_2} \right) \left(\frac{\partial \mu_1}{\partial n_1} \right) \nabla n_1 = -D_1 \nabla n_1 \qquad (16.4.10)$$

From this it follows that the relation between the phenomenological coefficient L_{11} and the diffusion coefficient is

$$L_{11} = \frac{D_1 T}{\left(1 + \dfrac{v_1 n_1}{v_2 n_2} \right) \left(\dfrac{\partial \mu_1}{\partial n_1} \right)} \qquad (16.4.11)$$

For diffusion of a solute in a solution, n_2 is the concentration of the solution and n_1 the concentration of the solute. For *dilute ideal solutions* recall that $\mu_1 = \mu_0(p, T) + RT \ln x_1$ in which $x_1 = n_1/(n_1 + n_2) \approx n_1/n_2$, and also that $n_1 \ll n_2$. These conditions simplify the relation between L_{11} and D to

$$L_{11} = \frac{D_1 n_1}{R} \qquad (16.4.12)$$

This is the relation we saw in Chapter 10 between the usual diffusion coefficient and the corresponding phenomenological coefficient.

To verify Onsager's reciprocal relations we need at least three components. For three- component isothermal diffusion, the entropy production per unit volume is

$$\sigma = -\frac{\mathbf{J}_1}{T} \cdot \nabla \mu_1 - \frac{\mathbf{J}_2}{T} \cdot \nabla \mu_2 - \frac{\mathbf{J}_3}{T} \cdot \nabla \mu_3 \qquad (16.4.13)$$

For a three-component system, the Gibbs–Duhem equation and the condition for no volume flow are as follows (equations 16.4.5 and 16.4.6):

$$n_1 \nabla \mu_1 + n_2 \nabla \mu_2 + n_3 \nabla \mu_3 = 0 \qquad (16.4.14)$$

$$\mathbf{J}_1 v_1 + \mathbf{J}_2 v_2 + \mathbf{J}_3 v_3 = 0 \qquad (16.4.15)$$

Let us assume that $\mathbf{J}_3$ and μ_3 are the variables for the solvent while $\mathbf{J}_1$, μ_1 and $\mathbf{J}_2$, μ_2 are the variables of two solutes whose diffusion cross effects are of interest. Using (16.4.14) and (16.4.15), $\mathbf{J}_3$ and μ_3 can be eliminated from the expression for the entropy production. The entropy production can then be written in terms of only the variables J_1, μ_1 and J_2, μ_2 of the two solutes (exc. 16.5):

$$\sigma = \mathbf{F}_1 \cdot \mathbf{J}_1 + \mathbf{F}_2 \cdot \mathbf{J}_2 \qquad (16.4.16)$$

in which the thermodynamic forces $\mathbf{F}_1$ and $\mathbf{F}_2$ are

$$\mathbf{F}_1 = -\frac{1}{T}\left[\nabla\mu_1 + \frac{n_1 v_1}{n_3 v_3}\nabla\mu_1 + \frac{n_2 v_1}{n_3 v_3}\nabla\mu_2\right] \qquad (16.4.17)$$

and

$$\mathbf{F}_2 = -\frac{1}{T}\left[\nabla\mu_2 + \frac{n_2 v_2}{n_3 v_3}\nabla\mu_2 + \frac{n_1 v_2}{n_3 v_3}\nabla\mu_1\right] \qquad (16.4.18)$$

The associated phenomenological laws then take the form

$$\mathbf{J}_1 = L_{11}\mathbf{F}_1 + L_{12}\mathbf{F}_2 \qquad (16.4.19)$$
$$\mathbf{J}_2 = L_{21}\mathbf{F}_1 + L_{22}\mathbf{F}_2 \qquad (16.4.20)$$

To verify the reciprocal relations, we must now relate L_{ik} and the experimentally measured diffusion coefficients D_{ik} of the generalized Fick's law:

$$\mathbf{J}_1 = -D_{11}\nabla n_1 - D_{12}\nabla n_2 \qquad (16.4.21)$$
$$\mathbf{J}_2 = -D_{21}\nabla n_1 - D_{22}\nabla n_2 \qquad (16.4.22)$$

If $\mathbf{J}_2 = 0$ note how these equations imply that a constant flow, $\mathbf{J}_1 = $ constant, due to a concentration gradient in n_1, will produce a concentration gradient in n_2. Let us assume that the flow and concentration gradients are along only one direction, say x, so that all the gradients correspond to derivatives with respect to x. (Extending the following calculation to three dimensions is straightforward.) We can write the forces F_k in terms of the gradients of the two diffusing components because the chemical potentials μ_k are functions of n_k. Thus

$$\frac{\partial\mu_1}{\partial x} = \frac{\partial\mu_1}{\partial n_1}\frac{\partial n_1}{\partial x} + \frac{\partial\mu_1}{\partial n_2}\frac{\partial n_2}{\partial x} \qquad (16.4.23)$$

A similar relation can be written for the gradient of μ_2. Using these relations in (16.4.17) and (16.4.18) and substituting them in (16.4.19) and (16.4.20), the flows J_k can be written in terms of the gradients of n_k. After some calculation (exc. 16.6) the following relations between the diffusion coefficients and the linear Onsager coefficients can be obtained:

$$L_{11} = T\frac{dD_{11} - bD_{12}}{ad - bc} \qquad L_{12} = T\frac{aD_{12} - cD_{11}}{ad - bc} \qquad (16.4.24)$$

$$L_{21} = T\frac{dD_{21} - bD_{22}}{ad - bc} \qquad L_{22} = T\frac{aD_{22} - cD_{21}}{ad - bc} \qquad (16.4.25)$$

in which

$$a = \left(1 + \frac{n_1 v_1}{n_3 v_3}\right)\left(\frac{\partial \mu_1}{\partial n_1}\right) + \frac{n_2 v_1}{n_3 v_3}\left(\frac{\partial \mu_2}{\partial n_1}\right) \quad b = \left(1 + \frac{n_2 v_2}{n_3 v_3}\right)\left(\frac{\partial \mu_2}{\partial n_1}\right) + \frac{n_1 v_2}{n_3 v_3}\left(\frac{\partial \mu_1}{\partial n_1}\right)$$

(16.4.26)

$$c = \left(1 + \frac{n_1 v_1}{n_3 v_3}\right)\left(\frac{\partial \mu_1}{\partial n_2}\right) + \frac{n_2 v_1}{n_3 v_3}\left(\frac{\partial \mu_2}{\partial n_2}\right) \quad d = \left(1 + \frac{n_2 v_2}{n_3 v_3}\right)\left(\frac{\partial \mu_2}{\partial n_2}\right) + \frac{n_1 v_2}{n_3 v_3}\left(\frac{\partial \mu_1}{\partial n_2}\right)$$

(16.4.27)

(Note that the only difference between (16.4.26) and (16.4.27) is that the derivative $\partial/\partial n_1$ is replaced by $\partial/\partial n_2$.) These relations can be written more compactly in matrix notation (exc. 16.7). From these relations it is easy to see that the implication of the reciprocal relations $L_{12} = L_{21}$ is

$$aD_{12} + bD_{22} = cD_{11} + dD_{21} \tag{16.4.28}$$

Experimental data for several three-component systems is summarized in Tables 16.3 and 16.4. Often the relations between the chemical potential and the concentration are not known precisely and accurate measurement of diffusion coefficients is rather difficult. Nevertheless, we see that within experimental error the reciprocal relations seem to hold very well.

Table 16.3 Experimental data on cross diffusion in molten silicates and verification of Onsager's reciprocal relations [1, 6, 7]

System	$D_{11}(m^2 s^{-1})$	$D_{12}(m^2 s^{-1})$	$D_{21}(m^2 s^{-1})$	$D_{22}(m^2 s^{-1})$	L_{12}/L_{21}	$T(K)$
CaO–Al$_2$O$_3$–SiO$_2$	6.8×10^{-11}	-2.0×10^{-11}	-3.3×10^{-11}	4.1×10^{-11}	1.46 ± 0.44	1723
CaO–Al$_2$O$_3$–SiO$_2$	1.0×10^{-10}	-2.8×10^{-11}	-4.2×10^{-11}	7.3×10^{-11}	1.46 ± 0.44	1773
CaO–Al$_2$O$_3$–SiO$_2$	1.8×10^{-10}	-4.6×10^{-11}	-6.4×10^{-11}	1.5×10^{-10}	1.29 ± 0.36	1823

Table 16.4 Experimental diffusion coefficients for the toluene–chlorobenzene–bromobenzene system at $T = 30\,°C$ and verification of Onsager's reciprocal relations [8]

X_1^*	X_2^*	$D_{11}/10^{-9}(m^2 s^{-1})$	$D_{12}/10^{-9}(m^2 s^{-1})$	$D_{21}/10^{-9}(m^2 s^{-1})$	$D_{22}/10^{-9}(m^2 s^{-1})$	L_{12}/L_{21}
0.25	0.50	1.848	-0.063	-0.052	1.797	1.052
0.26	0.03	1.570	-0.077	-0.012	1.606	0.980
0.70	0.15	2.132	0.051	-0.071	2.062	0.942
0.15	0.70	1.853	0.049	-0.068	1.841	0.915

$^*X_1 =$ mole fraction of toluene; $X_2 =$ mole fraction of chlorobenzene.

16.5 Chemical Reactions

In this section we shall look at the meaning of linear phenomenological laws in the context of chemical reactions. In a formalism in which the principle of detailed balance or microscopic reversibility is incorporated through the condition that forward rates of every elementary step balance the corresponding reverse rate, the Onsager reciprocity is implicit. No additional relations can be derived for the reaction rates if it is assumed that at equilibrium each elementary step is balanced by its reverse. Therefore, the main task in this section will be to relate the Onsager coefficients L_{ij} and the experimentally measured reaction rates. In our formalism the Onsager reciprocal relations will be automatically valid.

The entropy production due to chemical reactions is

$$\sigma = \sum_k \frac{A_k}{T} \frac{1}{V} \left(\frac{d\xi_k}{dt} \right) = \sum_k \frac{A_k}{T} v_k \qquad (16.5.1)$$

in which we have written v_k for the velocity of the kth reaction. In this case the thermodynamic forces are $F_k = (A_k/T)$ and the flows $J_k = v_k$. In Chapter 9 we have seen that *for a chemical reaction which can be identified as an elementary step*, the velocity v and the affinity A can be related to the forward and reverse reactions through the following relations:

$$v_k = (R_{kf} - R_{kr}) \qquad (16.5.2)$$

$$A_k = RT \ln \left(\frac{R_{kf}}{R_{kr}} \right) \qquad (16.5.3)$$

in which R_{kf} and R_{kr} are forward and reverse rates of the kth reaction and R is the gas constant. Using (16.5.3) in (16.5.2), we can write the velocity v_k as

$$v_k = R_{kf} \left(1 - e^{-A_k/RT} \right) \qquad (16.5.4)$$

a useful expression for discussing the linear phenomenological laws near thermodynamic equilibrium. It is important to keep in mind that (16.5.4) is valid only for an elementary step. Note that (16.5.3) incorporates the principle of detailed balance or microscopic reversibility according to which the forward and reverse reactions of every elementary step balance each other at equilibrium (which leads to the Onsager reciprocal relations). Also, the limit $A_k \to \infty$ implies that the velocity is entirely due to the forward reaction.

Equation (16.5.4) *does not* give the reaction velocity v_k as a function of the affinity A_k, because the term R_f has to be specified. There is no general thermodynamic expression relating velocities and affinities. Reaction velocities depend on many nonthermodynamic factors such as the presence of catalysts. (A catalyst does not have any effect on the state of equilibrium; also, because a catalyst changes the forward and reverse rate by the same factor, it does not alter the affinity either.) Close to thermodynamic equilibrium, however, there is a general linear relation between the two quantities. In this context, the general postulate of the linear phenomenological laws takes the form

$$v_k = \sum_j L_{kj} \frac{A_j}{T} \tag{16.5.5}$$

The coefficients L_{kj} can be related to the experimental quantities such as reaction rates as shown below.

SINGLE REACTION

For simplicity, let us consider a single reaction which is an elementary step. Then (16.5.4) becomes

$$v = R_f(1 - e^{-A/RT}) \tag{16.5.6}$$

At equilibrium $A = 0$. Let us denote the equilibrium value of the forward reaction rate by $R_{f,eq}$. Away from equilibrium, A has a nonzero value. By "close to equilibrium" we mean that

$$\frac{A}{RT} \ll 1 \tag{16.5.7}$$

When A is small compared to RT, and $R_f = R_{f,eq} + \Delta R_f$, we can expand (16.5.6) to obtain a linear relation between v and A:

$$v = R_{f,eq} \frac{A}{RT} + \cdots \tag{16.5.8}$$

to the leading order by ignoring smaller terms such as products of ΔR_f and A. Comparing (16.5.8) with the phenomenological law $v = LA/T$, we make the identification

$$L = \frac{R_{f,eq}}{R} = \frac{R_{r,eq}}{R} \tag{16.5.9}$$

where the last equality follows from the fact that the forward and reverse reaction rates of every elementary step are equal at equilibrium.

MANY REACTIONS

When the system consists of many reacting species and reactions, not all the reactions are independent. Take, for example, the following reactions:

$$O_2(g) + 2C(s) \rightleftharpoons 2CO(g) \tag{16.5.10}$$

$$O_2(g) + 2CO(g) \rightleftharpoons 2CO_2(g) \tag{16.5.11}$$

$$2O_2(g) + 2C(s) \rightleftharpoons 2CO_2(g) \tag{16.5.12}$$

The third reaction is the sum of the first two reactions. Therefore not all three are independent reactions. Thermodynamically this means that the affinity of the third reaction can be written as the sum of the first two. We have seen in Chapter 4 that the *affinity of a sum of reactions is the sum of the affinities*. Since the phenomenological relations are written in terms of independent thermodynamic forces, only the independent affinities are to be used. Also, without loss of generality we may consider affinities of elementary steps only because all reactions can be reduced to elementary steps.

If all the chemical reactions in the system are independent, then close to equilibrium, each velocity v_k is dependent on only the corresponding affinity and the equilibrium reaction rate is as in (16.5.8). There are no cross-coupling terms. In the general formalism, cross terms for chemical reactions appear when the total number of reactions is not the same as the number of independent reactions. In this case, some of the affinities can be expressed as linear functions of others. Let us take an example. For simplicity but without loss of generality, we consider a simple set of unimolecular reactions:

$$\begin{aligned} W \rightleftharpoons X \quad R_{1f}, \; R_{1r} \quad A_1 \quad v_1 \tag{16.5.13a} \\ X \rightleftharpoons Y \quad R_{2f}, \; R_{2r} \quad A_2 \quad v_2 \tag{16.5.13b} \\ W \rightleftharpoons Y \quad R_{3f}, \; R_{3r} \quad A_3 \quad v_3 \tag{16.5.13c} \end{aligned}$$

where the subscripts f and r stand for forward and reverse reaction rates. Only two out of the three reactions are independent, because the third can be expressed as the sum of the other two. Consequently, we have the relation

$$A_1 + A_2 = A_3 \tag{16.5.14}$$

The entropy production per unit volume due to these reactions is

$$\sigma = v_1 \frac{A_1}{T} + v_2 \frac{A_2}{T} + v_3 \frac{A_3}{T} \tag{16.5.15}$$

Using the relation between the affinities (16.5.14), this expression can be written in terms of two independent affinities A_1 and A_2:

$$\begin{aligned} \sigma &= (v_1 + v_3)\frac{A_1}{T} + (v_2 + v_3)\frac{A_2}{T} \\ &= v_1' \frac{A_1}{T} + v_2' \frac{A_2}{T} > 0 \end{aligned} \tag{16.5.16}$$

where $v'_1 = v_1 + v_3$ and $v'_2 = v_2 + v_3$. In terms of these independent velocities and affinities, the linear phenomenological laws may be written as

$$v'_1 = L_{11}\frac{A_1}{T} + L_{12}\frac{A_2}{T} \tag{16.5.17}$$

$$v'_2 = L_{21}\frac{A_1}{T} + L_{22}\frac{A_2}{T} \tag{16.5.18}$$

The relation between the phenomenological coefficients L_{ik} and the experimentally measured reaction rates can be obtained by using the general relation (16.5.4) between the velocities v_k and the affinities A_k. For example, close to equilibrium, i.e. when $(A_k/RT) \ll 1$, we can write v'_1 as

$$v'_1 = v_1 + v_3 = R_{1f}\left(1 - e^{-A_1/RT}\right) + R_{3f}\left(1 - e^{-A_3/RT}\right)$$

$$\approx R_{1f,eq}\frac{A_1}{RT} + R_{3f,eq}\frac{A_3}{RT} = \left(\frac{R_{1f,eq} + R_{3f,eq}}{R}\right)\frac{A_1}{T} + \frac{R_{3f,eq}}{R}\frac{A_2}{T} \tag{16.5.19}$$

using the fact that near equilibrium we have $R_{kf} \approx R_{kf,eq}$ the forward reaction rate at equilibrium. Comparing (16.5.19) with (16.5.17), we see that

$$L_{11} = \left(\frac{R_{1f,eq} + R_{3f,eq}}{R}\right) \quad \text{and} \quad L_{12} = \frac{R_{3f,eq}}{R} \tag{16.5.20}$$

Similarly, it is straightforward to show that

$$L_{22} = \left(\frac{R_{2f,eq} + R_{3f,eq}}{R}\right) \quad \text{and} \quad L_{21} = \frac{R_{3f,eq}}{R} \tag{16.5.21}$$

Thus one can relate the phenomenological coefficients L_{ik} to the reaction rates at equilibrium. We see that $L_{12} = L_{21}$. Since the principle of detailed balance or microscopic reversibility is incorporated into the formalism through $R_{3f} = R_{3r} = R_{3f,eq}$, the Onsager reciprocal relations are automatically valid.

ALTERNATIVE FORMS FOR σ

From the above considerations it is clear that the entropy production can be written in terms of A_2 and A_3 instead of A_1 and A_2. There is no unique way of writing the entropy production. In whatever way the independent affinities and velocities are chosen, the corresponding linear phenomenological coefficients can be obtained. The entropy production σ can be written in terms of different sets of independent reaction velocities and affinities:

$$\sigma = \sum_k v_k \frac{A_k}{T} = \sum_k v'_k \frac{A'_k}{T} > 0 \tag{16.5.22}$$

Equations (16.5.15) and (16.5.16) are examples. The number of independent reactions, and therefore the affinities, is constrained by the number of reacting species. In homogeneous systems in which the change in the concentrations of all the reacting species is only due to chemical reactions, we may choose the extents of reaction ξ_k to define the state of a system instead of the concentrations n_k. The chemical potentials μ_k are functions of ξ_k, p and T. However, since an extent of reaction relates the change in at least two reacting species, in a system consisting of r reacting species there are at most $(r-1)$ independent extents of reaction ξ_k. Thus all the chemical potentials can be expressed as $\mu_k(\xi_1, \xi_2, \xi_3, \dots, \xi_{r-1}, p, T)$. From this it is clear that, at any given pressure p and temperature T, there are only $(r-1)$ independent chemical potentials. Since the affinities A_k are linear functions of the chemical potentials, *in a system with r reacting species, there can be at most $(r-1)$ independent affinities*. (Sometimes this fact is derived using the "conservation of mass" in chemical reactions. Although this may be valid in ordinary chemical reactions, since mass is not conserved in nuclear reactions the argument is not general. In fact, mass is incidental to chemical reactions whose main consequence is the change in the number of molecules of the various reacting species.)

LINEARITY IN COUPLED REACTIONS

We have seen that the linear phenomenological laws are valid for chemical reactions with affinity A if the condition $A/RT \ll 1$ is satisfied. However, if the overall chemical reaction

$$X \longrightarrow Y \tag{16.5.23}$$

consists of m intermediates, $W_1, W_2, \dots, W_m$, one may still be justified in using the linearity even if $A/RT \ll 1$ is not valid. Let us suppose that the overall reaction (16.5.23) goes through the following series of reactions:

$$X \underset{}{\overset{(1)}{\rightleftharpoons}} W_1 \overset{(2)}{\rightleftharpoons} W_2 \overset{(3)}{\rightleftharpoons} W_3 \rightleftharpoons \cdots W_m \overset{(m+1)}{\rightleftharpoons} Y \tag{16.5.24}$$

The entropy production for this set of $(m+1)$ reactions is

$$T\sigma = A_1 v_1 + A_2 v_2 + \dots + A_{m+1} v_{m+1} \tag{16.5.25}$$

If the intermediate components W_k interconvert rapidly, then the reaction velocity of each of these reactions is essentially determined by the rate of the slowest reaction, which is called the **rate-determining step**. Let us assume that the last step $W_m \rightleftharpoons Y$ is the slow rate-determining step. The rate equations for this system are

$$\frac{d[\mathbf{X}]}{dt} = -v_1$$

$$\frac{d[\mathbf{W}_1]}{dt} = v_1 - v_2$$

$$\frac{d[\mathbf{W}_2]}{dt} = v_2 - v_3 \qquad (16.5.26)$$

$$\vdots$$

$$\frac{d[\mathbf{Y}]}{dt} = v_{m+1}$$

Because of the rapid interconversion, we may assume that a steady state is established for $[W_k]$ so that $d[W_k]/dt \approx 0$ (Such an assumption is used, for example, in obtaining the Michaelis–Menten rate law for enzyme kinetics.) This implies

$$v_1 = v_2 = \cdots = v_{m+1} = v \qquad (16.5.27)$$

Then the entropy production for the system becomes

$$T\sigma = (A_1 + A_2 + \ldots + A_{m+1})v = Av \qquad (16.5.28)$$

in which the overall affinity

$$A = A_1 + A_2 + \ldots + A_{m+1} \qquad (16.5.29)$$

Now if $A_k/RT \ll 1$ for each of the $(m+1)$ reactions, we are still in the region where the linear laws are valid, so from (16.5.8) we have

$$v_1 = R_{1f,eq}\frac{A_1}{RT} \quad v_2 = R_{2f,eq}\frac{A_2}{RT} \quad \cdots \quad v_{m+1} = R_{(m+1)f,eq}\frac{A_{m+1}}{RT} \quad (16.5.30)$$

in which $R_{1f,eq}$ is forward equilibrium reaction rate of reaction (1) in the scheme (16.5.24), etc.

In the above case, even if $A = \sum_{k=1}^{m+1} A_k \gg RT$ the linear phenomenological laws will be valid. A simple calculation (exc. 16.9) using (16.5.27), (16.5.28) and (16.5.30) shows that

$$v = \frac{R_{eff}}{RT}A \qquad (16.5.31)$$

in which the "effective reaction rate" R_{eff} is given by

$$\frac{1}{R_{eff}} = \frac{1}{R_{1f,eq}} + \frac{1}{R_{2f,eq}} + \frac{1}{R_{3f,eq}} + \ldots + \frac{1}{R_{(m+1)f,eq}} \qquad (16.5.32)$$

Since the overall reaction is not an elementary step but a result of many elementary steps, the relation $v = R_{eff}(1 - e^{-A/RT})$ is not valid.

Though we considered a coupled set of unimolecular reactions (16.5.24) to obtain (16.5.31), the result is more generally valid. Thus, *the linear phenomenological law is valid for an overall chemical reaction if $A/RT \ll 1$ for every elementary step in the reaction, and if concentrations of the reaction intermediates may be assumed to be in a steady state.*

16.6 Heat Conduction in Anisotropic Solids

In an anisotropic solid the flow of heat $\mathbf{J}_q$ may not be in the direction of the temperature gradient; a temperature gradient in one direction can cause the heat to flow in another direction. The entropy production is

$$\sigma = \sum_{i=1}^{3} \mathbf{J}_{qi} \frac{\partial}{\partial x_i}\left(\frac{1}{T}\right) \tag{16.6.1}$$

in which x_i are the Cartesian coordinates. The phenomenological laws for this system are

$$\mathbf{J}_{qi} = \sum_k L_{ik} \frac{\partial}{\partial x_k}\left(\frac{1}{T}\right) = \sum_k \left(\frac{-L_{ik}}{T^2}\right) \frac{\partial T}{\partial x_k} \tag{16.6.2}$$

For anisotropic solids the heat conductivity κ is a tensor of the second rank. The empirical Fourier law of heat conduction is then written as

$$\mathbf{J}_{qi} = -\sum_k \kappa_{ik} \frac{\partial T}{\partial x_k} \tag{16.6.3}$$

Comparison of (16.6.2) and (16.6.3) leads to

$$L_{ik} = T^2 \kappa_{ik} \tag{16.6.4}$$

Reciprocal relations $L_{ik} = L_{ki}$ then imply that

$$\kappa_{ik} = \kappa_{ki} \tag{16.6.5}$$

i.e. the heat conductivity is a symmetric tensor. However, for many solids, if the symmetry of the crystal structure itself implies that $\kappa_{ik} = \kappa_{ki}$, experimental verification of this equality would not confirm the reciprocal relations. On the other hand, solids with trigonal (C_3, C_{3i}), tetragonal (C_4, S_4, C_{4h}) and hexagonal (C_6, C_{3h}, C_{6h}) crystal symmetries imply that

$$\kappa_{12} = -\kappa_{21} \tag{16.6.6}$$

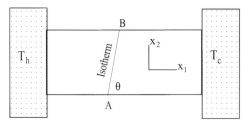

Figure 16.3 The method of Curie and Voigt to verify the reciprocal relations for anisotropic heat conduction. An anisotropic solid whose crystal symmetry implies $\kappa_{12} = -\kappa_{21}$ is placed in contact with two heat reservoirs of temperature T_h and T_c. If the reciprocal relations are valid then $\kappa_{12} = \kappa_{21} = 0$. If this is true, the isotherms should be perpendicular to the direction x_1, i.e. θ should be $90°$

If the reciprocal relations are valid, then

$$\kappa_{12} = \kappa_{21} = 0 \tag{16.6.7}$$

Equation (16.6.6) implies that a temperature gradient in the x-direction causes heat to flow in the positive y-direction but a gradient in the y-direction will cause heat to flow in the negative x-direction. Onsager's reciprocal relations imply that this is not possible. One method of experimental verification of this relation is due to Voigt and Curie (Fig. 16.3). Another method may be found in an article by Miller [1]. For crystals of apatite (calcium phosphate) and dolomite $(CaMg(CO_3)_2)$ it was found that $(\kappa_{12}/\kappa_{11}) < 0.0005$ [1], confirming the validity of reciprocal relations.

16.7 Electrokinetic Phenomena and the Saxen Relations

Electrokinetic phenomena are due to the coupling between electrical current and matter flow. Consider two chambers, 1 and 2, separated by a porous wall. If a voltage V is applied between the two chambers (Fig. 16.4), a current will flow until a pressure difference Δp is established at the steady state. This pressure difference is called the *electroosmotic pressure*. Conversely, if a fluid flow J from one chamber to another is achieved by a piston, an electric current I, called the *streaming current*, flows through the electrodes. As before, the thermodynamic description of these effects begins with the expression for the entropy production under the conditions specified above. In this case we essentially have

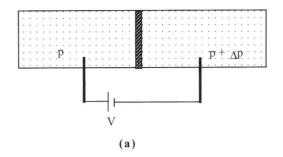

(a)

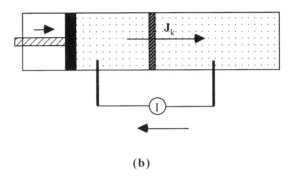

(b)

Figure 16.4 Electrokinetic phenomena. Two chambers containing electrolytes are separated by a porous wall or capillary. (a) An applied potential V generates a pressure difference Δp, called the *electroosmotic pressure*. (b) If the fluid is made to flow from one chamber to another by a piston, it generates an electrical current I, called the *streaming current*

a discontinuous system in which there are no gradients but differences in chemical potentials between the two chambers. For discontinuous systems the entropy production per unit volume σ is replaced by the total entropy production $d_i S/dt$. Furthermore, the entropy produced by the flow from chamber 1 to chamber 2 may be formally thought of as chemical reaction for which the difference in the electrochemical potential becomes the affinity. Thus we have

$$\frac{d_i S}{dt} = \sum_k \frac{\tilde{A}_k}{T} \frac{d\xi_k}{dt} \tag{16.7.1}$$

in which

$$\tilde{A}_k = \left(\mu_k^1 + z_k F \phi^1\right) - \left(\mu_k^2 + z_k F \phi^2\right) \tag{16.7.2}$$

$$d\xi_k = -dn_k^1 = dn_k^2 \tag{16.7.3}$$

In these equations the superscripts refer to the two chambers; z_k is the ion number of the component k, F is the Faraday constant, and ϕ is the electrical potential. For a relatively small difference in the pressure between the two chambers, since $(\partial \mu_k / \partial p) = v_k$, the partial molar volume, we may write

$$(\mu_k^1 - \mu_k^2) = v_k \Delta p \tag{16.7.4}$$

Equation (16.7.1) may now be written as

$$\frac{d_i S}{dt} = \frac{1}{T} \sum_k \left(-v_k \frac{dn_k^1}{dt} \right) \Delta p + \frac{1}{T} \sum_k (-I_k) \Delta \phi \tag{16.7.5}$$

in which $\Delta \phi = \phi^1 - \phi^2$ and $I_k = z_k F dn_k^1 / dt$, the electric current due to the flow of component k. Combining all the matter flow terms and the ion flow terms, equation (16.7.5) can now be written in the compact form

$$\frac{d_i S}{dt} = \frac{J \Delta p}{T} + \frac{I \Delta \phi}{T} \tag{16.7.6}$$

where

$$J = -\sum_k v_k \frac{dn_k^1}{dt} \quad \text{is the ``volume flow''} \tag{16.7.7}$$

$$I = -\sum_k I_k \quad \text{is the electric current} \tag{16.7.8}$$

The phenomenological equations that follow from (16.7.6) are

$$I = L_{11} \frac{\Delta \phi}{T} + L_{12} \frac{\Delta p}{T} \tag{16.7.9}$$

$$J = L_{21} \frac{\Delta \phi}{T} + L_{22} \frac{\Delta p}{T} \tag{16.7.10}$$

The reciprocal relations are

$$L_{12} = L_{21} \tag{16.7.11}$$

Experimentally, the following quantities can be measured:

• The streaming potential

$$\left(\frac{\Delta \phi}{\Delta p} \right)_{I=0} = -\frac{L_{12}}{L_{11}} \tag{16.7.12}$$

• Electroosmosis

$$\left(\frac{J}{I} \right)_{\Delta p=0} = \frac{L_{21}}{L_{11}} \tag{16.7.13}$$

- Electroosmotic pressure

$$\left(\frac{\Delta p}{\Delta \phi}\right)_{J=0} = -\frac{L_{21}}{L_{22}} \tag{16.7.14}$$

- Streaming current

$$\left(\frac{I}{J}\right)_{\Delta\phi=0} = \frac{L_{12}}{L_{22}} \tag{16.7.15}$$

As a consequence of the reciprocal relations $L_{12} = L_{21}$, we see from (16.7.12) to (16.7.15) that

$$\left(\frac{\Delta \phi}{\Delta p}\right)_{I=0} = -\left(\frac{J}{I}\right)_{\Delta p=0} \tag{16.7.16}$$

$$\left(\frac{\Delta p}{\Delta \phi}\right)_{J=0} = -\left(\frac{I}{J}\right)_{\Delta\phi=0} \tag{16.7.17}$$

These two relations, called the Saxen relations, were obtained originally by kinetic considerations for particular systems, but by virtue of the formalism of nonequilibrium thermodynamics we see their general validity.

16.8   Thermal Diffusion

The interaction between heat and matter flows produces two effects, the **Soret effect** and the **Dufour effect**. In the Soret effect, heat flow drives a flow of matter. In the Dufour effect, concentration gradients drive a heat flow. The reciprocal relations in this context can be obtained by writing the entropy production due to diffusion and heat flow:

$$\begin{aligned}
\sigma &= \mathbf{J}_u \cdot \nabla\left(\frac{1}{T}\right) - \sum_{k=1}^{w} \mathbf{J}_k \cdot \nabla\left(\frac{\mu_k}{T}\right) \\
&= \left(\mathbf{J}_u - \sum_{k=1}^{w} \mathbf{J}_k \mu_k\right) \cdot \nabla\left(\frac{1}{T}\right) - \sum_{k=1}^{w} \mathbf{J}_k \cdot \frac{1}{T}\nabla\mu_k
\end{aligned} \tag{16.8.1}$$

This expression, however, does not quite separate the thermal and concentration gradients as we would like, because the term $\nabla\mu_k$ contains the gradient of T (due to the fact that μ_k is a function of T, n_k and p). The explicit form of $\nabla\mu_k$ can be written using the relation

$$d\mu_k = (d\mu_k)_{p,T} + \left(\frac{\partial \mu_k}{\partial T}\right)_{n_k,p} dT + \left(\frac{\partial \mu_k}{\partial p}\right)_{n_k,T} dp \tag{16.8.2}$$

in which

$$(d\mu_k)_{p,T} = \left(\frac{\partial\mu_k}{\partial n}\right)_{p,T} dn_k \text{ variation due to concentration only.}$$

Now the term*

$$\left(\frac{\partial\mu_k}{\partial T}\right)_{n_k,p} = \frac{\partial}{\partial T}\left(\frac{\partial g}{\partial n_k}\right)_{p,T} = \left(\frac{\partial}{\partial n_k}\left(\frac{\partial g}{\partial T}\right)\right)_{p,T} = -\left(\frac{\partial s}{\partial n_k}\right)_{p,T}$$

Thus we see that (16.8.2) can be written as

$$d\mu_k = (d\mu_k)_{p,T} - s_k\,dT + \left(\frac{\partial\mu_k}{\partial p}\right)_{n_kT} dp \qquad (16.8.3)$$

in which the partial molar entropy $s_k = (\partial s/\partial n_k)_{p,T}$ (section 5.5). In this section we will consider systems in mechanical equilibrium for which $dp = 0$. Since the variation of any quantity Y with position can be written as $dY = (\nabla Y)\cdot d\mathbf{r}$, it follows that using (16.8.3) we can write

$$\nabla\mu_k = (\nabla\mu_k)_{p,T} - s_k\nabla T$$
$$= (\nabla\mu_k)_{p,T} + s_kT^2\nabla\frac{1}{T} \qquad (16.8.4)$$

Here we have used the fact that $dp = 0$ because the system is assumed to be in mechanical equilibrium. Substituting (16.8.4) into (16.8.1) we obtain

$$\sigma = \left(\mathbf{J}_u - \sum_{k=1}^{w}\mathbf{J}_k(\mu_k + Ts_k)\right)\cdot\nabla\left(\frac{1}{T}\right) - \sum_{k=1}^{w}\mathbf{J}_k\cdot\frac{1}{T}(\nabla\mu_k)_{p,T} \qquad (16.8.5)$$

Now, using the relation $g = h - Ts$, it is easily seen that $\mu_k + Ts_k = h_k$ where $h_k = (\partial h/\partial n_k)_{p,T}$ is the partial molar enthalpy. With this identification, a heat current that takes into account matter current can be defined as

$$\mathbf{J}_q \equiv \mathbf{J}_u - \sum_{k=1}^{w}h_k\mathbf{J}_k \qquad (16.8.6)$$

In a closed system under constant pressure, the change in enthalpy due to a change in composition is equal to the heat exchanged with the exterior. In an open system of a fixed volume, the heat exchanged is the difference between the change in energy and the change in enthalpy due to the matter flow. The vector $\mathbf{J}_q$ defined in (16.8.6) is called the **reduced heat flow**. In terms of $\mathbf{J}_q$ the entropy

---

* g and h are densities corresponding to G and H.

production may be written as

$$\sigma = \mathbf{J}_q \cdot \nabla \left(\frac{1}{T} \right) - \sum_{k=1}^{w} \mathbf{J}_k \cdot \frac{(\nabla \mu_k)_{T,p}}{T} \tag{16.8.7}$$

For simplicity, we shall consider a two-component system so that $w = 2$. As we noted in section 16.4 on diffusion, because of the Gibbs–Duhem relation at constant p and T, the chemical potentials are not independent. From (16.4.5) we have the following relation:

$$n_1 (\nabla \mu_1)_{p,T} + n_2 (\nabla \mu_2)_{p,T} = 0 \tag{16.8.8}$$

In addition, for no volume flow, we have from (16.4.6) the condition

$$\mathbf{J}_1 v_1 + \mathbf{J}_2 v_2 = 0 \tag{16.8.9}$$

As when obtaining (16.4.8), relations (16.8.8) and (16.8.9) can be used in (16.8.7) to give

$$\sigma = \mathbf{J}_q \cdot \nabla \left(\frac{1}{T} \right) - \frac{1}{T} \left(1 + \frac{v_1 n_1}{v_2 n_2} \right) \mathbf{J}_1 \cdot (\nabla \mu_1)_{p,T} \tag{16.8.10}$$

We can now write the phenomenological laws for the flows of heat and matter:

$$\mathbf{J}_q = L_{qq} \nabla \left(\frac{1}{T} \right) - L_{q1} \frac{1}{T} \left(1 + \frac{v_1 n_1}{v_2 n_2} \right) (\nabla \mu_1)_{p,T} \tag{16.8.11}$$

$$\mathbf{J}_1 = L_{1q} \nabla \left(\frac{1}{T} \right) - L_{11} \frac{1}{T} \left(1 + \frac{v_1 n_1}{v_2 n_2} \right) (\nabla \mu_1)_{p,T} \tag{16.8.12}$$

To relate the terms in this expression to Fourier law of heat conduction and Fick's law of diffusion, we write the gradients as $\nabla \mu_1 = (\partial \mu_1 / \partial n_1) \nabla n_1$ and $\nabla(1/T) = -(1/T^2) \nabla T$, so the two flows become:

$$\mathbf{J}_q = -\frac{L_{qq}}{T^2} \nabla T - L_{q1} \frac{1}{T} \left(1 + \frac{v_1 n_1}{v_2 n_2} \right) \frac{\partial \mu_1}{\partial n_1} \nabla n_1 \tag{16.8.13}$$

$$\mathbf{J}_1 = -\frac{L_{1q}}{T^2} \nabla T - L_{11} \frac{1}{T} \left(1 + \frac{v_1 n_1}{v_2 n_2} \right) \frac{\partial \mu_1}{\partial n_1} \nabla n_1 \tag{16.8.14}$$

We can now identify the diffusion coefficient and the heat conductivity.

$$D_1 = L_{11} \frac{1}{T} \left(1 + \frac{v_1 n_1}{v_2 n_2} \right) \frac{\partial \mu_1}{\partial n_1} \qquad \kappa = \frac{L_{qq}}{T^2} \tag{16.8.15}$$

and we have the reciprocal relations

$$L_{q1} = L_{1q} \tag{16.8.16}$$

The cross flow $-(L_{1q}/T^2)\nabla T$ is usually written as $-n_1 D_T \nabla T$, in which D_T is the **coefficient of thermal diffusion**, so that the flow of matter is proportional to n_1. The ratio of the thermal diffusion coefficient to the ordinary diffusion coefficient is the **Soret coefficient**.

$$s_T = \frac{D_T}{D_1} = \frac{L_{1q}}{D_1 T^2 n_1} \tag{16.8.17}$$

In a closed system with a temperature gradient (Fig. 16.5) a concentration gradient is set up due to the heat flow. The stationary state concentration gradient can be obtained by setting $\mathbf{J}_1 = 0$:

$$\mathbf{J}_1 = -\frac{L_{1q}}{T^2}\nabla T - D_1\nabla n_1 = 0 \tag{16.8.18}$$

Since $L_{1q}/T^2 = n_1 D_T$, the ratio of the two gradients is

$$\frac{\nabla n_1}{\nabla T} = -\frac{n_1 D_T}{D_1} = -n_1 s_T \tag{16.8.19}$$

The Soret coefficient has the dimensions of T^{-1}. It is generally small, in the range 10^{-2} to 10^{-3} K for electrolytes, nonelectrolytes and gases [10] but it might become larger in polymer solutions. Thermal diffusion has been utilized to separate isotopes [11].

The heat current carried by a flow of matter is identified by the **Dufour coefficient** D_d. Since the heat carried by the matter flow is proportional to the concentration n_1, the Dufour coefficient is defined by

$$n_1 D_d = L_{q1}\frac{1}{T}\left(1 + \frac{v_1 n_1}{v_2 n_2}\right)\frac{\partial \mu_1}{\partial n_1} \tag{16.8.20}$$

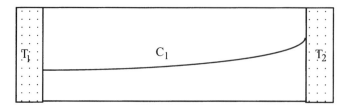

Figure 16.5 Thermal diffusion: a temperature gradient and the consequent flow of heat causes a concentration gradient

Since $L_{1q}/T^2 = n_1 D_T$, the Onsager reciprocal relations $L_{1q} = L_{q1}$, predict the relation

$$\frac{D_d}{D_T} = T\left(1 + \frac{v_1 n_1}{v_2 n_2}\right)\frac{\partial \mu_1}{\partial n_1} \qquad (16.8.21)$$

for the ratio of the Dufour and thermal diffusion coefficients. This has been confirmed experimentally.

Thus nonequilibrium thermodynamics gives a unified theory of irreversible processes. Onsager reciprocal relations are general, valid for all systems in which linear phenomenological laws apply.

References

1. Miller, D. G., *Chem. Rev.*, **60** (1960) 15–37.
2. Thomson, W., *Proc. R. Soc. (Edinburgh)*, **3** (1854), 225.
3. Onsager, L., *Phys. Rev.*, **37** (1931), 405–426.
4. Curie, P., *Oeuvres de Pierre Curie*. 1908, Paris: Gauthier-Villars.
5. Prigogine, I., *Etude Thermodynamique des Processus Irreversibles*. 1947, Liège: Desoer.
6. Sugawara, H., Nagata, K., and Goto, K., *Metall. Trans. B.*, **8** (1977), 605.
7. Spera, F. J. and Trial, A., *Science*, **259** (1993), 204–206.
8. Kett, T. K. and Anderson, D. K., *J. Phys. Chem.*, **73** (1969), 1268–1274.
9. Prigogine, I., *Introduction to Thermodynamics of Irreversible Processes*. 1967, New York: John Wiley.
10. Jost, J., *Diffusion in Solids, Liquids and Gases*. 1960, New York: Academic Press.
11. Jones, R. C. and Furry, W. H., *Rev. Mod. Phys.*, **18** (1946), 151–224.

Further Reading

- Katchalsky, A., *Nonequilibrium Thermodynamics and Biophysics*. 1965, Cambridge MA.: Harvard University Press.
- Jost, J., *Diffusion in Solids, Liquids and Gases*. 1960, New York: Academic Press.
- de Groot, S. R. and Mazur, P., *Non-Equilibrium Thermodynamics*. 1969, Amsterdam: North Holland.
- Forland, K. S., *Irreversible Thermodynamics: Theory and Applications*. 1988, New York: John Wiley.
- Haase, R., *Thermodynamics of Irreversible Processes*. 1990, New York: Dover.
- Kuiken, G. D. C., *Thermodynamics of Irreversible Processes*. 1994, New York: John Wiley.
- Landsberg, P. T., *Nature*, **238** (1972) 229–231.

- Samohyl, I., *Thermodynamics of Irreversible Processes in Fluid Mixtures: Approached by Rational Thermodynamics*. 1987, Leipzig: B. G. Teubner.
- Stratonovich, R. L., *Nonlinear Nonequilibrium Thermodynamics*. 1992, New York: Springer-Verlag.
- Wisniewski, S., *Thermodynamics of Nonequilibrium Processes*. 1976, Dordrecht: D. Reidel.
- The time evolution of the concentration to its steady state in the case of thermal diffusion was analyzed by S. R. de Groot in *Physica*, **9** (1952), 699.
- The steady states with thermal diffusion and chemical reactions were analyzed by I. Prigogine and R. Buess in *Acad. R. Belg.*, **38** (1952), 711.

Exercises

16.1 For a positive definite 2×2 matrix, show that (16.1.9) must be valid.

16.2 Give examples of the equality (16.2.11) hypothesized by Onsager. Give examples of situations in which it is not valid.

16.3 Estimate the cross-diffusion current of one component due to a gradient of another from the data given in Table 16.2 for reasonable gradients.

16.4 Obtain (16.4.7) from (16.4.1) and (16.4.5).

16.5 For diffusion in a three-component system, show that the entropy production is

$$\sigma = \mathbf{F}_1 \cdot \mathbf{J}_1 + \mathbf{F}_2 \cdot \mathbf{J}_2$$

in which the thermodynamic forces $\mathbf{F}_1$ and $\mathbf{F}_2$ are

$$\mathbf{F}_1 = -\frac{1}{T}\left[\nabla\mu_1 + \frac{n_1 v_1}{n_3 v_3}\nabla\mu_1 + \frac{n_2 v_1}{n_3 v_3}\nabla\mu_2\right]$$

and

$$\mathbf{F}_2 = -\frac{1}{T}\left[\nabla\mu_2 + \frac{n_2 v_2}{n_3 v_3}\nabla\mu_2 + \frac{n_1 v_2}{n_3 v_3}\nabla\mu_1\right]$$

16.6 For diffusion in a three-component system, show that the phenomenological coefficients are given by (16.4.24) to (16.4.27). (You can obtain this using Mathematica or Maple.)

16.7 For diffusion in a three-component system, write equations (16.4.17) to (16.4.27) in matrix notation.

16.8 For one of the reactions in Chapter 9, specify the conditions in which the linear phenomenological laws may be used.

16.9 Using (16.5.27), (16.5.28) and (16.5.30) show that a linear phenomenological relation $v = (R_{\text{eff}}/RT)A$ (equation 16.5.31) can be obtained in which the "effective reaction rate" R_{eff} is given by

$$\frac{1}{R_{\text{eff}}} = \frac{1}{R_{1\text{f,eq}}} + \frac{1}{R_{2\text{f,eq}}} + \frac{1}{R_{3\text{f,eq}}} + \ldots + \frac{1}{R_{(m+1)\text{f,eq}}}$$

17 NONEQUILIBRIUM STATIONARY STATES AND THEIR STABILITY: LINEAR REGIME

17.1 Stationary States Under Nonequilibrium Conditions

A system can be maintained in a nonequilibrium state through a flow of energy and matter. In the previous chapter we have seen some examples of non-equilibrium systems in the linear regime. In this section we will study some of these systems in more detail to understand the nature of the nonequilibrium states. In general, a system that is not in thermodynamic equilibrium need not be in a stationary (time-independent) state. Indeed, as we shall see in Chapters 18 and 19, systems that are far from equilibrium, for which the linear phenomenological laws are not valid, can exhibit very complex behavior such as concentration oscillations, propagating waves and even chaos. In the linear regime, however, all systems evolve to stationary states in which there is constant entropy production. Let us consider some simple examples to understand the entropy production and entropy flow in nonequilibrium stationary states in the linear regime.

THERMAL GRADIENTS

Let us consider a system of length L in contact with a hot thermal reservoir, at a temperature T_h, at one end and a cold thermal reservoirs, at temperature T_c, at the other (Fig. 17.1). In section 3.5, and in more detail in Chapter 16, we discussed the entropy production due to heat flow but we did not consider entropy balance in detail. We assume here that the conduction of heat is the only irreversible process. For this system, using Table 15.1 for the flows and forces, we see that the entropy production per unit volume is

$$\sigma = \mathbf{J}_q \cdot \nabla \frac{1}{T} \tag{17.1.1}$$

If we assume that the temperature gradient is only in the x direction, σ per unit length is given by

$$\sigma(x) = J_{qx} \frac{\partial}{\partial x} \frac{1}{T(x)} = -J_{qx} \frac{1}{T^2} \frac{\partial T(x)}{\partial x} \tag{17.1.2}$$

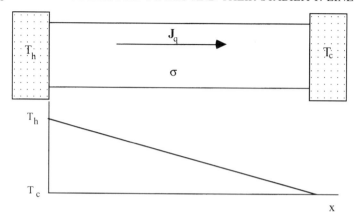

Figure 17.1 A simple thermal gradient maintained by a constant flow of heat. In the stationary state, the entropy current $J_{s,\text{out}} = d_iS/dt + J_{s,\text{in}}$. The stationary state can be obtained either as a solution of the Fourier equation for heat conduction or by using the theorem of minimum entropy production. Both lead to a temperature $T(x)$ that is a linear function of the position x

The total entropy production

$$\frac{d_iS}{dt} = \int_0^L \sigma(x)dx = \int_0^L J_q\left(\frac{\partial}{\partial x}\frac{1}{T}\right)dx \qquad (17.1.3)$$

Such a system reaches a state with stationary temperature distribution and a uniform heat flow $\mathbf{J}_q$. (A stationary temperature $T(x)$ implies that the heat flow is uniform; otherwise there will be an accumulation or depletion of heat, resulting in a time-dependent temperature.) The evolution of the temperature distribution can be obtained explicitly by using the Fourier laws of heat conduction:

$$C\frac{\partial T}{\partial t} = -\nabla \cdot \mathbf{J}_q \qquad \mathbf{J}_q = -\kappa\nabla T \qquad (17.1.4)$$

in which C is the heat capacity per unit volume and κ is the coefficient of heat conductivity. The first of these equations expresses the conservation of energy when the changes in energy are entirely due to heat flow. (For Fourier, who supported the caloric theory, this equation expressed the conservation of caloric.) For a one-dimensional system these two equations can be combined to obtain

$$C\frac{\partial T}{\partial t} = \kappa\frac{\partial^2 T}{\partial x^2} \qquad (17.1.5)$$

It is easy to see that the stationary state, $\partial T/\partial t = 0$, is one in which $T(x)$ is a linear function of x (Fig. 17.1) and $\mathbf{J}_q =$ constant. A stationary state also implies that all other thermodynamic quantities such as the total entropy S of the system are constant:

$$\frac{dS}{dt} = \frac{d_e S}{dt} + \frac{d_i S}{dt} = 0 \qquad (17.1.6)$$

The total entropy can be constant only when the entropy flowing out of the system is equal to the entropy entering the system plus the entropy produced in the system. This can be seen explicitly by evaluating the integral (17.1.3) (in which J_q is a constant):

$$\frac{d_i S}{dt} = \int_0^L J_q \left(\frac{\partial}{\partial x} \frac{1}{T} \right) dx = \frac{J_q}{T} \bigg|_0^L = \frac{J_q}{T_c} - \frac{J_q}{T_h} > 0 \qquad (17.1.7)$$

in which we can identify (J_q/T_h) as the entropy flowing into the system, $J_{s,\text{in}}$, and (J_q/T_c) as the entropy flowing out of the system, $J_{s,\text{out}}$. The entropy exchanged with the exterior is $d_e S/dt = [(J_q/T_h) - (J_q/T_c)]$. Note that the positivity of the entropy production requires that J_q be positive. Thus we have the entropy balance

$$\frac{d_i S}{dt} + (J_{s,\text{in}} - J_{s,\text{out}}) = \frac{d_i S}{dt} + \frac{d_e S}{dt} = 0 \qquad (17.1.8)$$

Since $d_i S/dt > 0$ the entropy exchanged with the exterior is $d_e S/dt = (J_{s,\text{in}} - J_{s,\text{out}}) < 0$. The nonequilibrium state is maintained through the exchange of negative entropy with the outside world; the system discards the entropy produced by the irreversible processes.

OPEN CHEMICAL SYSTEMS

In an open chemical system that exchanges matter and energy with the exterior, we can identify the energy and entropy flows associated with the exchange of matter and energy. Using the kinetic equations, we can obtain the stationary state. As an example, let us consider a chemical system undergoing a monomolecular reaction such as isomerization:

$$A \underset{k_{1r}}{\overset{k_{1f}}{\rightleftharpoons}} X \underset{k_{2r}}{\overset{k_{2f}}{\rightleftharpoons}} B \qquad (17.1.9)$$

The associated entropy production per unit volume is

$$\sigma = \frac{A_1}{T} v_1 + \frac{A_2}{T} v_2 > 0 \qquad (17.1.10)$$

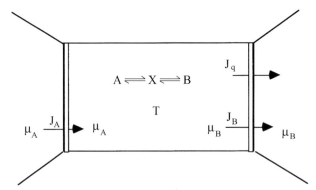

Figure 17.2 An open chemical system in which the chemical potentials μ_C and μ_B are maintained at a given nonequilibrium value by an inflow of component C and an outflow of component B. In this system the concentration of X is maintained at a nonequilibrium value. The system is also maintained at a constant temperature by removal of the heat of reaction

in which A_k and v_k $(k = 1, 2)$ are the affinities and velocities of the two reactions respectively. As we discussed in section 9.5, if R_{kf} is the forward reaction rate and R_{kr} is the reverse reaction rate, then

$$v_k = R_{kf} - R_{kr} \quad \text{and} \quad A_k = RT \ln\left(\frac{R_{kf}}{R_{kr}}\right) \tag{17.1.11}$$

We shall assume that the system is well mixed to maintain homogeneous concentrations and temperature. As illustrated in Fig. 17. 2, this system is in contact with reservoir with chemical potentials μ_A and μ_B, and the heat of reaction is compensated by a heat flow that keeps the system at constant temperature.

In a stationary state the total entropy of the system remains constant, i.e.

$$\frac{dS}{dt} = \frac{d_e S}{dt} + \frac{d_i S}{dt} = 0 \quad \text{where} \quad \frac{d_i S}{dt} = \int_V \sigma \, dV > 0 \tag{17.1.12}$$

which means the entropy exchange with the exterior must be negative:

$$\frac{d_e S}{dt} = -\frac{d_i S}{dt} < 0 \tag{17.1.13}$$

As in the case of thermal conduction, we can relate the entropy exchange to the inflow of A and outflow of B due to contact with the chemical reservoirs. Section

15.5 (equation 15.5.7) showed how the entropy current $\mathbf{J}_s$ is given by

$$\mathbf{J}_s = \frac{\mathbf{J}_u}{T} - \sum_k \frac{\mu_k \mathbf{J}_k}{T} \tag{17.1.14}$$

in which $\mathbf{J}_u$ is the energy flow. The total entropy flow into the system, d_eS/dt, is obtained by integrating $\mathbf{J}_s$ on the boundaries. Thus we can write

$$\frac{d_eS}{dt} = \frac{1}{T}\frac{d\Phi}{dt} - \frac{\mu_A}{T}\frac{d_eN_A}{dt} - \frac{\mu_B}{T}\frac{d_eN_B}{dt} < 0 \tag{17.1.15}$$

in which $d\Phi/dt$ is the total flow of energy (which must be distinguished from the rate of change of the total internal energy dU/dt) and dN_A/dt, dN_B/dt are the total flows of A and B. From (15.5.9) the entropy current can also be written as $\mathbf{J}_s = \mathbf{J}_q/T + \sum_k s_k \mathbf{J}_k$, in which the partial molar entropy is $s_k = (\partial s/\partial n_k)_T$ and $\mathbf{J}_q$ is the heat flow. If there is no flow of heat, the entropy flow becomes[*]

$$\frac{d_eS}{dt} = s_A\frac{d_eN_A}{dt} + s_B\frac{d_eN_B}{dt} \tag{17.1.16}$$

which, as noted above, must be negative for a nonequilibrium system. This means that the species flowing out of the system must carry more entropy than the species entering the system.

The stationary value of [X] is easily obtained from the kinetic equations:

$$\begin{aligned}
\frac{d[X]}{dt} &= v_1 - v_2 = (R_{1f} - R_{1r}) - (R_{2f} - R_{2r}) \\
&= k_{1f}[A] - k_{1r}[X] - k_{2f}[X] + k_{2r}[B]
\end{aligned} \tag{17.1.17}$$

Though it is more common to write the kinetic equations (17.1.17) in terms of concentrations, writing them in terms of velocities is more general—for it does not presume a rate law—and more convenient for formulating the thermodynamics of chemical reactions. The stationary state solution, $d[X]/dt = 0$, for (17.1.17) is simply

$$v_1 = v_2 \tag{17.1.18}$$

or $$[X] = \frac{k_{1f}[A] + k_{2r}[B]}{k_{1r} + k_{2f}} \tag{17.1.19}$$

---

[*] Note that the units of s_A and s_B are (entropy/mol).

If we have a series of coupled reactions:

$$X \overset{1}{\rightleftharpoons} W_1 \overset{2}{\rightleftharpoons} W_2 \rightleftharpoons \cdots \rightleftharpoons W_{n-1} \overset{n}{\rightleftharpoons} Y \tag{17.1.20}$$

with an inflow of M and an outflow of N, the above result for the steady state can be generalized (exc. 17.4) to

$$v_1 = v_2 = \ldots = v_n \tag{17.1.21}$$

in which the v_k are velocities of the indicated reactions.

ENTROPY PRODUCTION IN ELECTRICAL CIRCUIT ELEMENTS

The irreversible conversion of electrical energy into heat in electrical circuit elements, such as resistors, capacitors and inductances also leads to entropy production. The thermodynamic formalism of circuit elements can be developed by considering the changes in the energies associated with them. Section 10.1 showed that in the presence of a field we have

$$dU = T\,dS - p\,dV + \sum_k \mu_k \, dN_k + \sum_k Fz_k \phi_k \, dN_k \tag{17.1.22}$$

in which F is the Faraday constant and z_k the ion number; $Fz_k \, dN_k$ represents the amount of charge transferred dQ. If this charge is transferred from a potential ϕ_1 to a potential ϕ_2 by an irreversible process within a system, the entropy production is

$$\begin{aligned}
\frac{d_iS}{dt} &= \frac{1}{T}\sum_k \mu_k \frac{dN_k}{dt} - \frac{(\phi_2 - \phi_1)}{T}\sum_k Fz_k \frac{dN_k}{dt} \\
&= \frac{1}{T}\sum_k A_k v_k - \frac{(\phi_2 - \phi_1)}{T}\frac{dQ}{dt}
\end{aligned} \tag{17.1.23}$$

The first term is the entropy production due to chemical reactions, which can be dropped when considering only electrical circuit elements. For a resistor and a capacitor, $(\phi_1 - \phi_2)$ in the second term may be identified as the voltage V across the element and dQ/dt as the electric current I. If R is the resistance, according to Ohm's law, the voltage across the resistor $V_R = (\phi_1 - \phi_2) = IR$. The entropy production is

$$\frac{d_iS}{dt} = \frac{V_R I}{T} = \frac{RI^2}{T} > 0 \tag{17.1.24}$$

In this expression RI^2 is the well-known Ohmic heat produced per unit time by a current passing through a resistor. The entropy production is simply the rate of Ohmic heat generation divided by the temperature.

For a capacitor with capacitance C, the voltage decreases dV_C with transfer of charge dQ is given by $dV_C = -dQ/C$. The entropy production is therefore

$$\frac{d_i S}{dt} = \frac{V_C I}{T} = \frac{V_C}{T}\frac{dQ}{dt} = -\frac{C}{T}V_C\frac{dV_C}{dt}$$
$$= -\frac{1}{T}\frac{d}{dt}\left(\frac{CV_C^2}{2}\right) = -\frac{1}{T}\frac{d}{dt}\left(\frac{Q^2}{2C}\right) > 0 \tag{17.1.25}$$

where the term $CV_C^2/2 = Q^2/2C$ is the electrostatic energy stored in a capacitor. The entropy production is the rate of loss of this energy divided by its temperature. An ideal capacitor, once charged, will keep its charge indefinitely. Within such an ideal capacitor there is no dissipation of energy or entropy production. But all real capacitors will eventually lose their charge and reach equilibrium; equation (17.1.25) corresponds to the entropy production due to this irreversible process. (The internal discharging of a capacitor is the reaction $e^- + M^+ \rightarrow M$, in which M are the atoms that carry the charge. Note also that the flow of charge into a capacitor by the application of an external voltage corresponds to $d_e S$.)

The entropy production due to an inductance can be written in a similar manner, by noting that the energy stored in an inductance L carrying current I is equal to $LI^2/2$ and the voltage across it is $V_L = -L\,dI/dt$ (exc. 17.5). This energy is stored in the magnetic field. The entropy production associated with the dissipation of this energy is

$$\frac{d_i S}{dt} = -\frac{1}{T}\frac{d}{dt}\left(\frac{LI^2}{2}\right) = -\frac{LI}{T}\frac{dI}{dt} = \frac{V_L I}{T} > 0 \tag{17.1.26}$$

As in the case of an ideal capacitor, in an ideal inductance there is no loss of energy; a current once started will continue to exist indefinitely, as if in a perfect superconductor. In real inductances, however, the current decays with time. The entropy production for this irreversible process is given by (17.1.26).

The entropy production in circuit elements (equations 17.1.24 to 17.1.26) is in the form of a product of a thermodynamic force and a flow. In each case we can write the following linear phenomenological law relating the flows and the forces:

$$I = L_R\frac{V_R}{T} \tag{17.1.27}$$

$$I = L_C\frac{V_C}{T} \tag{17.1.28}$$

$$I = -L_L\frac{V_L}{T} \tag{17.1.29}$$

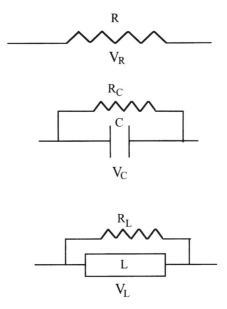

Figure 17.3 Elementary circuit elements, such as a resistor R, a capacitor C and an inductance L, also dissipate energy and produce entropy. In the thermodynamic formalism there are no ideal circuit elements with no dissipation of energy. Linear phenomenological laws give expressions for the rate of entropy production and dissipation of energy

in which L_R, L_C, L_L are linear phenomenological coefficients. In the case of the resistor, we identify (L_R/T) with the resistance $(1/R)$, in accordance with Ohm's law. For the capacitor we may think of an internal resistance $R_C = (T/L_C)$ that represents the slow dissipation of the charge. Equation (17.1.28) may be represented by an equivalent circuit (Fig. 17.3). By replacing I with dQ/dt in (17.1.28) we obtain a differential equation for the decay of the charge in the capacitor. Similarly, for the inductance, we identify the internal resistance by $R_L = (T/L_L)$. Equation (17.1.29) represents the irreversible decay of current in an inductance. In all three cases the entropy production is equal to the product of the voltage and the current divided by the temperature.

17.2 The Theorem of Minimum Entropy Production

In the previous section we have seen some examples of nonequilibrium stationary states in which one or more thermodynamic forces were maintained at a nonzero value. In the case of heat conduction, using Fourier's law of heat conduction (17.1.5), we found that the stationary state corresponded to a constant heat flow. In an open chemical system (17.1.9) in which the concentrations of A and B were maintained constant, using the kinetic equation (17.1.17) we found that in the stationary state the velocities of the two reactions were equal. This result could be extended to the case of many intermediates (17.1.20), in which case all the velocities of the reactions were equal in the stationary state.

And in the previous chapter we have seen how different flows J_k, $k = 1, 2, \ldots, n$, are coupled to the thermodynamic forces F_k. In such situations the system may be maintained away from equilibrium by constraining some forces F_k, $k = 1, 2, \ldots, s$, to be at a fixed nonzero value, while leaving the remaining forces F_k, $k = s + 1, \ldots, n$, free. In these cases, one often finds that the flows corresponding to the constrained forces reach a constant, $J_k = \text{constant}$, $k = 1, 2, \ldots, n$, whereas the unconstrained forces adjust so as to make their corresponding flows zero, $J_k = 0$, $k = s + 1, \ldots, n$. An example is thermal diffusion in which the stationary state corresponds to zero matter flow and constant heat flow (Fig. 16.5). In the linear regime, where the Onsager reciprocal relations are valid, all stationary states are characterized by the following general extremum principle [1,2]:

"In the linear regime, the total entropy production in a system subject to flow of energy and matter, $d_i S/dt = \int \sigma \, dV$, reaches a minimum value at the nonequilibrium stationary state."

Such a general criterion was sought by Lord Rayleigh, who suggested a "principle of least dissipation of energy". Lars Onsager (1903–1976), in his well-known article on the reciprocal relations, investigated this principle and suggested that "the rate of increase of entropy plays the role of a potential" [4]. The general formulation and the demonstration of the validity of this principle is due to Prigogine [1]. Let us look at some examples that demonstrate this principle.

For the case of coupled forces and flows, the principle of minimum entropy production can be demonstrated as follows. Consider a system with two forces and flows that are coupled. For notational convenience, we shall represent the total entropy production per unit time by P. Therefore,

$$P \equiv \frac{d_i S}{dt} = \int (F_1 J_1 + F_2 J_2) \, dV \qquad (17.2.1)$$

Let us assume that the force F_1 is maintained at a fixed value by a suitable nonequilibrium constraint (contact with a reservoirs, etc.). From kinetic considerations one generally finds that in the stationary state $J_1 = \text{constant}$ and $J_2 = 0$, i.e. for a fixed value of F_1, F_2 adjusts so that J_2 is zero. We now show that this stationary state corresponds to the state in which the entropy production P is minimized.

The linear phenomenological laws give

$$J_1 = L_{11}F_1 + L_{12}F_2 \quad \text{and} \quad J_2 = L_{21}F_1 + L_{22}F_2 \qquad (17.2.2)$$

Substituting (17.2.2) into (17.2.1) and using the Onsager reciprocal relations $L_{12} = L_{21}$, we obtain

$$P = \int (L_{11}F_1^2 + 2L_{12}F_1F_2 + L_{22}F_2^2)dV \qquad (17.2.3)$$

From (17.2.3) it follows that, for a fixed F_1, P as a function of F_2 is minimized when

$$\frac{\partial P}{\partial F_2} = \int 2(L_{22}F_2 + L_{21}F_1)dV = 0 \qquad (17.2.4)$$

Since this equation is valid for an arbitrary volume, the integrand is zero. By noting that $J_2 = L_{21}F_1 + L_{22}F_2$, we see at once that the entropy production is minimized when

$$J_2 = L_{21}F_1 + L_{22}F_2 = 0 \qquad (17.2.5)$$

That is, $P \equiv d_iS/dt$ is minimized when the flow J_2 corresponding to the unconstrained force F_2 vanishes. This result can easily be generalized to an arbitrary number of forces and flows. The stationary state is the state of minimum entropy production in which the flows J_k, corresponding to the unconstrained forces, are zero. Although nonequilibrium stationary states are generally obtained through kinetic considerations, minimization of entropy production provides an alternative way.

We shall now present several examples to illustrate the general validity of the principle of minimum entropy production.

EXAMPLE 1: STATIONARY STATES IN CHEMICAL SYSTEMS

Consider the chemical system (17.1.9) discussed in the previous section (Fig. 17.2):

$$A \underset{}{\overset{(1)}{\rightleftharpoons}} X \underset{}{\overset{(2)}{\rightleftharpoons}} B \qquad (17.2.6)$$

As before, the flows of A and B keep the chemical potentials μ_A and μ_B fixed, which implies that the sum of the affinities has a fixed value $\bar{A}$:

$$A_1 + A_2 = (\mu_A - \mu_X) + (\mu_X - \mu_B) = \mu_A - \mu_B \equiv \bar{A} \qquad (17.2.7)$$

In the previous section, by using kinetics, we have already seen that the nonequilibrium stationary state is completely specified by (17.1.18):

$$v_1 = v_2 \qquad (17.2.8)$$

We shall now show how this condition may also be obtained using the principle of minimum entropy production. The entropy production per unit volume for this system (which we assume is homogeneous) is

$$\frac{1}{V}\frac{d_i S}{dt} = \frac{P}{V} = \sigma = \frac{A_1}{T}v_1 + \frac{A_2}{T}v_2$$
$$= \frac{A_1}{T}v_1 + \frac{(\bar{A} - A_1)}{T}v_2 \qquad (17.2.9)$$

in which V is the system volume and in which we have included the constraint (17.2.7) explicitly. The value of the chemical potential (or concentration) of X in the stationary state determines the value of A_1, hence the value of the entropy production (17.2.9). Since we have already seen in the previous section that the stationary state is completely specified by $v_1 = v_2$, we will now show that, in the linear regime, the same result can be obtained by minimizing σ (hence P) as a function of A_1. In the linear regime, since the two reactions are independent, we have

$$v_1 = L_{11}\frac{A_1}{T} \qquad v_2 = L_{22}\frac{A_2}{T} = L_{22}\frac{(\bar{A} - A_1)}{T} \qquad (17.2.10)$$

in which we have used (17.2.7). Substituting (17.2.10) into (17.2.9) we obtain σ as a function of A_1:

$$\sigma(A_1) = L_{11}\frac{A_1^2}{T^2} + L_{22}\frac{(\bar{A} - A_1)^2}{T^2} \qquad (17.2.11)$$

This function reaches its minimum value when

$$\frac{\partial \sigma(A_1)}{\partial A_1} = \frac{L_{11}}{T^2}2A_1 - \frac{L_{22}}{T^2}2(\bar{A} - A_1) = 0 \qquad (17.2.12)$$

i.e.

$$\frac{L_{11}A_1}{T} - \frac{L_{22}A_2}{T} = v_1 - v_2 = 0 \qquad (17.2.13)$$

In the linear regime, the entropy production is minimized at the nonequilibrium stationary state.

We have expressed σ as a function of A_1. It is not necessary to express σ in terms of the affinities of the system, though it is convenient; σ can also be expressed in terms of the concentration [X]. The value of [X] that minimizes σ is the stationary state. We shall outline the main steps in this alternative demonstration of the principle, leaving some details as exercises.

In section 9.5 we have seen that the entropy production per unit volume for the two reactions (17.2.6) can also be written as

$$\frac{1}{V}\frac{d_i S}{dt} = \sigma = R\{(R_{1f} - R_{1r})\ln(R_{1f}/R_{1r}) + (R_{2f} - R_{2r})\ln(R_{2f}/R_{2r})\}$$

$$(17.2.14)$$

in which R_{kf} and R_{kr} are the forward and reverse reaction rates of reaction k and R is the gas constant. Now if these forward and reverse reaction rates are written in terms of the concentrations, we have an expression for σ in terms of the concentrations. Assuming the reactions in (17.2.6) are elementary steps, the rates may be written as

$$R_{1f} = k_{1f}[A] \qquad R_{1r} = k_{1r}[X] \qquad R_{2f} = k_{2f}[X] \qquad R_{2r} = k_{2r}[B] \quad (17.2.15)$$

At equilibrium each reaction is balanced by its reverse. The equilibrium concentrations of $[A]_{eq}$, $[X]_{eq}$ and $[B]_{eq}$ are easily evaluated using the principle of detailed balance:

$$[X]_{eq} = \frac{k_{1f}}{k_{1r}}[A]_{eq} = \frac{k_{2r}}{k_{2f}}[B]_{eq} \qquad (17.2.16)$$

We now define small deviations in concentrations from the equilibrium:

$$\delta_A = [A] - [A]_{eq} \quad \delta_X = [X] - [X]_{eq} \quad \delta_B = [B] - [B]_{eq} \qquad (17.2.17)$$

The deviations in [A] and [B] are due to the inflow of A and the outflow of B, so δ_A and δ_B are fixed by the flows. Only the concentration δ_X is determined by the chemical reactions. Using (17.2.17) in (17.2.14) the entropy production σ to the leading order in deviations (17.2.17) may be written (exc. 17.8) as

$$\sigma(\delta_X) = R\left\{\frac{(k_{1f}\delta_A - k_{1r}\delta_X)^2}{k_{1f}[A]_{eq}} + \frac{(k_{2f}\delta_X - k_{2r}\delta_B)^2}{k_{2f}[X]_{eq}}\right\} \qquad (17.2.18)$$

By setting $\partial\sigma/\partial\delta_X = 0$, the value of δ_X which minimizes σ (exc. 17.8) can easily be shown to be

$$\delta_X = \frac{k_{1f}\delta_A + k_{2r}\delta_B}{k_{1r} + k_{2f}} \qquad (17.2.19)$$

(This is another way of expressing equation 17.2.8). That δ_X given by (17.2.19) is identical to the stationary value can easily be verified. The kinetic equation for [X] that follows from the two reactions (17.2.6) is

$$\frac{d[X]}{dt} = k_{1f}[A] - k_{1r}[X] - k_{2f}[X] + k_{2r}[B] \qquad (17.2.20)$$

Substituting (17.2.17) into (17.2.20) gives the stationary state

$$\frac{d[X]}{dt} = k_{1f}\delta_A - k_{1r}\delta_X - k_{2r}\delta_X + k_{2r}\delta_B = 0 \qquad (17.2.21)$$

The solution δ_X of this equation is identical to (17.1.19). Thus the stationary value of δ_X is also the value for which the entropy production is minimized.

EXAMPLE 2: A SEQUENCE OF CHEMICAL REACTIONS

The principle of minimum entropy production can easily be demonstrated for more complex chemical systems. Example 1 can be generalized to an arbitrary number of intermediates.

$$X \overset{1}{\rightleftharpoons} W_1 \overset{2}{\rightleftharpoons} W_2 \rightleftharpoons \ldots \rightleftharpoons W_{n-1} \overset{n}{\rightleftharpoons} Y \qquad (17.2.22)$$

The entropy production in this case is

$$\frac{1}{V}\frac{d_iS}{dt} = \sigma = \frac{1}{T}(v_1A_1 + v_2A_2 + \ldots + v_nA_n) \qquad (17.2.23)$$

We assume the system is homogeneous, so we may assume the volume $V = 1$ without loss of generality. Since the affinity $\bar{A}$ of the net reaction $X \rightleftharpoons Y$ is the sum of the affinities of the constituent reactions:

$$\bar{A} = \sum_{k=1}^{n} A_k \qquad (17.2.24)$$

The inflow of X and outflow of Y keeps $\bar{A}$ at a fixed nonzero value, keeping the system away from thermodynamic equilibrium. This nonequilibrium constraint can be made explicit by writing $A_n = (\bar{A} - \sum_{k=1}^{n-1} A_k)$ and substituting it in (17.2.23). We then have σ as a function of $(n-1)$ independent affinities A_k:

$$\sigma = \frac{1}{T}\left(v_1A_1 + v_2A_2 + \ldots + v_{n-1}A_{n-1} + v_n\left(\bar{A} - \sum_{k=1}^{n-1}A_k\right)\right) \qquad (17.2.25)$$

Now, applying the linear phenomenological laws $v_k = L_{kk}(A_k/T)$ to this equation, we obtain

$$\sigma = \frac{1}{T^2}\left(L_{11}A_1^2 + L_{22}A_2^2 + \ldots + L_{(n-1)(n-1)}A_{n-1}^2 + L_{nn}\left(\bar{A} - \sum_{k=1}^{n-1}A_k\right)^2\right)$$

(17.2.26)

An elementary calculation shows that the condition for minimum entropy production $\partial\sigma/\partial A_k = 0$ leads to $v_k = v_n$. Since this is valid for all k, we have the following generalization of (17.2.8):

$$v_1 = v_2 = \ldots v_{n-1} = v_n$$

(17.2.27)

Since the kinetic equations for (17.2.22) are

$$\frac{d[W_k]}{dt} = v_k - v_{k+1}$$

(17.2.28)

it is clear that the stationary states $d[W_k]/dt = 0$ are identical to the states that minimize entropy production.

EXAMPLE 3: COUPLED CHEMICAL REACTIONS

As an example of a chemical reaction in which one of the affinities is unconstrained by the nonequilibrium conditions, let us consider the synthesis of HBr from H_2 and Br_2. In this case we expect the velocity of the unconstrained reaction to equal zero at the stationary state. We assume that the affinity of the net reaction

$$H_2 + Br_2 \rightleftharpoons 2HBr$$

(17.2.29)

is maintained at a fixed nonzero value by a suitable inflow of H_2 and Br_2 and removal of HBr. The intermediates of the reaction, H and Br, appear through the reactions

$$Br_2 \overset{1}{\rightleftharpoons} 2Br$$

(17.2.30)

$$Br + H_2 \overset{2}{\rightleftharpoons} HBr + H$$

(17.2.31)

$$Br_2 + H \overset{3}{\rightleftharpoons} HBr + Br$$

(17.2.32)

The affinity of the net reaction (17.2.29) is

$$A_2 + A_3 = \bar{A}$$

(17.2.33)

which we assume is kept at a nonzero value. The affinity A_1 of reaction (17.2.30) is not constrained. The entropy production per unit volume for this system is

$$
\begin{aligned}
\sigma &= \frac{1}{T}(v_1 A_1 + v_2 A_2 + v_3 A_3) \\
&= \frac{1}{T}(v_1 A_1 + v_2 A_2 + v_3(\bar{A} - A_2))
\end{aligned}
\tag{17.2.34}
$$

Again we shall assume a homogeneous system with $V = 1$, so that minimizing σ is equivalent to minimizing the total entropy production P. As was done above, using the phenomenological laws $v_k = L_{kk}(A_k/T)$ and setting $\partial\sigma/\partial A_k = 0$ for the two independent affinities A_1 and A_2, we see that the entropy production is extremized when

$$
v_1 = 0 \quad \text{and} \quad v_2 = v_3
\tag{17.2.35}
$$

This must also be the stationary state. Turning to the kinetic equations for H and Br, we have

$$
\frac{d[\mathrm{H}]}{dt} = v_2 - v_3
\tag{17.2.36}
$$

$$
\frac{d[\mathrm{Br}]}{dt} = 2v_1 - v_2 + v_3
\tag{17.2.37}
$$

The stationary states of these equations are the same as (17.2.35).

EXAMPLE 4: STATIONARY STATES IN THERMAL CONDUCTION

As an example of a continuous system, let us look at stationary states in heat conduction using the system we considered in Fig. 17.1. For a one-dimensional system the entropy production is

$$
P \equiv \frac{d_i S}{dt} = \int_0^L J_q\left(\frac{\partial}{\partial x}\frac{1}{T}\right) dx
\tag{17.2.38}
$$

Using the linear phenomenological law $J_q = L_{qq}\partial(1/T)/\partial x$, the above expression can be written as

$$
P = \int_0^L L_{qq}\left(\frac{\partial}{\partial x}\frac{1}{T}\right)^2 dx
\tag{17.2.39}
$$

Of all the possible functions $T(x)$, we want to obtain the function that minimizes the entropy production P. This can be done using the following basic result from the calculus of variations. The integral

$$I = \int_0^L \Lambda(f(x), \dot{f}(x)) dx \qquad (17.2.40)$$

in which the integrand $\Lambda(f(x), \dot{f}(x))$, a function of f and its derivative $\dot{f} \equiv \partial f/\partial x$ (for notational convenience we shall use $\dot{f}$ in place of $\partial f/\partial x$), is extremized when the function $f(x)$ is a solution of the following equation:

$$\frac{d}{dx}\frac{\partial \Lambda}{\partial \dot{f}} - \frac{\partial \Lambda}{\partial f} = 0 \qquad (17.2.41)$$

In applying this result to the entropy production (17.2.39), we identify f with (1/T) so that $\Lambda = L_{qq} \dot{f}^2$ Also, as was discussed in section 16.3 (16.3.6), in this calculation we assume $L_{qq} = \kappa T^2 \approx \kappa T_{avg}^2$ (in which κ is the thermal conductivity and T_{avg} is the average temperature) is approximately constant in accordance with the linear approximation. Then writing equation (17.2.41) for the entropy production, we obtain

$$\frac{d}{dx}L_{qq}\dot{f} = 0 \qquad (17.2.42)$$

Because we identified f with (1/T), this condition implies

$$L_{qq}\dot{f} = L_{qq}\frac{\partial}{\partial x}\frac{1}{T} = J_q = \text{constant} \qquad (17.2.43)$$

Since $L_{qq} \approx \kappa T_{avg}^2$ this condition can also be written as

$$\kappa\frac{\partial T}{\partial x} = \text{constant} \qquad (17.2.44)$$

Thus the function $T(x)$ that minimizes the entropy production P is linear in x, i.e. the entropy production is minimized when the heat current reaches a uniform value along the length of the system. This result has a formal similarity with the velocities of a sequence of coupled reactions all being constant along the reaction chain (Example 2). As expected, the stationary state obtained in the previous section using the heat conduction equation (17.1.5) is identical to (17.2.44).

EXAMPLE 5: STATIONARY STATES IN ELECTRICAL CIRCUIT ELEMENTS

In the previous section we have seen that the entropy production for electrical circuit elements is given by $Td_iS/dt = VI$ in which V is the voltage across the

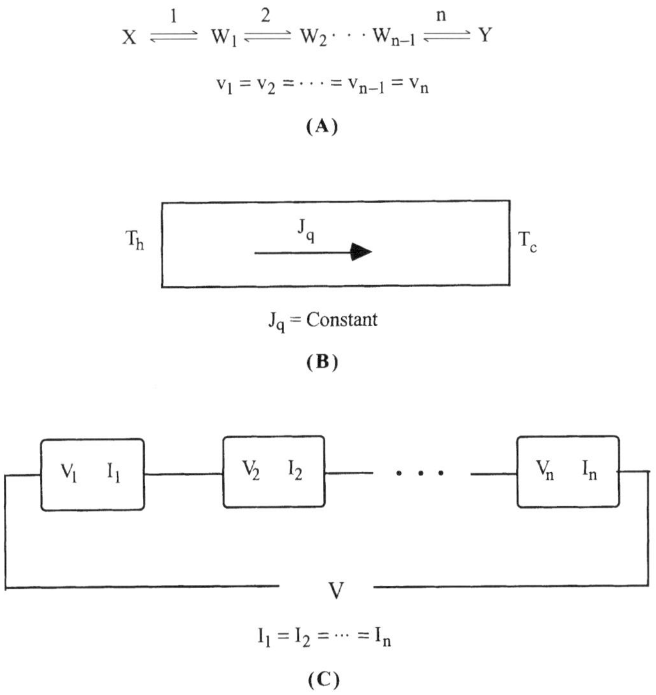

$$X \; \underset{\longleftarrow}{\overset{1}{\longrightarrow}} \; W_1 \; \underset{\longleftarrow}{\overset{2}{\longrightarrow}} \; W_2 \cdots W_{n-1} \; \underset{\longleftarrow}{\overset{n}{\longrightarrow}} \; Y$$

$$v_1 = v_2 = \cdots = v_{n-1} = v_n$$

(A)

T_h $\xrightarrow{\quad J_q \quad}$ T_c

$$J_q = \text{Constant}$$

(B)

$V_1 \quad I_1$ $V_2 \quad I_2$ $\cdots$ $V_n \quad I_n$

$$V$$

$$I_1 = I_2 = \cdots = I_n$$

(C)

Figure 17.4 For a nonequilibrium system consisting of a series of coupled subsystems, the entropy production in the linear regime is minimized when all the flows are equal. This state is the stationary state

circuit element and I the current passing through it. For each element k, the phenomenological laws imply that $I_k = L_{kk}(V_k/T)$. Let us consider n circuit elements connected in series, as shown in Fig. 17.4 (C). We assume that the total voltage drop V across the whole circuit is maintained at a constant value (just as a constant temperature was maintained for thermal conduction):

$$V = \sum_{k=1}^{n} V_k \tag{17.2.45}$$

The total entropy production for such a system will then be

$$P = \frac{d_i S}{dt} = \frac{1}{T}(V_1 I_1 + V_2 I_2 + \ldots + V_n I_n)$$

$$= \frac{1}{T^2}\left(L_{11}V_1^2 + L_{22}V_2^2 + \ldots + L_{(n-1)(n-1)}V_{n-1}^2 + L_{nn}\left(\sum_{k=1}^{n-1}(V - V_k)\right)^2 \right)$$

$$\tag{17.2.46}$$

in which we have used (17.2.45) to eliminate V_n. This equation is similar to (17.2.26), obtained for a sequence of chemical reactions in which the V_k take the place of the affinities A_k. We may now minimize the entropy production with respect to the $(n-1)$ independent V_k by setting $\partial P/\partial V_k = 0$. The result, as in the case of chemical reactions, is that the flows I_k must be equal:

$$I_1 = I_2 = \ldots = I_n \tag{17.2.47}$$

Thus, in a circuit element, the entropy production is minimized when the current is uniform along the circuit. (Feynman [5] indicates he has observed this relation between entropy production and uniformity of electric current.) In the analysis of electrical circuits, the condition that the current should be uniform is usually imposed on the system because we do not observe any charge accumulation in any part of the system. In electrical systems the relaxation to the stationary state of uniform I is extremely rapid, hence non-uniform or discontinuous I are not observed.

Examples 2, 4 and 5 illustrate a common feature implied by the principle of minimum entropy production (Fig. 17.4): in a series of coupled systems, entropy production is extremized when the flows are equal. In a chemical reaction it was the velocity v_k; for heat conduction it was the heat flow J_q; for an electric circuit it is the electric current I_k.

17.3 Time Variation of Entropy Production and the Stability of Stationary States

In the previous section we have seen that the stationary states in the linear regime are also states that extremize the internal entropy production. We shall now consider the stability of these states, and also show that the entropy production is *minimized*. In Chapter 14 we saw that the fluctuations near the equilibrium state decrease the entropy and that the irreversible processes drive the system back to the equilibrium state of maximum entropy. As the system approaches the state of equilibrium, the entropy production approaches zero. The approach to equilibrium can be described not only as a steady increase in entropy to its maximum value but also as a steady *decrease in entropy production to zero*. It is this latter approach that naturally extends to the linear regime, close to equilibrium.

Let us look at the time variation of the entropy production due to chemical reactions in an open system in the linear regime. As before, we assume homogeneity and unit volume. The entropy production is:

$$P \equiv \frac{d_iS}{dt} = \sum_k \frac{A_k}{T}\frac{d\xi_k}{dt} = \sum_k \frac{A_k}{T}v_k \tag{17.3.1}$$

In this equation, all the affinities A_k are functions of p, T and the extents of reaction ξ_k. In the linear regime, since $v_k = \sum_i L_{ki}(A_i/T)$, equation (17.3.1) becomes

$$P = \sum_{ik} \frac{L_{ik}}{T^2} A_i A_k \tag{17.3.2}$$

The time derivative of P can now be explicitly written by noting that at constant p and T,

$$\frac{dA_k}{dt} = \sum_j \left(\frac{\partial A_k}{\partial \xi_j}\right)_{p,T} \frac{d\xi_j}{dt} \tag{17.3.3}$$

Thus we find

$$\frac{dP}{dt} = \frac{1}{T^2} \sum_{ijk} L_{ik} \left[A_k \left(\frac{\partial A_i}{\partial \xi_j}\right) \frac{d\xi_j}{dt} + A_i \left(\frac{\partial A_k}{\partial \xi_j}\right) \frac{d\xi_j}{dt} \right] \tag{17.3.4}$$

By using the Onsager reciprocal relations $L_{ik} = L_{ki}$, and identifying $d\xi_k/dt \equiv v_k = \sum_i L_{ki}(A_i/T)$, equation (17.3.4) can be reduced to

$$\frac{dP}{dt} = \frac{2}{T} \sum_{ij} \left(\frac{\partial A_i}{\partial \xi_j}\right) v_i v_j \tag{17.3.5}$$

To see that (17.3.5) is negative, we turn to the stability conditions in Chapter 14, in particular to the following condition (14.1.9b) for stability with respect to fluctuations $\delta \xi_i$ in the extents of reaction:

$$\Delta_i S = \frac{1}{2T} \sum_{ij} \left(\frac{\partial A_i}{\partial \xi_j}\right)_{eq} \delta \xi_i \delta \xi_j < 0 \tag{17.3.6}$$

Since $\delta \xi_k$ can be positive or negative, condition (17.3.6) for stability of the equilibrium state implies that the matrix $(\partial A_i/\partial \xi_j)_{eq}$ must be negative definite. In a neighborhood of the equilibrium state, $(\partial A_i/\partial \xi_j)$ would retain its negative definiteness. In this neighborhood, expression (17.3.5) must also be negative definite. Hence in the neighborhood of equilibrium we have the inequalities

$$\boxed{P > 0} \tag{17.3.7}$$

$$\boxed{\frac{dP}{dt} = \frac{2}{T} \sum_{ij} \left(\frac{\partial A_i}{\partial \xi_j}\right) v_i v_j < 0} \tag{17.3.8}$$

These conditions ensure the stability of the nonequilibrium stationary states in

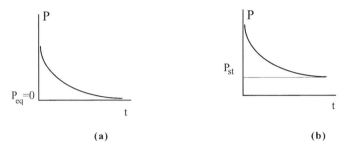

Figure 17.5 The time variation of the entropy production $P = d_iS/dt = \sum_k F_k J_k$ for equilibrium and near equilibrium states. (a) For a fluctuation from the equilibrium state, the initial nonzero value of P decreases to its equilibrium value of zero. (b) In the linear regime, a fluctuation from a nonequilibrium steady state can only increase the value of P above the stationary value P_{st}; irreversible processes drive P back to its minimum value P_{st}

the linear regime close to the equilibrium state (Fig. 17.5). At the stationary state, P has its minimum value. If a fluctuation drives P to a higher value, irreversible processes drive P back to its minimum stationary value. The result $dP/dt < 0$ for nonequilibrium states can be more generally proved [6]. The two conditions (17.3.7) and (17.3.8) constitute the "Lyapunov conditions" for the stability of a state, a topic we will discuss in detail in the next chapter.

References

1. Prigogine, I., *Etude Thermodynamique des Processus Irreversibles*, 1947, Liège: Desoer.
2. Prigogine, I., *Introduction to Thermodynamics of Irreversible Processes*, 1967, New York: John Wiley.
3. Rayleigh, L., *Proc. Math. Soc. London*, **4** (1873), 357–363.
4. Onsager, L., *Phys. Rev.*, **37** (1931), 405–426.
5. Feynman, R. P., Leighton, R. B., and Sands, M., *The Feynman Lectures on Physics*, Vol. II. 1964, Reading MA: Addison-Wesley, Ch. 19, p. 14.
6. Glansdorff, P. and Prigogine, I., *Thermodynamics of Structure Stability and Fluctuations*, 1971, New York: Wiley.

Exercises

17.1 (a) Using the Fourier law $\mathbf{J}_q = -\kappa \nabla T$, obtain the time-dependent equation for heat conduction:

$$C\frac{\partial T}{\partial t} = \kappa \nabla^2 T$$

in which C is the heat capacity per unit volume.

(b) For a one-dimensional system, show that the stationary state of the system leads to a linear temperature distribution.

(c) In planets the core is hotter than the surface. Consider a sphere of radius R whose core of radius R_1 is at a higher temperature T_1 than the surface temperature T_2. Obtain the stationary distribution of $T(r)$ and the heat flux $\mathbf{J}_q$ as a function of the radial distance r, using the Fourier law of heat conduction. (The conductivity of the earth cannot account for the heat flux measured at the surface of the earth. The transport of heat is therefore thought to be due to convective processes within the earth.)

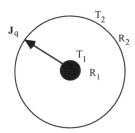

17.2 (a) Using the relation $s_m = s_{mo} + C_V \ln T$ for the molar entropy of a system, in which C_V is the molar heat capacity, obtain an expression for the total entropy of the system shown in Fig. 17.1. Let ρ be the density and M the molar mass. Assume that the distance between the hot and cold ends is L and that the area of cross section is unity. Also assume that the density ρ does not change much with T so that it is nearly uniform.

(b) Suppose the system were suddenly removed from being in contact with the reservoirs then insulated so that no heat escaped.

  (i) What would the final temperature of the system be when it reaches equilibrium?

  (ii) What would be the final entropy of the system?

(iii) What would be the increase in entropy compared to the initial nonequilibrium state?

17.3 Write a Mathematica code to simulate the reaction

$$\frac{d[X]}{dt} = k_{1f}[A] - k_{1r}[X] - k_{2f}[X] + k_{2r}[B]$$

Use it to study the relaxation of $[X]$ and the entropy production σ to its stationary value.

17.4 For a series of reactions

$$M \overset{1}{\rightleftharpoons} X_1 \overset{2}{\rightleftharpoons} X_2 \rightleftharpoons \cdots \rightleftharpoons X_{n-1} \overset{n}{\rightleftharpoons} N$$

with an inflow of M and an outflow of N, show that the steady state is given by

$$v_1 = v_2 = \cdots = v_n$$

in which the v_k are velocities of the indicated reactions.

17.5 Consider an ideal capacitor C in series with an inductance L.

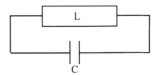

The voltage across the capacitor is $V_C = -Q/C$; the voltage across the inductance is $V_L = -L\,dI/dt$. For the illustrated circuit, the sum of these two voltages must be zero, i.e. $V_C + V_L = 0$. Using this fact, write a differential equation for Q and show that the quantity $(LI^2/2 + Q^2/2C)$ is constant in time. (The conservation of energy here, is similar to that of a simple harmonic oscillator.) If a resistor R is added to the circuit, show that the equation $dU/dt = -V_R I$ leads to the well-known equation $L\,d^2Q/dt + R\,dQ/dt + Q/C = 0$ of an LCR circuit.

17.6 Using equations (17.1.28) and (17.1.29) obtain the time variation of $I(t)$ and $Q(t)$ in a real capacitor and a real inductance. Using these expressions in (17.1.25) and (17.1.26) obtain the entropy production at any time t in these circuit elements with initial current I_0 and initial charge Q_0.

17.7 Demonstrate the theorem of minimum entropy production for an arbitrary number of constrained and unconstrained thermodynamic forces.

17.8 Consider the reaction $A \overset{(1)}{\rightleftharpoons} X \overset{(2)}{\rightleftharpoons} B$
(a) Show that the entropy production per unit volume is

$$\sigma(\delta_X) = R\left\{ \frac{(k_{1f}\delta_A - k_{1r}\delta_X)^2}{k_{1f}[A]_{eq}} + \frac{(k_{2f}\delta_X - k_{2r}\delta_B)^2}{k_{2f}[X]_{eq}} \right\}$$

in which $\delta_A = [A] - [A]_{eq}$, $\delta_X = [X] - [X]_{eq}$.
(b) Show that σ attains a minimum value for

$$\delta_X = \frac{k_{1f}\delta_A + k_{2r}\delta_B}{k_{1r} + k_{2f}}$$

PART V

ORDER THROUGH FLUCTUATIONS

18 NONLINEAR THERMODYNAMICS

18.1   Far-From-Equilibrium Systems

Systems, which are subject to a flow of energy and matter, can be driven far from thermodynamic equilibrium, into the "nonlinear" regime. In the nonlinear regime, the thermodynamic flows J_α are no longer linear functions of thermodynamic forces F_α. In the case of chemical reactions, we have seen that a system is in the linear regime if the affinities A_k are small compared to RT, i.e. $(A_k/RT) \ll 1$. The value of RT at $T = 300$ K is about 2.5 kJ/mol. Since the affinities of chemical reactions can easily reach the range 10–100 kJ/mol, the nonlinear regime is easily reached for chemical processes (exc. 18.1). It is more difficult to reach the nonlinear regime for transport processes such as heat conduction and diffusion.

In Nature, far from equilibrium systems are ubiquitous. The earth as a whole is an open system subject to the constant flow of energy from the sun. This influx of energy provides the driving force for the maintenance of life and is ultimately responsible for maintaining an atmosphere out of thermodynamic equilibrium (exc. 18.2). Every living cell lives through the flow of matter and energy.

As we shall see in the following sections, far-from-equilibrium states can lose their stability and evolve to one of the many states available to the system. The irreversible processes and the boundary conditions do not uniquely specify the nonequilibrium state to which the system will evolve; driven by internal fluctuations or other small influences, the system leaves the unstable state and evolves to one of the many possible new states. These new states can be highly organized states. In this realm of instability and evolution to new organized structures, very small factors, often beyond laboratory control, begin to decide the fate of a system. As for the certainty of Newtonian and Laplacian planetary motion and the uniqueness of equilibrium states, both begin to fade; we see instead a probabilistic Nature that generates new organized structures, a Nature that can create life itself.

18.2   General Properties of Entropy Production

In the linear regime we saw that the stationary states are those in which the total entropy production $P = \int_V \sigma \, dV$ reaches a minimum. This criterion also assured the stability of the stationary state. In the far-from-equilibrium nonlinear regime there is no such general principle for determining the state of the system.

Far-from-equilibrium states can become unstable and evolve to new organized states and we will identify the thermodynamic conditions under which this may happen. We begin by noting some general properties of the total entropy production P. These are statements regarding the time evolution of change δP due to small changes in the forces δF_k and the flows δJ_k.

Let P be the entropy production in a nonequilibrium stationary state. Since $P = \int_V \sigma \, dV = \int_V \sum_k F_k J_k \, dV$, the rate of change in P can be written as

$$\frac{dP}{dt} = \int_V \left(\frac{d\sigma}{dt} \right) dV = \int_v \left(\sum_k \frac{dF_k}{dt} J_k \right) dV + \int_V \left(\sum_k F_k \frac{dJ_k}{dt} \right) dV$$
$$\equiv \frac{d_F P}{dt} + \frac{d_J P}{dt} \tag{18.2.1}$$

in which $d_F P/dt$ is the change due to the changes in F_k and $d_J P/dt$ is the change due to the changes in J_k. Two general properties can now be stated [1–3]:

(i)   In the linear regime

$$\boxed{\frac{d_F P}{dt} = \frac{d_J P}{dt}} \tag{18.2.2}$$

(ii) For time-independent boundary conditions, even outside the linear regime

$$\boxed{\frac{d_F P}{dt} \leq 0} \tag{18.2.3}$$

($d_F P/dt = 0$ at the stationary state).

In contrast to the variation dG in the Gibbs free energy G, $d_F P$ is not a differential of a state function. Hence the fact that $d_F P$ can only decrease does not tell us how the state will evolve.

The first of the above relations follows from the linear relations $J_k = \sum_i L_{ki} F_i$ and the Onsager reciprocal relations $L_{ki} = L_{ik}$. First we note that

$$\sum_k dF_k J_k = \sum_{ki} dF_k L_{ki} F_i = \sum_{ki} (dF_k L_{ik}) F_i = \sum_i dJ_i F_i \tag{18.2.4}$$

Using this result in the definitions of $d_F P$ and $d_J P$ (18.2.1), we immediately see that

$$\frac{d_F P}{dt} = \int_V \left(\sum_k \frac{dF_k}{dt} J_k \right) dV = \int_V \left(\sum_k F_k \frac{dJ_k}{dt} \right) dV = \frac{d_J P}{dt} = \frac{1}{2} \frac{dP}{dt} \tag{18.2.5}$$

The general property (18.2.3) when applied to (18.2.5) gives us the result we have seen in the previous chapter:

$$\frac{dP}{dt} = 2\frac{d_F P}{dt} < 0 \quad \text{in the linear regime} \tag{18.2.6}$$

This shows, once again, that a perturbation in the total entropy production P from its stationary-state value monotonically decreases to its stationary state value, in accordance with the principle of minimum entropy production. A simple proof of (18.2.3) is given in Appendix 18.1.

We see that we now have two inequalities, $P \geq 0$ and $d_F P \leq 0$. The second inequality is an important evolution criterion. Let us indicate briefly two consequences. If only one concentration, say X, is involved in the evolution, $d_F P = v(X)(\partial A/\partial X)dX \equiv dW$. The variable W, thus defined, is then a "kinetic potential". But this is rather an exceptional case. The interesting consequence is that time-independent constraints may lead to states which are not stationary, states that oscillate in time. We shall see examples of such systems in Chapter 19, but let us consider here a simple example of a far-from-equilibrium chemical system where the dependence of velocities on affinities are antisymmetric i.e., $v_1 = lA_2, v_2 = -lA_1$ (Onsager's relations are not valid for systems far from equilibrium). The derivative $d_F P/dt$ in this case becomes:

$$\frac{d_F P}{dt} = v_1 \frac{dA_1}{dt} + v_2 \frac{dA_2}{dt} = lA_2 \frac{dA_1}{dt} - lA_1 \frac{dA_2}{dt} \leq 0 \tag{18.2.7}$$

By introducing the polar coordinates $A_1 = r\cos\theta$ and $A_2 = r\sin\theta$, it is easy to see that this equation can be written as:

$$\frac{d_F P}{dt} = -lr^2 \frac{d\theta}{dt} \leq 0 \tag{18.2.8}$$

The system rotates irreversibly in a direction determined by the sign of l. An example of such a system is the well-known Lotka–Volterra "prey–predator" interaction given as an exercise (exc. 18.9). We can also apply this inequality to derive a sufficient condition for the stability of a steady state. If all fluctuations $\delta_F P > 0$ then the steady state is stable. But here it is more expedient to use the Lyapunov theory of stability to which we turn now.

18.3 Stability of Nonequilibrium Stationary States

A very general criterion for stability of a state was formulated by Lyapunov [4]. We shall obtain the conditions for the stability of a nonequilibrium state using Lyapunov's theory.

LYAPUNOV'S THEORY OF STABILITY

Lyapunov's formulation gives conditions for stability in precise mathematical terms (with clear intuitive meaning). Let X_s be a stationary state of a physical system. In general, X may be an r-dimensional vector with components X_k, $k = 1, 2, \ldots, r$. We shall denote the components of X_s by X_{sk}. Let the time evolution of X be described by an equation

$$\frac{dX_k}{dt} = Z_k(X_1, X_2, \ldots X_r; \lambda_j) \qquad (18.3.1)$$

in which the λ_j are parameters that may or may not be independent of time. A simple example of such an equation is given in Box 18.1. In general, if the X_k are functions not only of time t, but also of positions $\mathbf{x}$, then (18.3.1) will be a partial differential equation in which Z_k will be a partial differential operator.

The stationary state X_{sk} is the solution to the set of coupled equations

$$\frac{dX_k}{dt} = Z_k(X_{s1}, X_{s2}, \ldots X_{sr}; \lambda_j) = 0 \qquad (k = 1, 2, \ldots, r) \qquad (18.3.2)$$

The stability of the stationary state can be understood by looking at the behavior of a small perturbation δX_k. To establish the stability of a state, first a *positive function $L(\delta X)$* of δX, which may be called a "distance," is defined in the space spanned by X_k. If this "distance" between X_{sk} and the perturbed state $(X_{sk} + \delta X_k)$ steadily decreases in time, the stationary state is stable. Thus state X_{sk} is stable if

$$\boxed{L(\delta X_k) > 0 \qquad \frac{dL(\delta X_k)}{dt} < 0} \qquad (18.3.3)$$

A function L that satisfies (18.3.3) is called a **Lyapunov function**. If the variables X_k are functions of position (as concentrations n_k in a nonequilibrium system can be), L is called a **Lyapunov functional**—a "functional" is a mapping of a set of functions to a number, real or complex. The notion of stability is not restricted to stationary states; it can also be extended to periodic states [4]. However, since we are interested in the stability of nonequilibrium stationary states, we shall not deal with the stability of periodic states at this point.

Box 18.1: Kinetic Equations and Lyapunov Stability Theory: An Example

Consider the open chemical system shown above with the following chemical reactions.

$$S + T \xrightarrow{k_1} A$$

$$S + A \xrightarrow{k_2} B$$

$$A + B \xrightarrow{k_3} P$$

For simplicity, we assume that the reverse reactions can be ignored. If the system is subject to an inflow of S and T and an outflow of P such that the concentrations of these species are maintained constant, we have the following kinetic equations for the concentration of A and B:

$$X_1 \equiv [A] \qquad X_2 \equiv [B]$$

$$\frac{dX_1}{dt} = k_1[S][T] - k_2[S]X_1 - k_3X_1X_2 \equiv Z_1(X_j, [S], [T])$$

$$\frac{dX_2}{dt} = k_2[S]X_1 - k_3X_1X_2 \equiv Z_2(X_j, [S], [T])$$

In this system, [S] and [T] correspond to the parameters λ_j in (18.3.1). For a given value of these parameters, stationary states, X_{s1} and X_{s2} are easily found by setting $dX_1/dt = dX_2/dt = 0$:

$$X_{s1} = \frac{k_1[T]}{2k_2} \qquad X_{s2} = \frac{k_2[S]}{k_3}$$

The stability of this stationary state is determined by examining the evolution of the perturbations δX_1 and δX_2 from this stationary state. A possible Lyapunov function L, for example, is

$$L(\delta X_1, \delta X_2) = \left[(\delta X_1)^2 + (\delta X_2)^2 \right] > 0$$

If it can be shown that $dL(\delta X_1, \delta X_2)/dt < 0$, then the stationary state (X_{s1}, X_{s2}) is stable.

SECOND VARIATION OF ENTROPY $-\delta^2 S$ AS A LYAPUNOV FUNCTIONAL

We have already seen that the second variation of entropy is a function that has a definite sign for any thermodynamic system. By considering the entropy density

$s(\mathbf{x})$ as a function of the energy density $u(\mathbf{x})$ and the concentrations $n_k(\mathbf{x})$, we can write ΔS, the change in entropy from the stationary value, in the form

$$
\begin{aligned}
\Delta S = & \int \left[\left(\frac{\partial s}{\partial u} \right)_{n_k} \delta u + \sum_k \left(\frac{\partial s}{\partial n_k} \right)_u \delta n_k \right] dV \\
& + \frac{1}{2} \int \left[\left(\frac{\partial^2 s}{\partial u^2} \right) (\delta u)^2 + 2 \sum_k \left(\frac{\partial^2 s}{\partial u \, \partial n_k} \right) \delta u \, \delta n_k + \sum_{ij} \left(\frac{\partial^2 s}{\partial n_i \, \partial n_j} \right) \delta n_i \, \delta n_j \right] dV \\
= & \; \delta S + \frac{1}{2} \delta^2 S
\end{aligned}
\tag{18.3.4}
$$

Since we are considering a nonequilibrium stationary state, the thermodynamic forces and the corresponding flows of energy, $\mathbf{J}_u$, and matter, $\mathbf{J}_k$, do not vanish. Hence the first variation $\delta S \neq 0$. The second variation $\delta^2 S$ has a definite sign because the integrand, which is the second variation of entropy of elemental volume that is locally in equilibrium, is negative (12.4.10):

$$
\frac{1}{2} \delta^2 S < 0
\tag{18.3.5}
$$

Appendix 18.2 contains the derivation of the following general result:

$$
\boxed{\frac{d}{dt} \frac{\delta^2 S}{2} = \sum_k \delta F_k \, \delta J_k}
\tag{18.3.6}
$$

In Chapter 14 (equation 14.1.16) we obtained the same equation for perturbations from the equilibrium state. Equation (18.3.6) shows that the time derivative of $\delta^2 S$ has the same form even under nonequilibrium conditions. The difference is that near equilibrium $\sum_k \delta F_k \delta J_k = \sum_k F_k J_k > 0$; this is not necessarily so far from equilibrium. We shall refer to this quantity as **excess entropy production**, but strictly speaking, it is the increase in entropy production only near the equilibrium state; for a perturbation from a nonequilibrium state, the increase in entropy production is equal to $\delta P = \delta_\mathrm{F} P + \delta_\mathrm{J} P$.

Expressions (18.3.5) and (18.3.6) would define a Lyapunov functional, $L = -\delta^2 S$ if the stationary state were such that $\sum_k \delta F_k \, \delta J_k > 0$. *Thus, a nonequilibrium stationary state is stable if*

$$
\frac{d}{dt} \frac{\delta^2 S}{2} = \sum_k \delta F_k \, \delta J_k > 0
\tag{18.3.7}
$$

If this inequality is violated, it only means that the system *may* be unstable, i.e. $\sum_k \delta F_k \, \delta J_k < 0$ is *a necessary but not a sufficient* condition for instability.

USING THE STABILITY CRITERION

Since $\delta^2 S < 0$ under both equilibrium and nonequilibrium conditions, the stability of a stationary state is assured if

$$\frac{d}{dt}\frac{\delta^2 S}{2} = \sum_k \delta F_k \, \delta J_k > 0 \qquad (18.3.8)$$

Let us apply this condition to simple chemical systems to understand when a nonequilibrium system may become unstable.

First, let us consider the following reaction:

$$A + B \underset{k_r}{\overset{k_f}{\rightleftharpoons}} C + D \qquad (18.3.9)$$

Assuming these reactions are elementary steps, we write the forward and reverse rates as

$$R_f = k_f[A][B] \quad \text{and} \quad R_r = k_r[C][D] \qquad (18.3.10)$$

We assume this system is maintained out of equilibrium by suitable flows. As we have seen in section 9.5, for a chemical reaction the affinity A and the velocity of reaction v are given by $A = RT \ln(R_f/R_r)$ and $v = (R_f - R_r)$. The time derivative of $\delta^2 S$, the "excess entropy production" (18.3.8), can be written in terms of $\delta F = \delta A/T$ and $\delta J = \delta v$. For a perturbation $\delta[B]$ from the stationary state, it is easy to show (exc. 18.4) that

$$\frac{1}{2}\frac{d\delta^2 S}{dt} = \sum_\alpha \delta J_\alpha \, \delta F_\alpha = \frac{\delta A}{T}\delta v = Rk_f \frac{[A]_s}{[B]_s}(\delta[B])^2 > 0 \qquad (18.3.11)$$

in which the subscript s indicates the nonequilibrium stationary-state values of the concentrations. Since $d\delta^2 S/dt$ is positive, the stationary state is stable.

The situation is different, however, for an *autocatalytic* reaction such as

$$2X + Y \underset{k_r}{\overset{k_f}{\rightleftharpoons}} 3X \qquad (18.3.12)$$

which appears in a reaction scheme called the "Brusselator" that we will consider in the next chapter. For this reaction, we can consider a nonequilibrium stationary state in which the concentrations are $[X]_s$ and $[Y]_s$ and a perturbation δX. Using the forward and reverse rates $R_f = k_f[X]^2[Y]$ and $R_r = k_r[X]^3$ in the expressions $A = RT \ln(R_f/R_r)$ and $v = (R_f - R_r)$, we can once again calculate

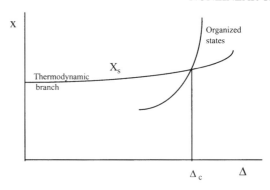

Figure 18.1 Each value of X represents a state of the system. The distance from equilibrium is represented by the parameter Δ. When $\Delta = 0$ the system is in a state of thermodynamic equilibrium. When Δ is small, the system is in a near-equilibrium state, which is an extrapolation of the equilibrium state; this family of states is called the thermodynamic branch. In some systems, such as those with autocatalysis, when Δ reaches a critical value Δ_c, the states belonging to the thermodynamic branch become unstable. When this happens, the system makes a transition to a new branch which may correspond to organized states

the excess entropy production to obtain

$$\frac{1}{2}\frac{d\delta^2 S}{dt} = \frac{\delta A}{T}\delta v = -R(2k_f[X]_s[Y]_s - 3k_r[X]_s^2)\frac{(\delta X)^2}{[X]_s} \qquad (18.3.13)$$

The excess entropy production can now become negative, particularly if $k_f \gg k_r$. Hence the stability is no longer assured and the stationary state *may* become unstable.

The above discussion can be summarized through a stability diagram as shown in Fig. 18.1. The value of the parameter Δ is a measure of the distance from equilibrium. For each value of Δ the system will relax to a stationary state, denoted by X_s. The equilibrium state corresponds to $\Delta = 0$; X_s is a continuous extension of the equilibrium state and it is called the **thermodynamic branch**. As long as condition (18.3.8) is satisfied, the thermodynamic branch is stable; if it is violated, the thermodynamic branch *may* become unstable. If it does become unstable, the system makes a transition to a new branch, which is generally an organized structure.

If the kinetic equations of the systems are known, there is a well-defined mathematical method to determine at what point the stationary state will become unstable. This is the linear stability analysis that we will discuss in the following section. Nonequilibrium instabilities give rise to a great variety of structures, which we will discuss in the next chapter

18.4   Linear Stability Analysis

In general, the rate equations of a chemical system take the form (18.3.1)

$$\frac{dX_k}{dt} = Z_k(X_1 \ldots X_n; \lambda_j) \tag{18.4.1}$$

where the X_k correspond to concentrations, such as [X] and [Y] in (18.3.12), and λ_j corresponds to concentrations that are maintained at a constant non-equilibrium value. We begin by assuming that a stationary solution X_k^0 of (18.4.1) is known. This means

$$Z_k(X_i^0, \ldots, X_n^0, \lambda) = 0 \tag{18.4.2}$$

We would like to know if this stationary solution will be stable to small perturbations x_i. Linear stability analysis provides the answer in the following way. Consider a small perturbation x_k:

$$X_k = X_k^0 + x_k(t) \tag{18.4.3}$$

Now the Taylor expansion of $Z_k(X_i)$ gives

$$Z_k(X_i^0 + x_i) = Z_k(X_i^0) + \sum_j \left(\frac{\partial Z_k}{\partial X_j}\right)_0 x_j + \ldots \tag{18.4.4}$$

in which the subscript 0 indicates that the derivative is evaluated at the stationary state X_i^0. In linear stability analysis, only the linear terms in x_j are retained; the higher-order terms are neglected by assuming the x_j are small. Substituting (18.4.4) into (18.4.1), since X_i^0 is a stationary state, we obtain for $x_k(t)$ the linear equation

$$\frac{dx_k}{dt} = \sum_j \Lambda_{kj}(\lambda) x_j \tag{18.4.5}$$

in which $\Lambda_{kj}(\lambda) = (\partial Z_k / \partial X_j)_0$ is a function of the parameter λ. In matrix notation, equation (18.4.5) can be written as

$$\frac{d\mathbf{x}}{dt} = \Lambda \mathbf{x} \tag{18.4.6}$$

in which the vector $\mathbf{x} = (x_1, x_2, x_3, \ldots, x_n)$ and Λ_{kj} are the elements of the matrix Λ. The matrix Λ is sometimes referred to as the **Jacobian matrix**.

The general solution of (18.4.6) can be written if the eigenvalues and the eigenvectors of the matrix Λ are known. Let ω_k be the eigenvalues and $\boldsymbol{\psi}_k$ the corresponding eigenvectors:

$$\Lambda \boldsymbol{\psi}_k = \omega_k \boldsymbol{\psi}_k \qquad (18.4.7)$$

In general, for an n-dimensional matrix there are n eigenvalues and n eigenvectors. (Note that $\boldsymbol{\psi}_k$ is a vector and the subscript k indicates different vectors.) If the eigenvalues ω_k and the eigenvectors $\boldsymbol{\psi}_k$ are known, it is easy to see that, corresponding to each eigenvector and its eigenvalue, we have the following solution to (18.4.6):

$$\mathbf{x} = e^{\omega_k t} \boldsymbol{\psi}_k \qquad (18.4.8)$$

This can be easily seen by substituting (18.4.8) into (18.4.6). Since a linear combination of solutions of a linear equation is also a solution, the general solution to (18.4.6) can be written as

$$\mathbf{x} = \sum_k c_k e^{\omega_k t} \boldsymbol{\psi}_k \qquad (18.4.9)$$

in which the coefficients c_k are determined by $\mathbf{x}$ at $t = 0$. Now the question of stability depends on whether the perturbation $\mathbf{x}$ will grow or decay with time. Clearly, this depends on the eigenvalues ω_k: if one or more of the eigenvalues have a positive real part, the associated solutions (18.4.8) will grow exponentially. The corresponding eigenvectors are called **unstable modes**. Since a random perturbation will be of the form (18.4.9) that includes the unstable modes, the existence of a single eigenvalue with positive real part is sufficient to make the perturbation grow with time. If all the eigenvalues have negative real parts, any small perturbation $\mathbf{x}$ in the vicinity of the stationary solution will exponentially decay or regress to zero. (This need not be true for large perturbations $\mathbf{x}$ for which the approximation (18.4.5) is not valid.)

Thus, the necessary and sufficient condition for the stability of a stationary state is that all the eigenvalues of the associated Jacobian matrix Λ have negative real parts. An eigenvalue with a positive real part implies instability.

The example given below illustrates the application of the linear stability theory to a chemical system. As we have seen in the previous section, thermodynamic considerations lead us to the conclusion that instability can arise only when the system is far from thermodynamic equilibrium and, generally, when autocatalysis is present.

The exponential growth of the perturbation does not continue indefinitely; the growth eventually stops due to the nonlinear terms. Through this process, the system makes a transition form an unstable state to a stable state. Thus, driven by instability, the system makes a transition to a new state. This new state is often an organized state, a state with lower entropy. These organized states can be maintained indefinitely as long as the flows are maintained.

AN EXAMPLE

We shall illustrate the use of linear stability theory with the following set of kinetic equations that we will study in more detail in the following chapter. Instead of using X_1 and X_2, we shall use concentrations [X] and [Y] for the system variables:

$$\frac{d[X]}{dt} = k_1[A] - k_2[B][X] + k_3[X]^2[Y] - k_4[X] = Z_1 \qquad (18.4.10)$$

$$\frac{d[Y]}{dt} = k_2[B][X] - k_3[X]^2[Y] = Z_2 \qquad (18.4.11)$$

Here [A] and [B] are the parameters (concentrations that are maintained at fixed values) corresponding to λ (18.4.1). One can easily obtain the stationary solutions to this equations (exc. 18.6):

$$[X]_s = \frac{k_1}{k_4}[A] \quad [Y]_s = \frac{k_4 k_2}{k_3 k_1} \frac{[B]}{[A]} \qquad (18.4.12)$$

The Jacobian matrix evaluated at the stationary state is

$$\begin{bmatrix} \dfrac{\partial Z_1}{\partial [X]} & \dfrac{\partial Z_1}{\partial [Y]} \\[2ex] \dfrac{\partial Z_2}{\partial [X]} & \dfrac{\partial Z_2}{\partial [Y]} \end{bmatrix} = \begin{bmatrix} k_2[B] - k_4 & k_3[X]_s^2 \\ -k_2[B] & -k_3[X]_s^2 \end{bmatrix} = \Lambda \qquad (18.4.13)$$

The stationary state (18.4.12) becomes unstable when the real parts of the eigenvalues of (18.4.13) become positive. The **eigenvalue equation** or the **characteristic equation** of a matrix Λ, whose solutions are the eigenvalues, is

$$\text{Det}[\Lambda - \lambda I] = 0 \qquad (18.4.14)$$

in which "Det" stands for the determinant. For a 2×2 matrix such as (18.4.13) it is easy to see that the characteristic equation is

$$\lambda^2 - (\Lambda_{11} + \Lambda_{22})\lambda + (\Lambda_{11}\Lambda_{22} - \Lambda_{21}\Lambda_{12}) = 0 \qquad (18.4.15)$$

in which Λ_{ij} are the elements of the matrix Λ. If all the matrix elements Λ_{ij} are real, as is the case for chemical systems, the solutions of the characteristic equation must be complex conjugate pairs because coefficients in the equation are real. For the matrix (18.4.13) we shall consider the case of a complex conjugate pair. We shall look at these solutions as functions of the concentration [B], and investigate whether their real parts, which are initially negative, can become positive due to an appropriate change in [B]. The point at which the real parts reach zero will be the point of transition from stability to instability.

For equation (18.4.15), since the coefficient of the linear term is the negative of the sum of the roots (exc. 18.7), if $\lambda_{\pm}$ are the two roots, we have

$$\lambda_+ + \lambda_- = (\Lambda_{11} + \Lambda_{22}) = k_2[B] - k_4 - k_3[X]_s^2 \qquad (18.4.16)$$

If the real parts of this complex conjugate pair, $\lambda_{\pm}$, are negative then $k_2[B] - k_4 - k_3[X]_s^2 < 0$; if they are positive then $k_2[B] - k_4 - k_3[X]_s^2 > 0^*$. Thus the condition that requires positive real parts for the onset of instability leads to

$$[B] > \frac{k_4}{k_2} + \frac{k_3}{k_2}[X]_s^2$$

or

$$[B] > \frac{k_4}{k_2} + \frac{k_3 k_1^2}{k_2 k_4^2}[A]^2 \qquad (18.4.17)$$

where we have used (18.4.12) for $[X]_s$. Thus, for a fixed value of [A], as the value of [B] increases, when condition (18.4.17) is satisfied, the stationary state (18.4.12) becomes unstable. In the next chapter we will see that this instability leads to oscillations.

Linear stability analysis does not provide a means of determining how the system will evolve when a state becomes unstable. To understand the system's behavior fully, the full nonlinear equation has to be considered. Often we encounter nonlinear equations for which solutions cannot be obtained analytically. However, with the availability of powerful desktop computers and software, numerical solutions can be obtained without much difficulty. To obtain numerical solutions to nonlinear equations considered in the following chapter, Mathematica codes are provided at the end of Chapter 19.

---

* If $\lambda_{\pm}$ are real roots, $\lambda_+ + \lambda_- > 0$ implies at least one of the roots is positive.

Appendix 18.1

In this appendix we show that, regardless of the distance from equilibrium,

$$\frac{d_F P}{dt} \leq 0 \qquad (A18.1.1)$$

The validity of (A18.1.1) depends on the validity of local equilibrium. In Chapter 12 we have seen that the second-order variation of entropy $\delta^2 S$ is negative because quantities such as the molar heat capacity C_V, isothermal compressibility κ_T, and $-\sum_{i,j}(\partial A_i/\partial \xi_j)\delta\xi_i\,\delta\xi_j$, are positive. This condition remains valid for an elemental volume δV which is in local equilibrium. We can see the relation between the derivative $d_F P/dt$ and quantities such as $-\sum_{i,j}(\partial A_i/\partial \xi_j)\delta\xi_i\,\delta\xi_j$, which have a definite sign, as follows.

CHEMICAL REACTIONS

Consider a closed homogeneous nonequilibrium system undergoing chemical reaction at uniform constant temperature. The affinities A_k are functions of the extents of reaction ξ_j and

$$\frac{\partial A_k}{\partial t} = \sum_j \left(\frac{\partial A_k}{\partial \xi_j}\right)\left(\frac{\partial \xi_j}{\partial t}\right) = \sum_j \left(\frac{\partial A_k}{\partial \xi_j}\right) v_j \qquad (A18.1.2)$$

Therefore:

$$\frac{d_F P}{dt} = \frac{1}{T}\sum_{k,j}\left(\frac{\partial A_k}{\partial \xi_j}\right) v_j v_k \leq 0 \qquad (A18.1.3)$$

which follows from the general relation $-\sum_{i,j}(\partial A_i/\partial \xi_j)\delta\xi_i\,\delta\xi_j \geq 0$ valid for a system in local equilibrium (12.4.5). This proof can be extended to open systems following along the lines of proof for isothermal diffusion given below.

ISOTHERMAL DIFFUSION

In this case we begin with

$$\frac{d_F P}{dt} = -\int \sum_k \mathbf{J}_k \cdot \frac{\partial}{\partial t}\nabla\left(\frac{\mu_k}{T}\right) dV = -\int \frac{1}{T}\sum_k \mathbf{J}_k \cdot \nabla\left(\frac{\partial \mu_k}{\partial t}\right) dV \qquad (A18.1.4)$$

Using the identity $\nabla \cdot (f\mathbf{J}) = f\nabla \cdot \mathbf{J} + \mathbf{J} \cdot \nabla f$, the right-hand side can be written as

$$-\int \frac{1}{T}\mathbf{J}_k \cdot \nabla\left(\frac{\partial \mu_k}{\partial t}\right) dV = -\int \frac{1}{T}\nabla \cdot \left[\mathbf{J}_k\left(\frac{\partial \mu_k}{\partial t}\right)\right] dV + \int \frac{1}{T}\left(\frac{\partial \mu_k}{\partial t}\right)\nabla \cdot \mathbf{J}_k\, dV$$

$$(A18.1.5)$$

Using Gauss's theorem, the first term can be converted into a surface integral. Since we assume the value of μ_k is time independent at the boundary, i.e. the boundary conditions are time independent, this surface integral vanishes. Using the relations

$$\frac{\partial \mu_k}{\partial t} = \sum_j \frac{\partial \mu_k}{\partial n_j}\frac{\partial n_j}{\partial t} \quad \text{and} \quad \frac{\partial n_k}{\partial t} = -\nabla \cdot \mathbf{J}_k \qquad (A18.1.6)$$

the second term can be written as

$$\int \frac{1}{T}\left(\frac{\partial \mu_k}{\partial t}\right)\nabla \cdot \mathbf{J}_k\, dV = \frac{-1}{T}\int \sum_j \frac{\partial \mu_k}{\partial n_j}\left(\frac{\partial n_j}{\partial t}\right)\left(\frac{\partial n_k}{\partial t}\right) dV \qquad (A18.1.7)$$

Combining (A18.1.7), (A18.1.5) and (A18.1.4), we arrive at

$$\frac{d_{\mathrm{F}}P}{dt} = \frac{-1}{T}\int \sum_{jk} \frac{\partial \mu_k}{\partial n_j}\left(\frac{\partial n_j}{\partial t}\right)\left(\frac{\partial n_k}{\partial t}\right) \leq 0$$

The right-hand side of this expression is negative because

$$-\sum_{jk} \frac{\partial \mu_k}{\partial n_j}\left(\frac{\partial n_j}{\partial t}\right)\left(\frac{\partial n_k}{\partial t}\right) \leq 0$$

is valid for systems in equilibrium (12.4.9). The general validity of (18.2.3) is proved in the literature [1]

Appendix 18.2

The relation

$$\frac{d}{dt}\frac{\delta^2 S}{2} = \sum_k \delta F_k\, \delta J_k \qquad (A18.2.1)$$

can be obtained as follows. We begin by taking the time derivative of $\delta^2 S/2$ as defined in (18.3.4). For notational simplicity, we shall denote the time derivatives of a quantity x by $\dot{x} \equiv \partial x/\partial t$. The time derivative of $\delta^2 S$ can be written as

$$\delta^2 \dot{S} = \int \left[\left(\frac{\partial^2 s}{\partial u^2} \right) 2\delta u(\delta \dot{u}) + 2 \sum_k \left(\frac{\partial^2 s}{\partial u \, \partial n_k} \right) (\delta \dot{u} \, \delta n_k + \delta u \, \delta \dot{n}_k) \right.$$
$$\left. + 2 \sum_{ik} \left(\frac{\partial^2 s}{\partial n_i \, \partial n_k} \right) \delta \dot{n}_i \, \delta n_k \right] dV \qquad \text{(A18.2.2)}$$

in which the factor 2 appears in the last term because we used the relation

$$\frac{\partial^2 s}{\partial n_i \, \partial n_k} = \frac{\partial^2 s}{\partial n_k \, \partial n_i}$$

Next, noting that $(\partial s/\partial u)_{n_k} = 1/T$ and $(\partial s/\partial n_k)_u = -\mu_k/T$, we can write (A18.2.2) as

$$\delta^2 \dot{S} = \int 2 \left[\left(\frac{\partial}{\partial u} \frac{1}{T} \right) \delta u(\delta \dot{u}) + \sum_k \left(\frac{\partial}{\partial n_k} \frac{1}{T} \right) \delta \dot{u} \, \delta n_k \right] dV$$
$$+ \int 2 \left[\sum_k \frac{\partial}{\partial u} \left(\frac{-\mu_k}{T} \right) \delta u \, \delta \dot{n}_k + \sum_{ik} \frac{\partial}{\partial n_i} \left(\frac{-\mu_k}{T} \right) \delta n_i \, \delta \dot{n}_k \right] dV \qquad \text{(A18.2.3)}$$

We now observe that, since u and n_k are the independent variables, we can write

$$\delta \left(\frac{1}{T} \right) = \sum_k \left(\frac{\partial}{\partial n_k} \frac{1}{T} \right) \delta n_k + \left(\frac{\partial}{\partial u} \frac{1}{T} \right) \delta u \qquad \text{(A18.2.4)}$$

$$\delta \left(\frac{\mu_i}{T} \right) = \sum_k \left(\frac{\partial}{\partial n_k} \frac{\mu_i}{T} \right) \delta n_k + \left(\frac{\partial}{\partial u} \frac{\mu_i}{T} \right) \delta u \qquad \text{(A18.2.5)}$$

Equations (A.18.2.4) and (A.18.2.5) enable us to reduce (A.18.2.3) to the simple form

$$\delta^2 \dot{S} = 2 \int \left[\delta \left(\frac{1}{T} \right) \delta \dot{u} + \sum_k \delta \left(\frac{-\mu_k}{T} \right) \delta \dot{n}_k \right] dV \qquad \text{(A18.2.6)}$$

This relation can be written in terms of the changes in thermodynamic forces $\delta \nabla(1/T)$ and $\delta \nabla(-\mu_k/T)$ and the corresponding flows $\delta \mathbf{J}_u$ and $\delta \mathbf{J}_k$, using the balance equations for energy density u and the concentrations n_k:

$$\frac{\partial u}{\partial t} = \dot{u} = -\nabla \cdot \mathbf{J}_u \tag{A18.2.7}$$

$$\frac{\partial n_k}{\partial t} = \dot{n}_k = -\nabla \cdot \mathbf{J}_k + \sum_i \nu_{ki} v_i \tag{A18.2.8}$$

in which ν_{ki} is the stoichiometric coefficient of reactant k in reaction i, and v_i is the velocity of reaction i. If we denote the stationary state densities and flows by u_s, n_{ks}, $\mathbf{J}_{us}$, $\mathbf{J}_{ks}$ and v_{is}, we have $\dot{u}_s = -\nabla \cdot \mathbf{J}_{us} = 0$ and $\dot{n}_{ks} = -\nabla \cdot \mathbf{J}_{ks} + \sum_i \nu_{ki} v_{is} = 0$. Consequently, for a perturbation $u = u_s + \delta u$, $\mathbf{J}_u = \mathbf{J}_{us} + \delta \mathbf{J}_u$, etc., from the stationary state, we have

$$\delta \dot{u} = -\nabla \cdot \delta \mathbf{J}_u \tag{A18.2.9}$$

$$\delta \dot{n}_k = -\nabla \cdot \delta \mathbf{J}_k + \sum_i \nu_{ki} \delta v_i \tag{A18.2.10}$$

We substitute these expressions for $\delta \dot{u}$ and $\delta \dot{n}_k$ into (A18.2.6) and use the identity

$$\nabla \cdot (f\mathbf{J}) = f\nabla \cdot \mathbf{J} + \mathbf{J} \cdot \nabla f \tag{A18.2.11}$$

in which f is a scalar function and $\mathbf{J}$ is a vector field, and we use Gauss's theorem

$$\int_V (\nabla \cdot \mathbf{J}) dV = \int_\Sigma \mathbf{J} \cdot d\mathbf{a} \tag{A18.2.12}$$

in which $\sum$ is the surface enclosing the volume V and $d\mathbf{a}$ is the element of surface area. All this allows (A.18.2.6) to be written as follows:

$$\begin{aligned}
\frac{1}{2}\delta^2 \dot{S} = &-\int_\Sigma \delta\left(\frac{1}{T}\right) \delta \mathbf{J}_u \cdot d\mathbf{a} + \int_V \delta\nabla\left(\frac{1}{T}\right) \delta \mathbf{J}_u \, dV \\
&+ \int_\Sigma \sum_k \delta\left(\frac{\mu_k}{T}\right) \delta \mathbf{J}_k \cdot d\mathbf{a} - \int_V \sum_k \delta\nabla\left(\frac{\mu_k}{T}\right) \delta \mathbf{J}_k \, dV \\
&+ \int_V \left[\sum_i \delta\left(\frac{A_i}{T}\right) \delta v_i\right] dV
\end{aligned} \tag{A18.2.13}$$

In obtaining this equation, we have used the relation $\sum_k \nu_{ki} \delta(\mu_k/T) = -\delta(A_i/T)$. The flows at the surface are fixed by the boundary conditions and are not subject to fluctuations, so the surface terms vanish. This leads us to the required result:

$$\frac{1}{2}\delta^2\dot{S} = \int_V \delta\nabla\left(\frac{1}{T}\right)\delta\mathbf{J}_u\,dV - \int_V \sum_k \delta\nabla\left(\frac{\mu_k}{T}\right)\delta\mathbf{J}_k\,dV + \int_V \left[\sum_i \delta\left(\frac{A_i}{T}\right)\delta v_i\right]dV$$

$$= \sum_\alpha \delta F_\alpha\,\delta J_\alpha \qquad\qquad\qquad\qquad (\text{A}18.2.14)$$

References

1. Glansdorff, P. and Prigogine, I., *Physica*, **20** (1954), 773.
2. Prigogine, I., *Introduction to Thermodynamics of Irreversible Processes*. 1967, New York: John Wiley.
3. Glansdorff, P. and Prigogine, I., *Thermodynamics of Structure, Stability and Fluctuations*. 1971, New York: Wiley
4. Minorski, N., *Nonlinear Oscillations*, 1962, Princeton NJ: Van Nostrand.

Exercises

18.1 Calculate the affinities of the following reaction systems for a range of concentrations (or partial pressures) of the reactants and the products and compare them with RT at $T = 298$ K. Determine the ranges in which the system is thermodynamically in the linear regime using appropriate data from tables.
(i) Racemization reaction $L \rightleftharpoons D$, (L and D are enantiomers)
(ii) Reaction $N_2O_4(g) \rightleftharpoons 2NO_2(g)$ (with partial pressures $P_{N_2O_4}$ and P_{NO_2})

18.2 (a) What factors would you note to conclude that the earth's atmosphere is not in thermodynamic equilibrium?
(b) Through an appropriate literature search, determine whether the atmospheres of Mars and Venus are in chemical equilibrium.

18.3 For the chemical reaction $A \rightleftharpoons B$, verify the general property $d_F P \leq 0$.

18.4 (a) Obtain inequality (18.3.11) for a perturbation $\delta[B]$ from the stationary states of reaction (18.3.9).
(b) Obtain the "excess entropy production" (18.3.13) for a perturbation $\delta[X]$ from the stationary states of reaction (18.3.12).

18.5 Obtain the excess entropy production and analyze the stability of the stationary states for the following reaction schemes:

(a) $W \rightleftharpoons X \rightleftharpoons Z$, in which the concentrations of W and Z are maintained fixed at a nonequilibrium value.

(b) $W + X \rightleftharpoons 2X, X \rightleftharpoons Z$, in which the concentrations of W and Z are maintained fixed at a nonequilibrium value

18.6 Show that the stationary states of (18.4.10) and (18.4.11) are (18.4.12).

18.7 For a polynomial equation of the type $\omega^n + A_1\omega^{n-1} + A_2\omega^{n-2} + \ldots + A_n = 0$ show that coefficient $A_1 = -(\lambda_1 + \lambda_2 + \lambda_3 + \ldots + \lambda_n)$ and coefficient $A_n = (-1)^n(\lambda_1\lambda_2\lambda_3\ldots\lambda_n)$, where λ_k are roots.

18.8 For the following equations, obtain the stationary states and analyze their stability as a function of the parameter λ:

(i)

$$\frac{dx}{dt} = -Ax^3 + C\lambda x$$

(ii)

$$\frac{dx}{dt} = -Ax^3 + Bx^2 + C\lambda x$$

(iii)

$$\frac{dx}{dt} = \lambda x - 2xy \qquad \frac{dy}{dt} = -y + xy$$

(iv)

$$\frac{dx}{dt} = -5x + 6y + x^2 - 3xy + 2y^2 \qquad \frac{dy}{dt} = \lambda x - 14y + 2x^2 - 5xy + 4y^2$$

18.9 Consider the reaction scheme:

$$A + X \rightleftharpoons 2X$$
$$Y + X \rightleftharpoons 2Y$$
$$Y \rightleftharpoons E$$

Far from equilibrium, we only keep the forward reactions. Using the linear stability theory, show that the perturbations around the non-equilibrium steady state lead to oscillations in [X] and [Y], as was discussed in section 18.2. This model was used by Lotka and Volterra to describe the "struggle of life" (see V. Volterra, *Theorie mathématique de la Lutte pour la Vie*, Paris: Gauthier Villars, 1931). Here X is the prey (lamb) and Y is the predator (wolf). This model of the prey–predator interaction shows that the populations X and Y will exhibit oscillations.

19 DISSIPATIVE STRUCTURES

19.1 The Constructive Role of Irreversible Processes

One of the most profound lessons of nonequilibrium thermodynamics is the dual role of irreversible processes: as destroyers of order near equilibrium and as creators of order far from equilibrium. For far from equilibrium systems, there are no general extremum principles that predict the state to which it will evolve. The lack of extremum principles that uniquely predict the state to which a nonequilibrium system will evolve is a fundamental aspect of nonequilibrium systems. In stark contrast to equilibrium systems which evolve to a state that minimizes a free energy, nonequilibrium systems can evolve unpredictably; their state cannot always be uniquely specified by macroscopic rate equations. This is because, for a given set of nonequilibrium conditions, it is often possible to have more than one state. As a result of random fluctuations, or other random factors such as small inhomogeneities or imperfections, the system evolves to one of the many possible states. Which one of these states a particular system will evolve to is, in general, not predictable. The new states thus attained are often "ordered states" that possess spatiotemporal organization. Patterns in fluid flow, inhomogeneities in concentrations exhibiting geometrical patterns with great symmetry, or periodic variations of concentrations are examples of such ordered states. Because of its fundamental character, we shall refer to the general phenomenon of a nonequilibrium system evolving to an ordered state as a result of fluctuations as "**order through fluctuations**" [1, 2].

In nonequilibrium systems, oscillating concentrations and geometrical concentration patterns can be a result of chemical reactions and diffusion, the same dissipative processes that, in a closed system, wipe out inhomogeneities and drive the system to a stationary, timeless homogeneous state of equilibrium. Since the creation and maintenance of organized nonequilibrium states are due to dissipative processes, they are called **dissipative structures** [3].

The two concepts of *dissipative structures* and *order through fluctuations* encapsulate the main aspects of nonequilibrium order that we describe in this chapter.

19.2 Loss of Stability, Bifurcation and Symmetry Breaking

In the previous chapter we have seen that the stability of the thermodynamic branch is no longer assured when a system is driven far from equilibrium. In section 18.3 we have seen how a necessary condition (18.3.7) for a system to become unstable can be obtained by using the second variation of entropy, $\delta^2 S$. Beyond this point, we are confronted with a multiplicity of states and unpredictability. To understand the precise conditions for instability and the subsequent behavior of a system, we need to use the specific features of the system, such as the rates of chemical reactions and the hydrodynamic equations. There are, however, some general features of far-from-equilibrium systems that we will summarize in this section. A detailed discussion of dissipative structures will be presented in the following sections.

The loss of stability of a nonequilibrium state can be analyzed using the general theory of stability for solutions of a nonlinear differential equation. Here we encounter the basic relationship between the loss of stability, multiplicity of solutions and symmetry. We also encounter the phenomenon of "bifurcation" or "branching" of new solutions of a differential equation from a particular solution. We shall first illustrate these general features for a simple nonlinear differential equation and then show how they are used to describe far-from-equilibrium systems.

AN ELEMENTARY EXAMPLE OF BIFURCATION AND SYMMETRY BREAKING

Consider the equation

$$\frac{d\alpha}{dt} = -\alpha^3 + \lambda\alpha \tag{19.2.1}$$

in which λ is a parameter. Our objective is to study the stationary solutions of this equation as a function of λ. Equation (19.2.1) possesses a simple twofold symmetry: it remains invariant when α is replaced by $-\alpha$. This means that if $\alpha(t)$ is a solution, then $-\alpha(t)$ is also solution. If $\alpha(t) \neq -\alpha(t)$, then there are two solutions to the equation. In this way, symmetry and multiplicity of solutions are related.

The stationary states of this differential equation are

$$\alpha = 0 \qquad \alpha = \pm\sqrt{\lambda} \tag{19.2.2}$$

Note the multiplicity of solutions related to symmetry. When a solution does not possess the symmetries of the differential equation, i.e. when $\alpha \neq -\alpha$, it is said to be a solution with a **broken symmetry** or a solution that has broken the symmetry. In this case the solution $\alpha = 0$ is invariant when α is replaced by $-\alpha$,

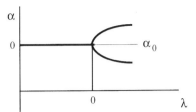

Figure 19.1 The bifurcation of solutions $\alpha = 0$ and $\alpha = \pm\sqrt{\lambda}$ to (19.2.1) as a function of the parameter λ. Thin lines represent solutions that are unstable

but the solution $\alpha = \pm\sqrt{\lambda}$ is not. Hence $\alpha = \pm\sqrt{\lambda}$ is said to have broken the symmetry of the differential equation. (Though this idea may seem rather trivial in this simple case, it has a rather important and nontrivial significance for nonequilibrium systems.)

Let us assume that, for physical reasons, we are seeking only real solutions of (19.2.1). When $\lambda < 0$ there is only one real solution, but when $\lambda > 0$ there are three solutions, as shown in Fig. 19.1. The new solutions for $\lambda > 0$ *branch* or *bifurcate* from the solution $\alpha = 0$. The value of λ at which new solutions bifurcate is called the **bifurcation point**. In Fig. 19.1 $\lambda = 0$ is the bifurcation point. Similar bifurcation of new solutions from a given solution occurs generally in nonlinear equations, be they a simple algebraic equation as above, a set of coupled ordinary differential equations or more complex partial differential equations.

Turning to the question of stability, we shall now see that solution $\alpha = 0$ becomes unstable precisely at the point where new solutions $\alpha = \pm\sqrt{\lambda}$ emerge. As we have seen earlier, a stationary solution α_S is locally stable if a small perturbation $\delta(t)$ from the solution decays to the stationary state. Thus we must look at the time evolution of $\alpha = \alpha_S + \delta(t)$ to determine if α_S is stable or not. Substituting $\alpha = \alpha_S + \delta(t)$ into (19.2.1), and keeping only terms of the first order in δ, we obtain

$$\frac{d\delta}{dt} = -3\alpha_S^2\delta + \lambda\delta \tag{19.2.3}$$

For the stationary state $\alpha_S = 0$, we see that the solution is stable if $\lambda < 0$, because $\delta(t)$ decays exponentially. On the other hand, if $\lambda > 0$ the solution is locally unstable because $\delta(t)$ grows exponentially. At the same time, if we use equation (19.2.3) to analyze the stability of the stationary states $\alpha_S = \pm\sqrt{\lambda}$, we find that they are stable. These stability properties of the stationary states mean that, as λ moves from a value less than zero to a value greater than zero, the solution $\alpha = 0$ becomes unstable and the system makes a transition to one of the two new solutions that bifurcate at $\lambda = 0$. To which of the two possible states the system will evolve is not deterministic; it depends on the random fluctuations in the system. The loss of stability implies that a random fluctuation will grow and drive the system to one of the two states, $\alpha_S = +\sqrt{\lambda}$ or $\alpha_S = -\sqrt{\lambda}$.

The bifurcation of new solutions at exactly the point where one solution loses stability is not a coincidence, it is a general property of the solutions of nonlinear equations. (This general relation between bifurcation and stability of solutions of nonlinear equations can be explained using *topological degree theory*, which is beyond the scope of this discussion.)

GENERAL THEORY OF BIFURCATION

In far-from-equilibrium systems the loss of stability of the thermodynamic branch and the transition to a dissipative structure follows the same general features shown in our simple example. The parameter such as λ corresponds to constraints—e.g. flow rates or concentrations maintained at a nonequilibrium value — that keep the system away from equilibrium. When λ reaches a particular value, the thermodynamic branch becomes unstable but at the same time new solutions now become possible; driven by fluctuations, the system makes a transition to one of the new states. As we did in section 18.4, let us specify the state of the system by $X_k, k = 1, 2, \ldots, n$ which in general may be functions of both position $\mathbf{r}$ and time t. Let the equation that describes the spatiotemporal evolution of the system be

$$\frac{\partial X_k}{\partial t} = Z_k(X_i, \lambda) \tag{19.2.4}$$

Here λ is the nonequilibrium constraint. If the system under consideration is a homogeneous chemical system, then Z_k is specified by the rates of chemical reactions. For an inhomogeneous system, Z_k may contain partial derivatives to account for diffusion and other transport processes. It is remarkable that, whatever the complexity of Z_k, the loss of stability of a solution of (19.2.4) at a particular value of λ and bifurcation of new solutions at this point are similar to those of (19.2.1). As in the case of (19.2.1), the symmetries of (19.2.4) are related to the multiplicity of solutions. For example, in an isotropic system, the equations should be invariant under the inversion $\mathbf{r} \rightarrow -\mathbf{r}$. In this case, if $X_k(\mathbf{r}, t)$ is a solution then $X_k(-\mathbf{r}, t)$ will also be a solution; if $X_k(\mathbf{r}, t) \neq X_k(-\mathbf{r}, t)$ then there are two distinct solutions which are mirror images of each other.

Let X_{sk} be a stationary solution of (19.2.4). The stability of this state can be analyzed as before by considering the evolution of $X_k = X_{sk} + \delta_k$ where δ_k is a small perturbation. If δ_k decays exponentially, then the stationary state is stable. This generally happens when λ is less than a "critical value" λ_c. When λ exceeds λ_c it may happen that the perturbations δ_k, instead of decaying exponentially, grow exponentially, thus making the state X_{sk} unstable. Precisely at λ_c, new solutions to (19.2.4) will appear. As we will see in detail in the following sections, in the vicinity of λ_c, the new solutions often take the form

$$X_k(\mathbf{r},\, t;\lambda) = X_{sk}\,(\lambda_c) + \alpha_k\,\psi_k(\mathbf{r}, t) \qquad (19.2.5)$$

in which, $X_{sk}(\lambda_c)$ is the stationary state when $\lambda = \lambda_c$, α_k are a set of "amplitudes" that are to be determined and $\psi_k(r, t)$ are functions that can be obtained from Z_k in (19.2.4). The general theory of bifurcation provides a means of obtaining the time evolution of the amplitudes α_k through a set of equations of the type

$$\frac{d\alpha_k}{dt} = G(\alpha_k,\, \lambda) \qquad (19.2.6)$$

These are called the **bifurcation equations**. In fact, though (19.2.1) is an equation in its own right, it is also a bifurcation equation for systems that break a two-fold symmetry. The multiplicity of solutions to (19.2.6) corresponds to the multiplicity of solutions to the original equation (19.2.4).

In this manner, instability, bifurcation, multiplicity of solutions and symmetry are all interrelated. We shall now give a few detailed examples of instability of the thermodynamic branch leading to dissipative structures.

19.3 Chiral Symmetry Breaking and Life

The chemistry of life as we know it is founded on a remarkable asymmetry. A molecule whose geometrical structure is not identical to its mirror image is said to possess **chirality**, or handedness. Mirror-image structures of a chiral molecule are called **enantiomers**. Just as we distinguish the left and the right hand, the two mirror-image structures are identified as L- and D-enantiomers (L for "levo" and D for "dextro"; R and S is another convention of identifying the two enantiomers). Amino acids, the building blocks of proteins, and deoxyribose in DNA are chiral molecules. From bacteria to man, nearly all amino acids that take part in the chemistry of life are L-amino acids (Fig. 19.2) and the riboses in DNA and RNA are D-riboses (Fig. 19.3). As Francis Crick notes, "The first great unifying principle of biochemistry is that the key molecules have the same hand in all organisms." This is all the more remarkable because chemical reactions show equal preference for the two mirror-image forms (except for very small differences due to parity-nonconserving electroweak interactions [4–6]).

Figure 19.2 Proteins are made exclusively of L-amino acids. The amino acid shown in L-alanine. In other L-amino acids, different groups of atoms take the place of CH_3

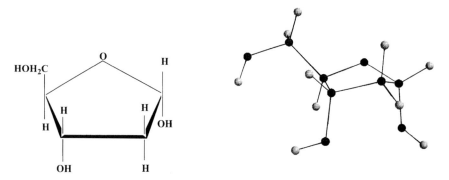

Figure 19.3 2-deoxy-D-ribose shown above is a basic chiral building block of DNA. Its mirror image structure, 2-deoxy-L-ribose, is excluded from the chemistry of life

Biochemistry's hidden asymmetry was discovered by Louis Pasteur in 1857. Nearly 150 years later, its true origin remains an unsolved problem, but we can see how such a state might be realized in the framework of dissipative structures. First, we note that such an asymmetry can arise only under far-from-equilibrium conditions; at equilibrium the concentrations of the two enantiomers will be equal. The maintenance of this asymmetry requires constant catalytic production of the preferred enantiomer in the face of interconversion between enantiomers, called **racemization**. (Racemization drives the system to the equilibrium state in which the concentrations of the two enantiomers will become equal.) Second, following the paradigm of order through fluctuations, we will presently see how, in systems with appropriate chiral autocatalysis, the thermodynamic branch, which contains equal amounts of L- and D-enantiomers, can become unstable. The instability is accompanied by the bifurcation of asymmetric states, or states of broken symmetry, in which one enantiomer dominates. Driven by random fluctuations, the system makes, a transition to one of the two possible states.

In 1953 F. C. Frank [7] devised a simple model reaction scheme with chiral autocatalysis that could amplify a small initial asymmetry. We shall modify this

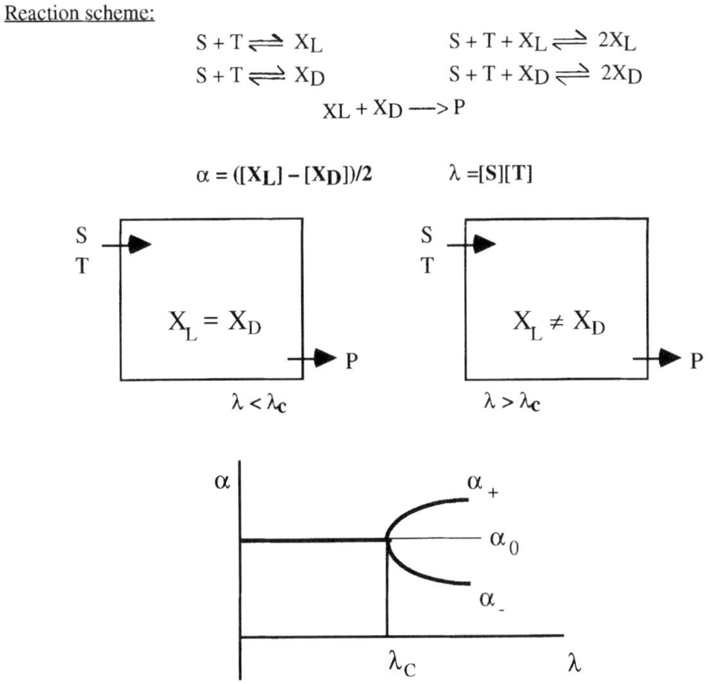

Reaction scheme:
S + T ⇌ XL
S + T + XL ⇌ 2XL
S + T ⇌ XD
S + T + XD ⇌ 2XD
XL + XD —> P
α = ([XL] – [XD])/2
λ =[S][T]
S
T
XL = XD
P
λ < λc
S
T
XL ≠ XD
P
λ > λc
α
α+
α0
α-
λC
λ

Figure 19.4 A simple autocatalytic reaction scheme in which X_L and X_D are produced with equal preference. However, in an open system, this leads to a dissipative structure in which $X_L \neq X_D$, a state of broken symmetry. A bifurcation diagram shows some general features of transitions to dissipative structures

reaction scheme so that its nonequilibrium aspects, instability and bifurcation of symmetry breaking states can be clearly seen (Fig. 19.4). It includes chirally autocatalytic reactions:

$$S + T \underset{k_{1r}}{\overset{k_{1f}}{\rightleftharpoons}} X_L \tag{19.3.1}$$

$$S + T + X_L \underset{k_{2r}}{\overset{k_{2f}}{\rightleftharpoons}} 2X_L \tag{19.3.2}$$

$$S + T \underset{k_{1r}}{\overset{k_{1f}}{\rightleftharpoons}} X_D \tag{19.3.3}$$

$$S + T + X_D \underset{k_{2r}}{\overset{k_{2f}}{\rightleftharpoons}} 2X_D \tag{19.3.4}$$

$$X_L + X_D \xrightarrow{k_3} P \tag{19.3.5}$$

Each enantiomer of X is produced directly from the achiral* reactants S and T, as shown in (19.3.1) and (19.3.3) and autocatalytically, as shown in (19.3.2) and (19.3.4). In addition, the two enantiomers react with one another and turn into an inactive compound, P. Due to symmetry, the rate constants for the direct reactions, (19.3.1) and (19.3.3), as well as the autocatalytic reactions, (19.3.2) and (19.3.4), must be equal. It is easy to see that at equilibrium the system will be in a *symmetric state*, i.e. $[X_L] = [X_D]$ (exc. 19.3). Now let us consider an open system into which S and T are pumped and from which P is removed. For mathematical simplicity, we assume that the pumping is done in such a way that the concentrations [S] and [T] are maintained at a fixed level, and that due to removal of P the reverse reaction in (19.3.5) may be ignored. The kinetic equation, of this system are

$$\frac{d[X_L]}{dt} = k_{1f}[S][T] - k_{1r}[X_L] + k_{2f}[X_L][S][T] - k_{2r}[X_L]^2 - k_3[X_L][X_D] \quad (19.3.6)$$

$$\frac{d[X_D]}{dt} = k_{1f}[S][T] - k_{1r}[X_D] + k_{2f}[X_D][S][T] - k_{2r}[X_D]^2 - k_3[X_L][X_D] \quad (19.3.7)$$

To make the symmetric and asymmetric states explicit, it is convenient to define the following variables:

$$\lambda = [S][T] \qquad \alpha = \frac{[X_L] - [X_D]}{2} \qquad \beta = \frac{[X_L] + [X_D]}{2} \quad (19.3.8)$$

When equations (19.3.6) and (19.3.7) are rewritten in terms of α and β (exc. 19.4), we have

$$\frac{d\alpha}{dt} = -k_{1r}\alpha + k_{2f}\lambda\alpha - 2k_{2r}\alpha\beta \quad (19.3.9)$$

$$\frac{d\beta}{dt} = k_{1f}\lambda - k_{1r}\beta + k_{2f}\lambda\beta - k_{2r}(\beta^2 + \alpha^2) - k_3(\beta^2 - \alpha^2) \quad (19.3.10)$$

The stationary states of this equation can be obtained after a little calculation (by setting $d\alpha/dt = d\beta/dt = 0$). A complete analytic study of the solutions of (19.3.9) and (19.3.10) and their stability is somewhat lengthy and it can be found in the literature [8]. We shall only state the main results of this analysis. With the Mathematica code provided in Appendix 19.1, the reader can explore the properties of the equation quite easily and verify the phenomenon of chiral symmetry breaking in this system (Fig. 19.5).

---

* Objects that do not possess a sense of handendness are called achiral. The molecule NH_3 is an example of an achiral molecule.

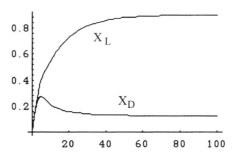

Figure 19.5 Time evolution of X_L and X_D obtained using Mathematica code A given in Appendix 19.1. For $\lambda > \lambda_c$ a small initial fluctuation in X_L grows to establish a state of broken symmetry in which the concentrations of X_L and X_D are unequal

For small values of λ the stable stationary state is

$$\alpha_s = 0 \tag{19.3.11}$$

$$\beta_s = \frac{2k_{2r}\beta_a + \sqrt{(2k_{2r}\beta_a)^2 + 4(k_{2r} + k_3)k_{1f}\lambda}}{2(k_{2r} + k_3)} \tag{19.3.12}$$

in which

$$\beta_a = \frac{k_{2f}\lambda - k_{1r}}{2k_{2r}}$$

This is a *symmetric* solution in which $[X_L] = [X_D]$ (as indicated by the subscript s). Using the stability analysis described in the previous chapter, it can be shown that this symmetric solution becomes unstable when λ greater than a critical value λ_c. The value of λ_c is given by

$$\lambda_c = \frac{s + \sqrt{s^2 - 4k_{2f}^2 k_{1r}^2}}{2k_{2f}^2} \tag{19.3.13}$$

where

$$s = 2k_{2f}k_{1r} + \frac{4k_{2r}^2 k_{1r}}{k_3 - k_{2r}} \tag{19.3.14}$$

For the system of equations (19.3.9) and (19.3.10) it is possible to obtain an asymmetric stationary solution analytically:

$$\alpha_a = \pm\sqrt{\beta_a^2 - \frac{k_{1f}\lambda}{k_3 - k_{2r}}} \qquad (19.3.15)$$

$$\beta_a = \frac{k_{2f}\lambda - k_{1r}}{2k_{2r}} \qquad (19.3.16)$$

in which the subscript a stands for *asymmetric*. We recommend the reader to use Mathematica code A in Appendix 19.1 to verify all these properties of the system.

To date, no chemical reaction has produced chiral asymmetry in this simple manner. However, symmetry breaking does occur in the crystallization of $NaClO_3$ [9, 10] in far from equilibrium conditions. The simple model however leads to interesting conclusions regarding the sensitivity of bifurcation discussed below.

NONEQUILIBRIUM SYMMETRY BREAKING AND THE ORIGIN OF BIOMOLECULAR ASYMMETRY

The above example shows how a far-from-equilibrium chemical system can generate and maintain chiral asymmetry, but it only provides a general framework in which we must seek the origins of biomolecular handedness. The origin of biomolecular handedness, or life's **homochirality**, remains to be explained [11, 12]. Here we shall confine our discussion to how the theory of nonequilibrium symmetry breaking contributes to this important topic. We cannot yet say with confidence whether chiral asymmetry arose in a prebiotic (i.e. before life) process and facilitated the evolution of life, or whether some primitive form of life that incorporated both L- and D-amino acids arose first and subsequent evolution of this life form led to the homochirality of L-amino acids and D-sugars. Both views have their proponents.

A related question is whether the dominance of L-amino acids in biochemistry was a matter of chance or whether it was a consequence of the extremely small but systematic chiral asymmetry due to electroweak interactions that is known to exist at the atomic and molecular levels [4–6, 13, 14]. Theories that support both views have been put forward but there is no general consensus on this matter either, mainly because there is a dearth of persuasive experimental evidence. However, the theory of nonequilibrium symmetry breaking provides a valuable means of assessing the plausible role of different models. For example, if we consider a prebiotic symmetry-breaking process that might have occurred in the oceans, it is possible to develop a general theory of symmetry breaking. A parameter λ, similar to the one in the above model, can be defined for any

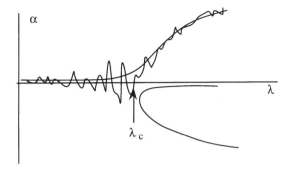

Figure 19.6 A symmetry-breaking transition or bifurcation in the presence of a small bias that favors one of the bifurcating branches. It can be analyzed through the general equation (19.3.17) and the probability of the system making a transition to the favored branch is given by equation (19.3.18)

symmetry-breaking system. When $\lambda < \lambda_c$, the system will be in a symmetric state; for $\lambda > \lambda_c$ the symmetric state will become unstable and evolve to an asymmetric state. Furthermore, regardless of the complexities of the reaction scheme, based on symmetry considerations only, it is possible to describe the bifurcation of the chiral symmetry-breaking states with an equation of the following type [15, 16]:

$$\frac{d\alpha}{dt} = -A\alpha^3 + B(\lambda - \lambda_c)\alpha + Cg + \sqrt{\varepsilon}f(t) \qquad (19.3.17)$$

in which the coefficients A, B, and C depend on the concentrations of the reactants and on the reaction rates (Fig. 19.6). The parameter g, is a small systematic bias, such as due to the electroweak force [5, 14] or other systematic chiral influences such as spin-polarized electrons that emerge from radioactive decay [17], or circularly polarized electromagnetic radiation emitted by certain stars that might fill large regions of space for long periods of time [11]. The systematic influence appears in the form of the rates of production or destruction of one enantiomer being larger than that of the other. The term $\sqrt{\varepsilon}f(t)$ represents random fluctuations with root-mean-square value $\sqrt{\varepsilon}$. Since the assumptions about rates of production for biomolecules, their catalytic activities and their concentrations determine the coefficients A, B and C, rather than the details of the chemical reaction scheme, the model is constrained by our general understanding of the prebiotic chemistry. Equation (19.3.17) provides a way to assess whether a given prebiotic model can produce and maintain the required asymmetry in a reasonable amount of time.

Furthermore, equation (19.3.17) can also give us a quantitative measure for the systematic chiral influence g. Detailed analysis, [16, 18] of this equation has shown that the sensitivity of the bifurcation to the systematic influence depends on the rate at which the system moves through the critical point λ_c, i.e. we assume that $\lambda = \lambda_0 + \gamma t$, so that the initial value of λ_0 is less than λ_c, but that λ gradually increases to a value larger than λ_c at an average rate γ. This process may correspond, for example, to a slow increase in the concentrations of biomolecules in the oceans. It has been shown, [16, 18] that the probability P of the system making a symmetry-breaking transition to the asymmetric state favored by the systematic chiral influence g is given by

$$P = \frac{1}{2\pi} \int_{-\infty}^{N} e^{-x^2/2} \, dx \quad \text{where} \quad N = \frac{Cg}{\sqrt{\varepsilon/2}} \left(\frac{\pi}{B\gamma} \right)^{1/4} \qquad (19.3.18)$$

Although *derived in the context of biomolecular handedness, this formula is generally valid for any system that breaks a two-fold symmetry, such as mirror inversion.* Using this formula, it is possible to understand the extraordinary sensitivity of bifurcation to small systematic biases that favor one enantiomer by increasing its production rate. For example, it can be estimated that the chiral asymmetry of the electroweak interaction can create differences of the order of one part in 10^{17} between the enantiomers. Application of the above theory shows that if the autocatalytic production rate of the chiral molecules is faster than the racemization rates, then for a period in the range 10^4 to 10^5 years, the enantiomer favored by the electroweak force will dominate [16]. For such a scenario, there is currently no experimental evidence to show us how chiral autocatalysis with the required properties can originate in prebiotic chiral molecules.

Many different scenarios have been suggested for the possible origins of biomolecular handedness. An extensive review can be found in the literature [19]. Note that, even if one is considering a process of chiral asymmetry generation after "life" arose, equations of the type (19.3.17) can still be used to describe the symmetry-breaking process, but this time the model will contain as "reactants" the self-replicating unit of life.

19.4   Chemical Oscillations

Our next example of a dissipative structure illustrates how the breaking of time-translation symmetry leads to oscillatory behavior. Some early reports of concentration oscillations were discounted because it was widely believed that such behavior was not consistent with thermodynamics. That is why the report on oscillating reactions by Bray in 1921 and Belousov in 1959 were met with skepticism [20]. Although it is true that oscillations of the extent of reaction ξ about its equilibrium value will violate the Second Law, oscillations of concentration about a nonequilibrium value of ξ do not violate the Second Law. When it was realized that systems far from thermodynamic equilibrium could exhibit oscillations, interest in these and other oscillating reactions rose sharply and gave rise to a rich study of dissipative structures in chemical systems.

Developments in the theoretical understanding of instability for non-equilibrium states in the 1960s [3] stimulated the experimental study of autocatalytic chemical kinetics that could give rise to concentration oscillations through the phenomenon of bifurcation. In 1968 Prigogine and Lefever [21] developed a simple model that not only demonstrated clearly how a non-equilibrium system can become unstable and make a transition to an oscillatory state, but also proved to be a rich source for theoretical understanding of propagating waves and almost every other phenomenon observed in real chemical systems, most of which are extremely complex to study. Due to its impact on the study of dissipative structures, it is often called the **Brusselator** (after its place of origin, the Brussels School of Thermodynamics) or the "trimolecular model" due to the trimolecular autocatalytic step in the reaction scheme. Because of its theoretical simplicity, we shall first discuss this reaction. Here is the reaction scheme:

$$A \xrightarrow{k_1} X \tag{19.4.1}$$

$$B + X \xrightarrow{k_2} Y + D \tag{19.4.2}$$

$$2X + Y \xrightarrow{k_3} 3X \tag{19.4.3}$$

$$X \xrightarrow{k_4} E \tag{19.4.4}$$

The net reaction of this scheme is $A + B \rightarrow D + E$. We assume the concentrations of the reactants A and B are maintained at a desired nonequilibrium value through appropriate flows. The products D and E are removed as they are formed. We also assume the reaction occurs in a solution that is well stirred, hence homogeneous. If we further assume that all the reverse reactions are so slow they can be neglected, we have the following rate equations for the species X and Y:

$$\frac{d[X]}{dt} = k_1[A] - k_2[B][X] + k_3[X]^2[Y] - k_4[X] \equiv Z_1 \tag{19.4.5}$$

$$\frac{d[Y]}{dt} = k_2[B][X] - k_3[X]^2[Y] \equiv Z_2 \tag{19.4.6}$$

One can easily verify (exc. 19.5) that the stationary solutions to these equations are

$$[X]_s = \frac{k_1}{k_4}[A] \qquad [Y]_s = \frac{k_4 k_2}{k_3 k_1} \frac{[B]}{[A]} \tag{19.4.7}$$

As was explained in section 18.4, the stability of the stationary state depends on the eigenvalues of the Jacobian matrix

$$
\begin{bmatrix}
\dfrac{\partial Z_1}{\partial [X]} & \dfrac{\partial Z_1}{\partial [Y]} \\[2ex]
\dfrac{\partial Z_2}{\partial [X]} & \dfrac{\partial Z_2}{\partial [Y]}
\end{bmatrix}
\tag{19.4.8}
$$

evaluated at the stationary state (19.4.7). The explicit form of the Jacobian matrix is

$$
\begin{bmatrix}
k_2[B] - k_4 & k_3[X]_s^2 \\[1ex]
-k_2[B] & -k_3[X]_s^2
\end{bmatrix}
\tag{19.4.9}
$$

The example in section 18.4 shows how the stationary state (19.4.7) becomes unstable when a complex conjugate pair of eigenvalues cross the imaginary axis; for the Brusselator this happens when:

$$
[B] > \frac{k_4}{k_2} + \frac{k_3 k_1^2}{k_2 k_4^2}[A]^2
\tag{19.4.10}
$$

The system makes a transition to an oscillatory state and the resulting oscillations are shown in Fig. 19.7. The steady states and the transition to oscillations can easily be investigated using the Mathematica codes provided in Appendix 19.1.

THE BELOUSOV–ZHABOTINSKY REACTION

Once it became clear that concentration oscillations are not inconsistent with the laws of thermodynamics (as the theoretical models of oscillating reactions showed), interest grew in the neglected 1959 report by Belousov and the later experiments of Zhabotinsky reported in a 1964 article [22]. These experimental studies of Belousov and Zhabotinsky were conducted in the Soviet Union and made known to the Western world through the Brussels School of Thermodynamics. In the United States, the study of the Belousov–Zhabotinsky oscillations was taken up by Field, Körös and Noyes [23], who performed a through-study of the reaction mechanism in the early 1970s. This was an important landmark in the study of oscillating reactions. Field, Körös and Noyes identified the key steps in the rather complex **Belousov–Zhabotinsky reaction**, and developed a model — which we shall refer to as the **FKN model**— consisting of only three variables that showed how the oscillations arise.

The Belousov–Zhabotinsky reaction is basically catalytic oxidation of an organic compound such as malonic acid, $CH_2(COOH)_2$. The reaction occurs in an aqueous solution and is easily performed in a beaker by simply adding the

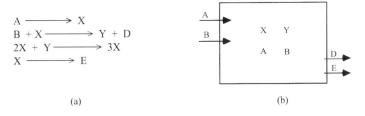

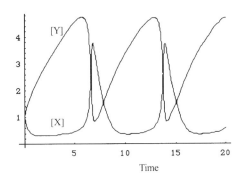

Figure 19.7 Oscillations in [X] and [Y] of the Brusselator obtained using Mathematica code B in Appendix 19.1

following reactants in the concentrations shown:

$$[H^+] = 2.0\,M \qquad [CH_2(COOH)_2] = 0.28\,M$$
$$[BrO_3^-] = 6.3 \times 10^{-2}\,M \qquad [Ce^{4+}] = 2.0 \times 10^{-3}\,M$$

After an initial "induction" period, the oscillatory behavior can be seen in the variation of the concentration of the Ce^{4+} ion, due to which there is a change in color from colorless to yellow. Many variations of this reaction — with more dramatic variations of color—are known today. (A wealth of detail about the Belousov–Zhabotinsky may be found in the literature [24].)

Box 19.1 presents a simplified version of the reaction mechanism from which the FKN model was developed. (Later models of the Belousov–Zhabotinzky reactions have included as many as 22 reaction steps.) The FKN model of the Belousov–Zhabotinsky reaction makes the following identifications: $A = BrO_3^-$, $X = HBrO_2$, $Y = Br^-$, $Z = Ce^{4+}$, $P = [HBrO]$ and $B = [Org]$, organic species that is oxidized. In modeling the reaction, $[H^+]$ is absorbed in the definition of the rate constant. The reaction scheme consists of the following steps:

- Generation of $HBrO_2$

$$A + Y \to X + P \qquad k_1[A][Y] \qquad (19.4.11)$$

- Autocatalytic production of $HBrO_2$

$$A + X \rightarrow 2X + 2Z \qquad k_2[A][X] \qquad (19.4.12)$$

- Consumption of $HBrO_2$

$$X + Y \rightarrow 2P \qquad k_3[X][Y] \qquad (19.4.13)$$

$$2X \rightarrow A + P \qquad k_4[X]^2 \qquad (19.4.14)$$

- Oxidation of the organic reactants

$$B + Z \rightarrow (f/2)Y \qquad k_5[B][Z] \qquad (19.4.15)$$

Box 19.1 The Belousov–Zhabotinsky Reaction and the FKN Model

The Field–Köros–Noyes (FKN) model of the Belousov–Zhabotinsky reaction consists of the following steps with $A = BrO_3^-$, $X = HBrO_2$, $Y = Br^-$, $Z = Ce^{4+}$, $P = HBrO$ and $B = Org$. In modeling the reaction, $[H^+]$ is absorbed in the definition of the rate constant.

- Generation of $HBrO_2$: $A + Y \rightarrow X + P$

$$BrO_3^- + Br^- + 2H^+ \rightarrow HBrO_2 + HBrO \qquad (1)$$

- Autocatalytic production of $HBrO_2$: $A + X \rightarrow 2X + 2Z$

$$BrO_3^- + HBrO_2 + H^+ \rightarrow 2BrO_2^{\cdot} + H_2O \qquad (2)$$

$$BrO_2^{\cdot} + Ce^{3+} + H^+ \rightarrow HBrO_2 + Ce^{4+} \qquad (3)$$

The net reaction, $(2) + 2(3)$, is autocatalytic in $HBrO_2$. Since the rate-determining step is (2), the reaction is modeled as $BrO_3^- + HBrO_2 \xrightarrow{H^+,Ce^{3+}} 2Ce^{4+} + 2HBrO_2$

- Consumption of $HBrO_2$: $X + Y \rightarrow 2P$ and $2X \rightarrow A + P$

$$HBrO_2 + Br^- + H^+ \rightarrow 2HBrO \qquad (4)$$

$$2HBrO_2 \rightarrow BrO_3^- + HBrO + H^+ \qquad (5)$$

- Oxidation of the organic reactants: $B + Z \rightarrow (f/2)Y$

$$CH_2(COOH)_2 + Br_2 \rightarrow BrCH(COOH)_2 + H^+ + Br^- \qquad (7)$$

$$Ce^{4+} + \tfrac{1}{2}[CH_2(COOH)_2 + BrCH(COOH)_2 \rightarrow \tfrac{f}{2}Br^- + Ce^{3+} + products \qquad (8)$$

The oxidation of the organic species is a complex reaction. It is approximated by a single rate determining step (8). In the FKN model, concentration [B] of the organic species is assumed to be constant. The value of the effective stoichiometric coefficient f is a variable parameter. Oscillations occur if f is in the range 0.5–2.4.

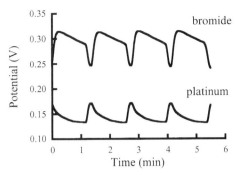

Figure 19.8 Experimentally observed oscillations in [Br$^-$] for the Belousov–Zhabotinsky reaction; the concentrations were measured using electrodes (Courtesy John Pojman)

The corresponding rate equations are

$$\frac{d[X]}{dt} = k_1[A][Y] + k_2[A][X] - k_3[X][Y] - 2k_4[X]^2 \qquad (19.4.16)$$

$$\frac{d[Y]}{dt} = -k_1[A][Y] - k_3[X][Y] + \frac{f}{2}k_5[B][Z] \qquad (19.4.17)$$

$$\frac{d[Z]}{dt} = 2k_2[A][X] - k_5[B][Z] \qquad (19.4.18)$$

Stationary states of this equation can be found after a little calculation (exc. 19.7). To study its stability involves analyzing the roots of a third-degree equation. There are many analytical methods [25] to analyze the oscillatory behavior of such a system, but these details are outside our main objective of giving examples of oscillating chemical systems. The oscillatory behavior of these equations may be numerically studied quite easily using Mathematica code C in Appendix 19.1 (Figs. 19.8 and 19.9). For numerical solutions, one may use the following data [25]:

$$k_1 = 1.28 \, \text{mol}^{-1} \, \text{L} \, \text{s}^{-1} \quad k_2 = 8.0 \, \text{mol}^{-1} \, \text{L} \, \text{s}^{-1} \quad k_3 = 8.0 \times 10^5 \, \text{mol}^{-1} \, \text{L} \, \text{s}^{-1}$$

$$k_4 = 2.0 \times 10^3 \, \text{mol}^{-1} \, \text{L} \, \text{s}^{-1} \quad k_5 = 1.0 \, \text{mol}^{-1} \, \text{L} \, \text{s}^{-1}$$

$$[B] = [\text{Org}] = 0.02 \, \text{M} \quad [A] = [\text{BrO}_3^-] = 0.06 \, \text{M} \quad 0.5 < f < 2.4$$

$$(19.4.19)$$

The Belousov–Zhabotinsky reaction shows oscillations of great variety and complexity; it even exhibits chaos. In chaotic systems arbitrarily close initial conditions diverge exponentially; the system exhibits aperiodic behavior. A

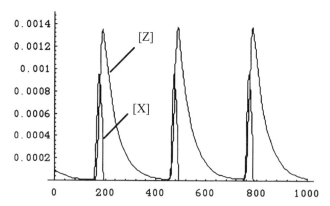

Figure 19.9 Oscillatory solutions obtained numerically using the FKN model of the Belousov–Zhabotinsky reaction: $[X] = [HBrO_2]$ and $[Z] = [Ce^{4+}]$. The plots were obtained using Mathematica code C in Appendix 19.1

review by Epstein and Showalter [33] summarizes these aspects. It also produces propagating waves and multistability. A large number of very interesting phenomena have been studied using this reaction [24, 25].

OTHER OSCILLATING REACTIONS

During the last two decades, many more oscillating chemical reactions have been discovered. Indeed, Irving Epstein and coworkers in the United States [26–28] and De Kepper and Boissonade in France [32] developed a systematic way of designing oscillating chemical reactions. In biochemical systems, one of the most interesting oscillating behaviors is found in the glycolytic reaction. A recent monograph by Albert Goldbeter [29] summarizes the vast amount of study on oscillatory biochemical systems.

19.5   Turing Structures and Propagating Waves

From the delicate beauty of the butterfly to the "fearful symmetry" of the tiger, Nature if full of wondrous patterns, both animate and the inanimate. How do these patterns arise? Dissipative process in systems far from thermodynamic equilibrium may provide at least a partial answer.

The emergence of biological morphology during embryonic development — with hands and feet and eyes all in the right place — is a fascinating subject (a popular account of this subject is Lewis Wolperts's *Triumph of the Embryo*, 1991, Oxford University Press)). What mechanism produces the morphology of living organisms? In 1952 the British mathematician Alan Turing suggested a mechanism based on the processes of chemical reactions and diffusion [30]. He showed, by devising a simple model, how chemical reactions and diffusion can work in consonance to produce stable stationary patterns of concentrations. Turing proposed it to explain biological morphogenesis. Today we know that biological morphogenesis is a very complex process, too complex to be explained entirely by the processes of diffusion and chemical reactions. However, Turing's observation has gained much attention since the 1970s due to the great interest in theoretical and experimental study of far-from-equilibrium chemical systems. In this section we will briefly describe a **Turing structure**, or a stationary **spatial dissipative structure**, using the Brusselator of section 19.4.

For simplicity, we shall consider a system with one spatial dimensional, coordinate r, in which diffusion occurs (Fig. 19.10). We assume the system extends form $-L$ to $+L$. We must also specify spatial boundary conditions; the usual boundary conditions are that either the concentrations of the reactants or their flows are maintained at a constant value at the boundaries (or even a combination of both). For our example, we shall assume that the flows of the reactants are zero at the boundaries. Since diffusion flow is proportional to the derivative $\partial C/\partial r$, (in which C is the concentration), the no-flow boundary conditions imply that the derivatives of the concentrations are zero at the boundaries.

When diffusion is included as a transport process, the kinetic equations (19.4.5) and (19.4.6) become

$$\frac{\partial[X]}{\partial t} = D_X \frac{\partial^2[X]}{\partial r^2} + k_1[A] - k_2[B][X] + k_3[X]^2[Y] - k_4[X] \qquad (19.5.1)$$

$$\frac{\partial[Y]}{\partial t} = D_Y \frac{\partial^2[Y]}{\partial r^2} + k_2[B][X] - k_3[X]^2[Y] \qquad (19.5.2)$$

The boundary conditions are

$$\left.\frac{\partial[X]}{\partial r}\right|_{r=-L} = \partial\left.\frac{[X]}{\partial r}\right|_{r=+L} = 0$$

in which D_X and D_Y are the diffusion coefficients and r is the spatial coordinate. As before, we assume that [A] and [B] are maintained at a fixed uniform value along the entire system (an assumption that simplifies the mathematics but which is difficult to achieve in practice). Diffusion usually homogenizes the concentration in a system, but when coupled with autocatalytic chemical reactions under far-from-equilibrium conditions, it actually generates inhomogeneities or patterns. For pattern formation, the diffusion coefficients must be different. If the diffusion coefficients are nearly equal, then diffusion does not cause an instability; diffusion only tends to homogenize the instability that

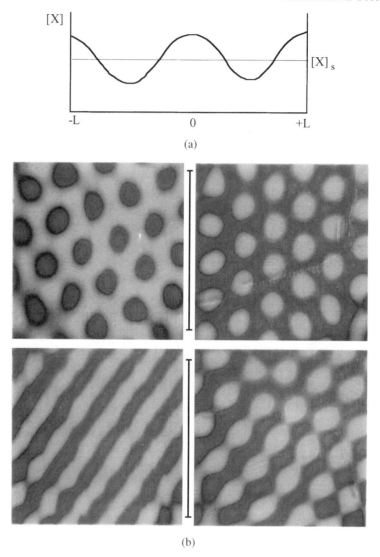

(a)

(b)

Figure 19.10 (a) Turing structure in a one-dimensional Brusselator model. (b) Turing structures observed in chlorite-iodide-malonic acid reaction in an acidic aqueous solution (Courtesy Harry L. Swinney). The size of each square is nearly 1 mm.

already exists. This can be seen as follows. We begin by considering the stability of the stationary state (19.4.7), the concentrations being homogeneous in the entire system:

$$[X]_s = \frac{k_1}{k_4}[A] \qquad [Y]_s = \frac{k_4 k_2}{k_3 k_1}\frac{[B]}{[A]} \qquad (19.5.3)$$

The stability of this solution depends on the behavior of a small perturbation. If δX and δY are the small perturbation from $[X]_s$ and $[Y]_s$, it is easy to see that equations linearized about the steady state (19.5.3) are

$$\frac{\partial}{\partial t}\begin{pmatrix}\delta X\\\delta Y\end{pmatrix} = \begin{bmatrix}D_X\dfrac{\partial^2}{\partial r^2} & 0 \\ 0 & D_Y\dfrac{\partial^2}{\partial r^2}\end{bmatrix}\begin{pmatrix}\delta X\\\delta Y\end{pmatrix} + \begin{bmatrix}k_2[B]-k_4 & k_3[X]_s^2 \\ -k_2[B] & -k_3[X]_s^2\end{bmatrix}\begin{pmatrix}\delta X\\\delta Y\end{pmatrix}$$

$$(19.5.4)$$

If we assume $D_X = D_Y = D$, equation (19.5.4) can be written as

$$\frac{\partial}{\partial t}\begin{pmatrix}\delta X\\\delta Y\end{pmatrix} = D\frac{\partial^2}{\partial r^2}I\begin{pmatrix}\delta X\\\delta Y\end{pmatrix} + M\begin{pmatrix}\delta X\\\delta Y\end{pmatrix} \qquad (19.5.5)$$

in which M is the matrix in the second term of (19.5.4) and I is the identity matrix. For a *linear equation* of this type, the spatial part of the solutions can always be written as combinations of $\sin Kx$ and $\cos Kx$, in which the wavenumbers K are chosen so that the boundary conditions are satisfied. This means that if we understand the behavior of a perturbation of the type

$$\begin{pmatrix}\delta X(t)\\\delta Y(t)\end{pmatrix}\sin Kr \quad \text{and} \quad \begin{pmatrix}\delta X(t)\\\delta Y(t)\end{pmatrix}\cos Kr \qquad (19.5.6)$$

in which the spatial part is separated, then the behavior of all linear combinations of these basic solutions can be deduced. If we substitute (19.5.6) into (19.5.5), we obtain

$$\frac{\partial}{\partial t}\begin{pmatrix}\delta X(t)\\\delta Y(t)\end{pmatrix} = (-DK^2\,I + M)\begin{pmatrix}\delta X(t)\\\delta Y(t)\end{pmatrix} \qquad (19.5.7)$$

From this expression it is clear that if λ_+ and λ_- are the eigenvalues of M, the addition of diffusion will only change the eigenvalues to $(\lambda_+ - DK^2)$ and $(\lambda_- - DK^2)$. Since it is the positivity of the real part of the eigenvalue that indicates instability, we see that (in this case, where $D_X = D_Y$), diffusion does not generate a new instability; it only makes steady states more stable to perturbations with $K \neq 0$. So the solution to (19.5.7) with $K = 0$ is the least stable state because its eigenvalues will have the largest real parts.

For the emergence of spatial patterns, the diffusion coefficients must be *unequal*. In a small region, if one species diffuses out more rapidly than the other, the growth of one species may be facilitated by the depletion of the other. If this happens, the homogeneous state will no longer be stable and inhomogeneities will begin to grow. When the diffusion coefficients are unequal, it is easy to see that in place of the matrix $(-K^2DI + M)$ we have the matrix

$$\begin{bmatrix} k_2[B] - k_4 - K^2D_X & k_3[X]_s^2 \\ -k_2[B] & -k_3[X]_s^2 - K^2D_Y \end{bmatrix} \qquad (19.5.8)$$

For an instability to produce stationary **spatial structures**, the two eigenvalues of this matrix must be real, and at least one must become positive. If the eigenvalues are real, and one becomes positive due to the variations in the parameters [B] and [A], then the unstable perturbation will be of the form

$$\begin{pmatrix} c_1 \\ c_2 \end{pmatrix} \sin(Kr)e^{\lambda_+ t} \quad \text{or} \quad \begin{pmatrix} c_1 \\ c_2 \end{pmatrix} \cos(Kr)e^{\lambda_+ t} \qquad (19.5.9)$$

in which λ_+ is the eigenvalue with positive real part. This indicates a growth of a spatial pattern $\sin Kr$ or $\cos Kr$ without any temporal oscillations; it will evolve to a stationary pattern or a Turing structure.

On the other hand, if the eigenvalues are a complex-conjugate pair, then the solutions to the perturbation equation (19.5.4) will be of the form

$$\begin{pmatrix} c_1 \\ c_2 \end{pmatrix} \sin(Kr)e^{(\lambda_{re} \pm i\lambda_{im})t} \quad \text{or} \quad \begin{pmatrix} c_1 \\ c_2 \end{pmatrix} \cos(Kr)e^{(\lambda_{re} \pm i\lambda_{im})t} \qquad (19.5.10)$$

in which $\begin{pmatrix} c_1 \\ c_2 \end{pmatrix}$ is the eigenvector with the eigenvalue $\lambda = \lambda_{re} \pm i\lambda_{im}$, with its real and imaginary parts as shown. If the real part λ_{re} is positive, the perturbation (19.5.10) will grow. The unstable perturbation contains oscillations in time, due to the factor $e^{i\lambda_{im}t}$ as well as variations in space due to the factor $\sin Kr$ or $\cos Kr$. Such a perturbation corresponds to a **propagating wave**.

For matrix (19.5.8) the condition for one of its two real eigenvalues to cross zero can be obtained as follows. First we note that the determinant, Det, of a matrix is the product of the eigenvalues. If the eigenvalues are λ_+ and λ_-, we have

$$(\lambda_+ \lambda_-) = \text{Det} = (k_2[B] - k_4 - K^2D_X)(-k_3[X]_s^2 - K^2D_Y)$$
$$+ (k_2[B])(k_3[X]_s^2) \qquad (19.5.11)$$

Before the onset of the instability, both eigenvalues are negative, hence Det > 0. Let us assume that when the parameter [B] is varied, λ_+ crosses zero and becomes positive. Then, at the point where $\lambda_+ = 0$, we have Det = 0 and when $\lambda_+ > 0$ we have Det < 0. Thus the condition for the instability may be stated

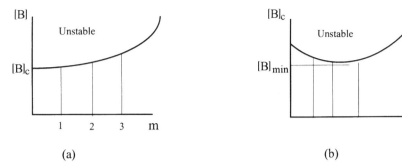

Figure 19.11 (a) Stability diagram showing the value of [B] for which modes of wavenumber m become unstable when D_X and D_Y are nearly equal. The instability occurs when $[B] = [B]_c$ at which $K = 0$, leading to homogeneous oscillations. (b) Stability diagram showing the value of [B] for which modes of wavenumber m become unstable when the difference between D_X and D_Y is large. As [B] increases, when it just above $[B]_{min}$, the pattern with the value of K^2 that is consistent with the boundary conditions will become unstable and grow

as

$$\text{Det} = (k_2 [B] - k_4 - K^2 D_X)(-k_3 [X]_s^2 - K^2 D_Y)$$
$$+ (k_2 [B])(k_3 [X]_s^2) < 0 \tag{19.5.12}$$

Using $[X]_s = (k_1/k_4) [A]$ this inequality can be rewritten as

$$[B] > \frac{1}{k_2} [k_4 + K^2 D_X] \left[1 + \frac{k_3 k_1^2 [A]^2}{k_4^2} \frac{1}{K^2 D_Y} \right] \tag{19.5.13}$$

This then is the condition under which a Turing structure will arise in the Brusselator model. As [B] increases, the lowest value $[B]_c$ for which (19.5.13) is satisfied will trigger an instability. The value of $[B]_c$ can be found by plotting

$$[B]_c = \frac{1}{k_2} [k_4 + K^2 D_X] \left[1 + \frac{k_3 k_1^2 [A]^2}{k_4^2} \frac{1}{K^2 D_Y} \right] \tag{19.5.14}$$

as a function of K^2. As shown in Fig. 19.11 (b), this plot has a minimum. When [B] reaches this minimum value, the corresponding K_{min} will be the wavenumber of the stationary pattern. The minimum occurs at the following values (exc. 19.9):

$$K_{min}^2 = A \sqrt{\frac{k_3 k_1^2}{k_4 D_X D_Y}} \quad \text{and} \quad [B]_c = [B]_{min} = \frac{1}{K_2} \left[\sqrt{k_4} + A \sqrt{\frac{D_X k_3 k_1^2}{D_Y k_4^2}} \right]^2 \tag{19.5.15}$$

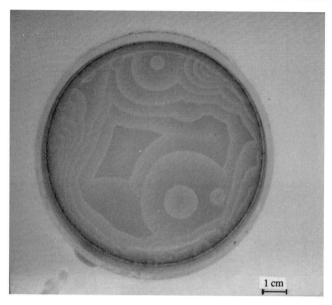

Figure 19.12   Traveling waves in the Belousov–Zhabotinsky reaction

Experimentally, traveling waves have been observed in the Belousov–Zhabotinsky reaction (Fig. 19.12) but only recently have the Turing patterns been realized in the laboratory [31].

The examples shown in this chapter are only a small part of the rich variety of behavior encountered in far-from-equilibrium chemical systems. Here our objective is only to show a few examples; an extensive description would form a book in itself! At the end of the chapter there is a list of monographs and conference proceedings that give a detailed descriptions of oscillations, propagating waves, Turing structures, pattern formation on catalytic surfaces, multistability and chaos (both temporal and spatiotemporal). Dissipative structures have also been found in other fields such as hydrodynamics and optics.

19.6   Structural Instability and Biochemical Evolution

We conclude this chapter with a few remarks on another kind of instability often called "structural instability" and its relevance to biochemical evolution. In the previous sections we have seen instabilities giving rise to organized states. These instabilities arose in a given set of chemical reactions. In nonequilibrium

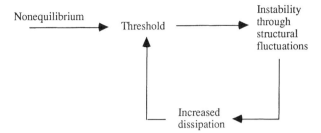

Figure 19.13 Structural instabilities during molecular evolution give rise to new processes that tend to increase entropy production.

chemical systems, instability may also arise by the introduction of a new chemical species which gives rise to new reactions; these new reactions may destabilize the system and drive it to a new state of organization. In this case the "structure" of the chemical reaction network is itself subject to changes. Each new species alters the reaction kinetics and this may drastically alter the state of the system, i.e., due to the appearance of a new chemical species the system may become unstable and evolve to a new state.

This type of structural instability can be seen most easily in the evolution of self-replicating molecules with a steady supply of monomers. Let us consider a set of autocatalytic polymers that are capable of self-replication through a template mechanism. In this case, each new polymer is a new autocatalytic species. Let us further assume that this self-replication is subject to random errors or mutations. Each mutation of a self-replicating molecule introduces a new species and new chemical reactions. Thus if we write a set of kinetic equations for such a system, each time a random mutation occurs, the set of equations itself will change. Under a given set of nonequilibrium conditions or "environment" some (or perhaps most) of the mutations may not produce a polymer whose rate of self-replication is larger than those of others. The appearance of such a new species may cause a small change in the population of various polymers but no significant change will arise. However, some of the mutations might give rise to a polymer with a high rate of self-replication. This would correspond to fluctuation to which the system is unstable. The new polymer may now dominate the system and alter the population significantly. This of course corresponds to Darwinian evolution at the molecular level, the paradigm of the survival of the fittest. Many detailed studies of such structural instabilities and molecular evolution have been conducted [34–37]. These models are beyond the scope of this text but we will note an interesting thermodynamic feature summarized in Fig. 19.13. Each new structural instability generally increases the dissipation or entropy production in the system because it increases the number of reactions. This is in contrast to the near-equilibrium situations discussed in Chapter 17 in which the entropy production tends to a minimum. Structural instability may progressively drive far-from-equilibrium systems to higher states of entropy production and higher states of order. Needless to say, biochemical evolution and the origin of life is a very complex process which we are only beginning to understand. But now we see instability, fluctuation and evolution to organized states as a general nonequilibrium process whose most spectacular manifestation is the evolution of life.

Appendix 19.1: Mathematica Codes

CODE A: MATHEMATICA CODE FOR SOLVING THE SET OF EQUATION (19.3.6) AND (19.3.7)

```
(* Chemical kinetics showing chiral symmetry breaking *)
k1f = 0.5; k1r = 0.1; k2f = 0.1; k2r = 0.2; k3 = 0.5; S = 0.5;
T = 0.5; Soln1 = NDSolve[{
XL'[t] == k1f*S*T - k1r*XL[t] + k2f*S*T*XL[t]-k2r*[XL]^2
- k3*XL[t]*XD[t],
XD'[t] == k1f*S*T - k1r*XD[t] + k2f*S*T*XD[t]-k2r*[XD]^2
- k3*XL[t]*XD[t],
XL[0] == 0.002, XD[0] == 0.0},{XL,XD},{t,0,100},
MaxSteps - > 500]
```

The solution can be plotted using the following command:
```
Plot[Evaluate[{XL[t],XD[t]}/.Soln1],{t,0,100}]
```

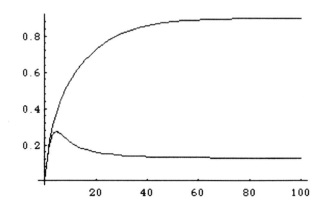

CODE B: MATHEMATICA CODE FOR THE BRUSSELATOR

```
(* Chemical kinetics: the Brusselator *)
k1 = 1.0; k2 = 1.0; k3 = 1.0; k4 = 1.0; A = 1.0; B = 3.0;
Soln2 = NDSolve[{ X'[t] == k1*A - k2*B*X[t]
+ k3*(X[t]^2)*Y[t] - k4*X[t],
Y'[t] == k2*B*X[t] - k3*(X[t]^2)*Y[t],
X[0] == 1.0,Y[0] == 1.0},{X,Y},{t,0,20},
MaxSteps - > 500]
```

The solution can be plotted using the following commands:

```
Plot[Evaluate[{X[t]}/.Soln2],{t,0,20}]
```

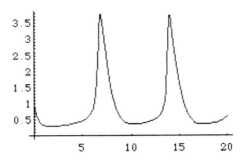

```
Plot[Evaluate[{X[t],Y[t]}/.Soln2],{t,0,20}]
```

CODE C: MATHEMATICA CODE FOR FKN MODEL OF THE BELOUSOV–
ZHABOTINSKY REACTION

```
(* Chemical kinetics: the Belousov-Zhabotinsky reaction/
FKN *)
(* X = HBrO2  Y = Br-  Z = Ce4+  B = Org  A = BrO3- *)
k1 = 1.28; k2 = 8.0; k3 = 8.0*10^5; k4 = 2*10^3; k5 = 1.0;
A = 0.06; B = 0.02;f = 1.5;
```

```
Soln3 = NDSolve[{
X'[t] == k1*A*Y[t] + k2*A*X[t] - k3*X[t]*Y[t]
      - 2*k4*X[t]^2,
Y'[t] == -k1*A*Y[t] - k3*X[t]*Y[t] + (f/2)*k5*B*Z[t],
Z'[t] == 2*k2*A*X[t] - k5*B*Z[t],
X[0] == 2*10^-7,Y[0] == 0.00002,Z[0] == 0.0001},{X,Y,Z},
{t,0,500}, MaxSteps -> 1000]
```

The solution can be plotted using the following commands:

```
Plot[Evaluate[{X[t]}/.Soln3],{t,0,500},PlotRange ->
{0.0,10^-4}]
```

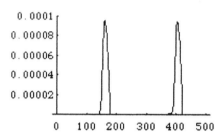

```
Plot[Evaluate[{Y[t]}/.Soln3],{t,0,500},PlotRange ->
{0.0,10^-4}]
```

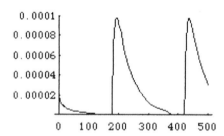

Similarly the Z[t] can also be plotted with PlotRange->{0.0,2*10^-3}.

References

1. Prigogine, *From Being to Becoming*, 1980, San Francisco: W. H. Freeman.
2. Prigogine, I. and Stengers I., *Order Out of Chaos*, 1984, New York: Bantam.
3. Prigogine, I., *Introduction to Thermodynamics of Irreversible Processes*, 1967, New York: John Wiley.
4. Hegstrom, R. and Kondepudi, D. K., *Sci. Am.*, Jan 1990, pp. 108–115.
5. Mason, S. F. and Tranter G. E., *Chem. Phys. Lett.*, **94** (1983), 34–37.
6. Hegstrom, R. A., Rein, D. W., and Sandars, P. G. H., *J. Chem. Phys.*, **73** (1980), 2329–2341.
7. Frank, F. C., *Biochem. Biophys. Acta*, **11** (1953), 459.
8. Kondepudi, D. K. and Nelson, G. W., *Physica A*, **125** (1984), 465–496.

9. Kondepudi, D. K., Kaufman R., and Singh N., *Science*, **250** (1990), 975–976.
10. Kondepudi, D. K. et al., *J. Am. Chem. Soc.*, **115** (1993), 10211–10216.
11. Bonner, W. A., *Origins of Life*, **21** (1992), 407–420.
12. Bonner, W. A., *Origins of Life and Evol. Biosphere*, **21** (1991), 59–111.
13. Bouchiat, M.-A. and Pottier, L., *Science*, **234** (1986), 1203–1210.
14. Mason, S. F. and Tranter, G. E., *Proc. R. Soc. London*, **A397** (1985), 45–65.
15. Kondepudi, D. K. and Nelson, G. W., *Physica A*, **125** (1984), 465–496.
16. Kondepudi, D. K. and Nelson, G. W., *Nature*, **314** (1985), 438–441.
17. Hegstrom, R., *Nature*, **315** (1985) 749.
18. Kondepudi, D. K., *BioSystems*, **20** (1987), 75–83.
19. Cline, D. B. (ed.), *Physical Origin of Homochirality in Life*, 1996, New York: American Institute of Physics.
20. Winfree, A. T., *J. Chem. Ed.*, **61** (1984), 661–663.
21. Prigogine, I. and Lefever, R., *J. Chem. Phys.*, **48** (1968), 1695–1700.
22. Zhabotinsky, A. M., *Biophysika*, **9** (1964), 306.
23. Field, R. J., Körös, E. and Noyes, R. M., *J. Am. Chem. Soc.*, **94** (1972), 8649–8664.
24. Field, R. J. and Burger, M. (eds), *Oscillations and Traveling Waves in Chemical Systems*, 1985, New York: Wiley.
25. Gray, P. and Scott, K. S., *Chemical Oscillations and Instabilities*, 1990, Oxford: Clarendon Press.
26. Epstein, I. R. and Orban, M., in *Oscillations and Traveling Waves in Chemical Systems*, R. J. F. Fiel and M. Burger (eds). 1985, New York: Wiley.
27. Epstein, I., K. Kustin, De Kepper, P. and Orbán M., *Sci. Am.*, Mar 1983, pp. 96–108.
28. Epstein, I. R., *J. Chem. Ed.*, **69** (1989), 191.
29. Goldbeter, A., *Biochemical Oscillations and Cellular Rhythms: the molecular bases of periodic and chaotic behaviour*, 1996, Cambridge: Cambridge University Press.
30. Turing, A., *Phil. Trans. R. Soc. London*, **B237** (1952), 37.
31. Kapral, R. and Showalter, K. (eds), *Chemical Waves and Patterns*, 1994, New York: Kluwer.
32. Boissonade, J. and Dekepper, P., *J. Phys. Chem.*, **84** (1980), 501–506.
33. Epstein, I. R. and Showalter, K., *J. Phys. Chem.*, **100** (1996), 13132–13143.
34. Prigogine, I., Nicolis, G. and Babloyantz, A., *Physics Today*, **25** (1972), No. 11, p. 23; No. 12, p. 38.
35. Eigen, M. and Schuster, P., *The Hypercycle – A Principle of Natural Self-organization*, 1979, Heidelberg: Springer.
36. Nicolis, G. and Prigogine, I., *Self-organization in Nonequilibrium Systems*, 1977, New York: Wiley.
37. Küppers, B-O., *Molecular Theory of Evolution*, 1983, Berlin: Springer.

Further Reading

- Nicolis, G. and Prigogine, I., *Self-Organization in Nonequilibrium Systems*. 1977, New York: Wiley.
- Vidal, C. and Pacault, A., (eds), *Non-Linear Phenomenon in Chemical Dynamics*. 1981, Berlin: Springer Verlag.
- Epstein, I., Kustin, K., De Kepper, P. and Orbán, M., *Sci. Am.*, Mar 1983, pp. 112.
- Field, R. J. and Burger, M. (eds), *Oscillations and Traveling Waves in Chemical Systems*. 1985, New York: Wiley.

- State-of-the-art symposium: self-organization in chemistry. *J. Chem. Ed.*, **66** (1989) No. 3; articles by several authors.
- Gray, P. and Scott, K. S., *Chemical Oscillations and Instabilities*. 1990, Oxford: Clarendon Press.
- Manneville, P., *Dissipative Structures and Weak Turbulence*. 1990, San Diego CA: Academic Press.
- Baras, F. and Walgraef, D. (eds), Nonequilibrium chemical dynamics: from experiment to microscopic simulation. *Physica A*, **188** (1992), No. 1–3; special issue.
- Ciba Foundation Symposium 162, *Biological Asymmetry and Handedness*, 1991, London: John Wiley.
- Kapral, R. and Showalter, K. (eds), *Chemical Waves and Patterns*. 1994, New York: Kluwer.

Exercises

19.1 Analyze the stability of solutions $\alpha = 0$ and $\alpha = \pm\sqrt{\lambda}$ for equation (19.2.1) and show explicitly that when $\lambda > 0$ the solution $\alpha = 0$ becomes unstable whereas the solutions $\alpha = \pm\sqrt{\lambda}$ are stable.

19.2 Write a Mathematica or Maple code to obtain the solutions of the equations in Exercise 19.1. Plot these solutions as a function of time for various initial conditions and show explicitly that the solutions evolve to stable stationary states.

19.3 For the reaction scheme (19.3.1) to (19.3.5), using the principle of detailed balance, verify that the concentrations of X_L and X_D will be equal at equilibrium.

19.4 Using the variables α, β and λ defined in (19.3.8), show that the kinetic equations (19.3.6) and (19.3.7) can be written in the forms (19.3.9) and (19.3.10).

19.5 Show that (19.4.7) are the stationary states of the kinetic equations of the Brusselator (19.4.5) and (19.4.6).

19.6 (a) Write the kinetic equations for [X] and [Y], assuming that [A] and [B] are fixed, for the following scheme (called the Lotka–Volterra model):

$$A + Y \rightarrow 2X \qquad X + Y \rightarrow 2Y \qquad Y \rightarrow B$$

(b) Obtain its steady states and analyze its stability as a function of the parameters [A] and [B].

19.7 (a) Using the dimensionless variables defined by

$$x = \frac{[X]}{X_0} \qquad y = \frac{[Y]}{Y_0} \qquad z = \frac{[Z]}{Z_0} \qquad \tau = \frac{t}{T_0}$$

in which

$$X_0 = \frac{k_2[A]}{2k_4} \qquad Y_0 = \frac{k_2[A]}{k_3} \qquad Z_0 = \frac{(k_2[A])^2}{k_4 k_5[B]} \qquad T_0 = \frac{1}{k_5[B]}$$

show that the kinetic equations (19.4.16) to (19.4.18) can be written as

$$\varepsilon \frac{dx}{d\tau} = qy - xy + x(1 - x)$$

$$\varepsilon' \frac{dy}{d\tau} = -qy - xy + fz$$

$$\frac{dz}{d\tau} = x - z$$

in which

$$\varepsilon = \frac{k_5[B]}{k_2[A]} \qquad \varepsilon' = \frac{2k_5 k_4[B]}{k_3 k_2[A]} \quad \text{and} \quad q = \frac{2k_1 k_4}{k_3 k_2}$$

(See Tyson, J. J., Scaling and reducing the Field–Körös–Noyes mechanism of the Belousov–Zhabotinsky reaction, *J. Phys. Chem.*, **86**, 1982, 3006–3012.)
(b) Find the stationary states of this set of equations.

19.8 Using Mathematica code C in Appendix 19.1, obtain the range of values for the parameter *f* in which oscillations occur. Also plot the relation between the period of oscillations and the value of *f*.

19.9 Show that the minimum of (19.5.14) occurs at the values given by (19.5.15)

20 WHERE DO WE GO FROM HERE?

In the introduction we emphasized that there is no final formulation of science; this also applies to thermodynamics. The method based on local equilibrium that we have followed appears to be satisfactory in a large domain of experimentation and observation. But there remain situations where some extension and modification are necessary. Let us enumerate a few of them.

First of all we have the case of rarefied media, where the idea of local equilibrium fails. The average energy at each point depends on the temperature at the boundaries. Important astrophysical situations belong to this catagory.

We then have the case of strong gradients, where we expect the failure of linear laws such as the Fourier law for heat conduction. Not much is known either experimentally or theoretically. Attempts to introduce such nonlinear outcomes into the thermodynamic description have led to "extended thermodynamics" [1] already mentioned in the text.

Finally, we have very interesting memory effects which appear for long times (as compared to characteristic relaxation times). This field started with important numerical simulations by Alder and Wainright [2], who showed that nonequilibrium processes may have "long-time tails." In other words, the approach to equilibrium is not exponential, as was generally believed, but polynomial (e.g. $t^{-3/2}$), which is much slower. To explain this effect, consider a molecule we set in motion with respect to the medium; its momentum is transmitted to the medium, which in turn reacts back on the molecule. This leads to memory effects which are discussed in many papers [3,4]. As a result, Nature has a much longer memory of irreversible processes than it was thought before. Again this shows that local equilibrium is an approximation, albeit a very good one.

However, the formulation of nonequilibrium thermodynamics as used in this book has already led to innumerable applications in very diverse fields. To whet the appetite, we shall quote just a few of them.

The first example is in *materials science*. Concepts such as fluctuations, dissipative structure and self-organization play an essential role in the true revolution that is occurring in this field. A good introduction is given by Walgraef [5]. Through new techno-logies (laser and particle irradiation, ion implantation, ultrafast quenches) it is now possible to produce materials in highly nonequilibrium conditions—thereby escaping the tyranny of the equilibrium phase diagram. Here are some examples from Walgraef's book:

- Materials such as quasicrystals, high-temperature superconductors, semiconductor heterostructures and superlattices are typical examples of materials produced in nonequilibrium conditions.

- It is now possible to produce complex structures or composites that simultaneously satisfy very diverse requirements. To do so, one has to control the material on length scales that vary from the atomic to the micrometer level. Selforganization is a precious ally for the design of such materials.

- Many materials are used in very demanding conditions. Submitted to deformation, corrosion, irradiation, etc., their defect populations acquire complex behaviors, well described by reaction diffusion equations, and may therefore become organized in very regular structures that affect their physical properties. It is also clear now that instabilities and patterns occur all the time in materials science. They affect the properties of the materials, hence they need to be understood and controlled.

- It is well known that defects play an important role in determining material properties. Point defects play a major role in all macroscopic material properties that are related to atomic diffusion mechanisms and to electronic properties in semiconductors. Line defects, or dislocations, are unquestionably recognized as the basic elements that lead to plasticity and fracture (Fig. 20.1). Although the study of individual solid-state defects has reached an advanced level, investigations into the collective behavior of defects under nonequilibrium conditions remain in their infancy. Nonetheless, significant progress has been made in dislocation dynamics and plastic instabilities over the past several years, and the importance of nonlinear phenomena has also been assessed in this field. Dislocation structures have been observed experimentally.

Curiously, the instabilities and self-organization that occur in far-from-equilibrium systems as a result of basic physical processes, such as chemical reactions and diffusion, also occur at the much more complex level of living systems. Mathematical modeling of these complex systems also consists of irreversible nonlinear equations. A basic feature in all these systems is the possibility of amplifying small fluctuations under certain conditions; this is what makes the systems unstable. The reason they undergo instabilities is often due to autocatalytic processes, which cause them to make a transition to states with distinctly different organization. Thus the paradigm of "order through fluctuations" also holds here.

One example of pattern formation in complex systems occurs in the life cycle of the cellular slime mold *Dictyostelium discoideum*. Figure 20.2 describes the life cycle of this species. Beginning as isolated amebas at the unicellular stage (a), they move in the surrounding medium, feed on such nutrients such as bacteria and proliferate by cell division. Globally speaking, they constitute a uniform system, inasmuch as their density (number of cells per square centimeter) is essentially constant. Suppose now that the amebas are starved; in the laboratory this is induced deliberately, in Nature it may happen because of

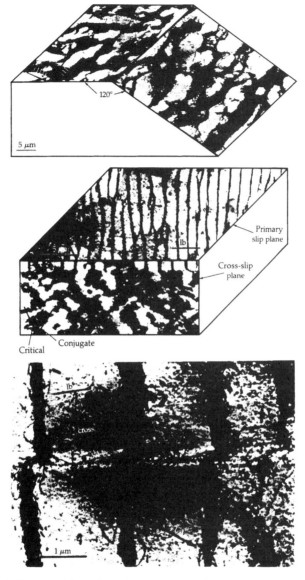

Figure 20.1 Three-dimensional dislocation structures in a periodically stressed single crystal of copper (Courtesy D. Walgraef)

less favorable ambient conditions. This is the analog of applying a constraint in a physical or chemical experiment. Interestingly, the individual cells do not die. Rather, they respond to the constraint by aggregating (b) toward a center of attraction. The initial homogeneity is broken; space becomes structured. The resulting multicellular body, the plasmodium (c), is capable of moving, presumably to seek more favorable conditions of temperature and moisture.

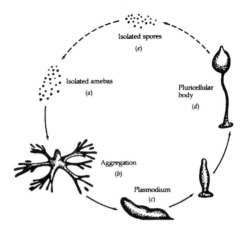

Figure 20.2 Life cycle of the cellular
slime mold *Dictyostelium discoideum*
(Courtesy A. Goldbeter)

After this migration the body differentiates (d) and gives rise to two kinds of
cells, one constitutes the stalk and the other constitutes a fruiting body within
which spores are formed. Eventually the spores are disseminated (e) in the
ambient medium, and if the conditions are favorable, they germinate to become
amebas and the life cycle begins again.

Let us investigate the aggregation stage, (Fig. 20.3) in more detail. After
starvation some of the cells begin to synthesize and release signals of a chemical
substance known as cyclic adenosine monophosphte (cAMP) in the extracellular
medium. The synthesis and release are periodic, just as in the chemical clock of
the BZ system, with a well-defined period for given experimental conditions.
The cAMP released by the "pioneer" cells diffuses into the extracellular
medium and reaches the surface of the neighboring cells. Two types of events
are then switched on. First, these cells perform an oriented movement called
chemotaxis toward the regions of higher concentration of cAMP, i.e. toward the
pioneer cells. This motion gives rise to density patterns among the cells that look
very much like the wave patterns in the BZ reagent (Fig. 19.12). Second, the
process of aggregation is accelerated by the ability of sensitized cells to amplify
the signal and to relay it in the medium. This enables the organism to control a
large territory and form a multicellular body comprising some 10^5 cells.

Thus, the response to the starvation constraint gives rise to a new level of
organization, resulting from the concerted behavior of a large number of cells
and enabling the organism to respond flexibly to a hostile environment. What are
the mechanisms mediating this transition? Let us first observe that the process of
chemotaxis leads to an amplification of the heterogeneity formed initially, when
the pioneer cells begin to emit pulses of cAMP. Because it enhances the density

Figure 20.3 Concentric and spiral waves of aggregation cell populations of *Dictyostelium discoideum* on an agar surface. The bands of cells moving toward the center appear bright (Courtesy A. Goldbeter)

of cells near the emission center, chemotaxis enhances movement of the other cells toward it. This constitutes what one usually calls a feedback loop, very similar to chemical autocatalysis.

As it turns out, a second feedback mechanism is present in *Dictyostelium discoideum* which operates at the subcellular level and is responsible for both the periodic emission of cAMP and the relay of the chemotactic signal. This mechanism is related to the synthesis of cAMP by the cell. The cAMP arises from transformation of another important cellular constituent, adenosine triphosphate (ATP), which (through its phosphate bond) is one of the principal carriers of energy within living cells. But the ATP $\rightarrow$ cAMP transformation is not spontaneous; a catalyst is needed to accelerate it to a level compatible with vital requirements. In biological systems the tasks of catalysis are assumed by special molecules called enzymes. Some enzymes have a single active site which the reactants must reach in order to transform into products. But in many cases there are cooperative enzymes, which have several sites; some of the sites are catalytic and others are regulatory. When special effector molecules bind to the regulatory sites, the catalytic function is considerably affected. In some cases the molecules reacting with or produced from the catalytic site may also act as effector molecules. This will switch on a feedback loop, which will be positive (activation) if the result is the enhancement of the rate of catalysis, or negative (inhibition) otherwise. The enzyme that catalyzes ATP $\rightarrow$ cAMP conversion is called adenylate cyclase and is fixed at the interior of the cell membrane. It interacts with a receptor fixed at the exterior phase of the membrane in a

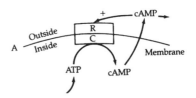

Figure 20.4 Oscillatory sysnthesis of cAMP in the slime mold *Dictyostelium discoideum*

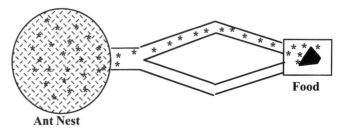

Figure 20.5 Bifurcation in the behavior of social insects such as ants can be seen in their choice of path to a food source (Deneubourg et al. [7])

cooperative fashion, whose details are not completely elucidated. The resulting cAMP diffuses into the extracellular medium through the cell membrane and can bind to the receptor and activate it (Fig. 20.4). In this way it enhances its own production, thereby giving rise to a feedback loop capable of amplifying signals and of inducing oscillatory behavior.

Many other examples may be found in the literature. Bifurcations can be found in the behavior of social insects as well [7]. Imagine an ant nest connected to a food source by two paths, identical except for the directions of their two limbs (Fig. 20.5). At first, equal numbers of ants are traveling on the two paths. After some time, practically all the ants are found on the same path due to the catalytic effects of chemical substances called "pheromones," produced by the ants. Note that which bridge will be used is unpredictable. This corresponds to a typical symmetry-breaking bifurcation.

There have been promising applications of nonequilibrium physics to geology. Numerous geological deposits exhibit spectacular regular mineralization structures at a variety of space scales. Among them are metamorphic layers of thickness a few millimeters to several meters, centimeter-scale structure in granite, and agates with bands several millimeters or centimeters wide. Figure 20.6 shows two examples. The traditional interpretation attributes these structures to sequential phenomena, tracing the effect of successive environmental or climatic changes. It appears, however, that a more satisfactory

Figure 20.6 (a) Skarn from Sa Leone, Sardinia. The light bands, 1–2 mm thick, consist of andraditic garnet (calcium and ferricion bearing). The dark bands, 5–8 thick, consist of magnetite and quartz. The white rectangle is 1 cm long. (b) Orbicular diorite from Epoo, Finland. The concentric shells are alternately richer in biotite (dark) and plagioclase (light). The radius of the orbicule is 10 cm (Photographs courtesy of (a) B. Guy and (b) E. Merino)

interpretation would be to attribute them to symmetry-breaking transitions induced by nonequilibrium constraints.

The climatic conditions that prevailed in the last 200–300 million years were extremely different from those of the present day. During this period, with the exception of the Quaternary era (our era, which began about 2 million years ago), there was practically no ice on the continents and the sea level was about 80 m higher than at present. Climate was particularly mild, and the temperature differences between equatorial (25–30 °C) and polar (8–10 °C) regions were relatively lower.

It was during the Teritary era, some 40 million years ago, that a sharper contrast between equatorial and polar temperatures began to develop. In the relatively short time of 100 000 years, the sea temperature south of New Zealand dropped by several degrees. This was probably the beginning of the Antarctic current, which reduces the exchange of heat between high and low latitudes and contributes to a further cooling of the masses of water "trapped" in this way near the polar regions. Once again, we see a feedback mechanism in action.

At the beginning of the Quaternary era this difference was sufficiently important to allow for the formation and maintenance of continental ice. A series of intermittent glaciations took place in, the northern hemisphere, sometimes pushing the glaciers as far as the middle latitudes. These climatic episodes present an average periodicity of about 100 000 years, though with considerable random-looking variation (Fig. 20.7).

The last advance of continental ice in the northern hemisphere attained its maximum some 18 000 years ago, and its relics are still with us. Although the amount of continental ice today is about 30 million cubic kilometers, confined essentially to Antarctica and Greenland, there was at that time about 70–80 million cubic kilometers covering, in addition, much of North America and northern Europe. Because of the huge quantities of water trapped in the glaciers,

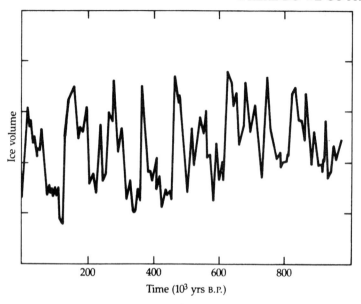

Figure 20.7 Variation of global ice volume during the last 1 million years, inferred from the isotope record of deep-sea cores (After ref. [6])

the sea level was some 120 m lower than today. Since then a large part of the ice has melted, thus defining the coastlines and most of the other features of the present-day landscape. The fact that our ecosystem is unstable makes it difficult to separate the "anthropic signal" from the spontaneous evolution of the system.

In conclusion, we cannot escape the feeling that we live in an age of transition, an age that demands a better understanding of our environment. We must find and explore new resources, and achieve a less destructive coexistence with Nature. We cannot anticipate the outcome of this period of transition, but it is clear that science and especially nonequilibrium physics, is bound to play an increasingly important role in our effort to meet the challenge of understanding and reshaping our global environment.

References

1. Jou, D., *Extended Irreversible Thermodynamics*, 1996, New York: Springer-Verlago.
2. Alder, B. and Wainwright, T. *Phys. Rev. A*, **1** (1970) 18.
3. Resibois, P. and de Leener, M. *Classical Kinetic Theory of Fluids*, 1977, New York: Wiley.
4. Petrosky, T. and Prigogine, I. forthcoming
5. Walgraef, D., *Spatio-Temporal Pattern Formation*, 1997, New York: Springer-Verlago.
6. Nicolis, G. and Prigogine, I. *Exploring Complexity*, 1989, New York: W. H. Freeman.
7. Deneubourg, J. L., Pasteels, J. and Verhaege, J. C., *J. Theor. Biol.*, **105** (1983), 259.

POSTFACE

Nature has a history—for a long time the ideal of physics was geometry, as implied in Einstein's general relatively. Relativity is certainly one of the great achievement of the human mind. But the geometrical view is incomplete. Now we see that narrative elements also play a basic role. This leads to a different concept of nature in which the arrow of time is essential. After all, this arrow appears as the feature which is common to all objects in the expanding bubble which is our universe. We all age in the same direction; all stars, all rocks age in the same direction even if the mechanism of aging is different in each case.

Time, better the direction of time, is the fundamental existential dimension of human life. We discover now that the flow of time is universal. Time is no more separating men from nature.

STANDARD THERMODYNAMIC PROPERTIES

The properties listed are:

$\Delta_f H°$ Standard enthalpy of formation at 298.15 K in J/mol

$\Delta_f G°$ Standard Gibbs energy of formation at 298.15 K in J/mol

$S°$ Standard entropy at 298.15 K in J/mol K

C_p Heat capacity at constant pressure at 298.15 K in J/mol K

The standard state pressure is 100 kPa (1 bar). An entry of 0.0 for $\Delta_f H°$ for an element indicates the reference state of that element.

Molecular formula	Name	State	$\Delta_f H°$ kJ/mol	$\Delta_f G°$ kJ/mol	$S°$ J/mol K	C_p J/mol K
Compounds not containing carbon:						
Ac	Actinium	gas	406.0	366.0	188.1	20.8
Ag	Silver	cry	0.0	0.0	42.6	25.4
AgBr	Silver bromide	cry	−100.4	−96.9	107.1	52.4
AgBrO₃	Silver bromate	cry	−10.5	71.3	151.9	
AgCl	Silver chloride	cry	−127.0	−109.8	96.3	50.8
AgClO₃	Silver chlorate	cry	−30.3	64.5	142.0	
Al	Aluminum	cry	0.0	0.0	28.3	24.4
		gas	330.0	289.4	164.6	21.4
AlB₃H₁₂	Aluminium borohydride	liq	−16.3	145.0	289.1	194.6
AlBr	Aluminum bromide (AlBr)	gas	−4.0	−42.0	239.5	35.6
AlCl	Aluminum chloride (AlCl)	gas	−47.7	−74.1	228.1	35.0
AlCl₃	Aluminum trichloride	cry	−704.2	−628.8	110.7	91.8
AlF	Aluminum fluoride (AlF)	gas	−258.2	−283.7	215.0	31.9
AlF₃	Aluminum trifluoride	cry	−1510.4	−1431.1	66.5	75.1
AlI₃	Aluminum triiodide	cry	−313.8	−300.8	159.0	98.7
AlO₄P	Aluminum phosphate (AlPO₄)	cry	−1733.8	−1617.9	90.8	93.2
AlS	Aluminum sulfide (AlS)	gas	200.9	150.1	230.6	33.4
Al₂O	Aluminum oxide (Al₂O)	gas	−130.0	−159.0	259.4	45.7
Al₂O₃	Aluminum oxide (Al₂O₃)	cry	−1675.7	−1582.3	50.9	79.0
Ar	Argon	gas	0.0		154.8	20.8
As	Arsenic (gray)	cry	0.0		35.1	24.6
AsBr₃	Arsenic tribromide	gas	−130.0	−159.0	363.9	79.2
AsCl₃	Arsenic trichloride	gas	−261.5	−248.9	327.2	75.7
AsF₃	Arsenic trifluoride	liq	−821.3	−774.2	181.2	126.6
As₂	Arsenic (As₂)	gas	222.2	171.9	239.4	35.0
Au	Gold	cry	0.0	0.0	47.4	25.4
AuH	Gold hydride (AuH)	gas	295.0	265.7	211.2	29.2
B	Boron	cry (rhombic)	0.0	0.0	5.9	11.1
BCl	Chloroborane (BCl)	gas	149.5	120.9	213.2	31.7

Molecular formula	Name	State	$\Delta_f H°$ kJ/mol	$\Delta_f G°$ kJ/mol	$S°$ J/mol K	C_p J/mol K
BCl_3	Boron trichloride	liq	−427.2	−387.4	206.3	106.7
BF	Fluoroborane (BF)	gas	−122.2	−149.8	200.5	29.6
BH_3O_3	Boric acid (H_3BO_3)	cry	−1094.3	−968.9	88.8	81.4
BH_4K	Potassium borohydride	cry	−227.4	−160.3	106.3	96.1
BH_4Li	Lithium borohydride	cry	−190.8	−125.0	75.9	82.6
BH_4Na	Sodium borohydride	cry	−188.6	−123.9	101.3	86.8
BN	Boron nitride (BN)	cry	−254.4	−228.4	14.8	19.7
B_2	Boron (B_2)	gas	830.5	774.0	201.9	30.5
Ba	Barium	cry	0.0	0.0	62.8	28.1
		gas	180.0	146.0	170.2	20.8
$BaBr_2$	Barium bromide	cry	−757.3	−736.8	146.0	
$BaCl_2$	Barium chloride	cry	−858.6	−810.4	123.7	75.1
BaF_2	Barium fluoride	cry	−1207.1	−1156.8	96.4	71.2
BaO	Barium oxide	cry	−553.5	−525.1	70.4	47.8
BaO_4S	Barium sulfate	cry	−1473.2	−1362.2	132.2	101.8
Be	Beryllium	cry	0.0	0.0	9.5	16.4
$BeCl_2$	Beryllium chloride	cry	−490.4	−445.6	82.7	64.8
BeF_2	Beryllium fluoride	cry	−1026.8	−979.4	53.4	51.8
BeH_2O_2	Beryllium hydroxide	cry	−902.5	−815.0	51.9	
BeO_4S	Beryllium sulfate	cry	−1205.2	−1093.8	77.9	85.7
Bi	Bismuth	cry	0.0	0.0	56.7	25.5
$BiCl_3$	Bismuth trichloride	cry	−379.1	−315.0	177.0	105.0
Bi_2O_3	Bismuth oxide (Bi_2O_3)	cry	−573.9	−493.7	151.5	113.5
Bi_2S_3	Bismuth sulfide (Bi_2S_3)	cry	−143.1	−140.6	200.4	122.2
Br	Bromine	gas	111.9	82.4	175.0	20.8
BrF	Bromine fluoride	gas	−93.8	−109.2	229.0	33.0
BrH	Hydrogen bromide	gas	−36.3	−53.4	198.7	29.1
BrH_4N	Ammonium bromide	cry	−270.8	−175.2	113.0	96.0
BrK	Potassium bromide	cry	−393.8	−380.7	95.9	52.3
$BrKO_3$	Potassium bromate	cry	−360.2	−217.2	149.2	105.2
BrLi	Lithium bromide	cry	−351.2	−342.0	74.3	
BrNa	Sodium bromide	cry	−361.1	−349.0	86.8	51.4
Br_2Ca	Calcium bromide	cry	−682.8	−663.6	130.0	
Br_2Hg	Mercury bromide ($HgBr_2$)	cry	−170.7	−153.1	172.0	
Br_2Mg	Magnesium bromide	cry	−524.3	−503.8	117.2	
Br_2Zn	Zinc bromide	cry	−328.7	−312.1	138.5	
Br_4Ti	Titanium bromide ($TiBr_4$)	cry	−616.7	−589.5	243.5	131.5
Ca	Calcium	cry	0.0	0.0	41.6	25.9
$CaCl_2$	Calcium chloride	cry	−795.4	−748.8	108.4	72.9
CaF_2	Calcium fluoride	cry	−1228.0	−1175.6	68.5	67.0
CaH_2	Calcium hydride (CaH_2)	cry	−181.5	−142.5	41.4	41.0
CaH_2O_2	Calcium hydroxide	cry	−985.2	−897.5	83.4	87.5
CaN_2O_6	Calcium nitrate	cry	−938.2	−742.8	193.2	149.4
CaO	Calcium oxide	cry	−634.9	−603.3	38.1	42.0
CaO_4S	Calcium sulfate	cry	−1434.5	−1322.0	106.5	99.7
CaS	Calcium sulfide	cry	−482.4	−477.4	56.5	47.4
$Ca_3O_8P_2$	Calcium phosphate	cry	−4120.8	−3884.7	236.0	227.8
Cd	Cadmium	cry	0.0	0.0	51.8	26.0
CdO	Cadmium oxide	cry	−258.4	−228.7	54.8	43.4
CdO_4S	Cadmium sulfate	cry	−933.3	−822.7	123.0	99.6
Cl	Chlorine	gas	121.3	105.3	165.2	21.8
ClCu	Copper chloride (CuC1)	cry	−137.2	−119.9	86.2	48.5

Molecular formula	Name	State	$\Delta_f H°$ kJ/mol	$\Delta_f G°$ kJ/mol	$S°$ J/mol K	C_p J/mol K
ClF	Chlorine fluoride	gas	−50.3	−51.8	217.9	32.1
ClH	Hydrogen chloride	gas	−92.3	−95.3	186.9	29.1
ClHO	Hypochlorous acid (HOCl)	gas	−78.7	−66.1	236.7	37.2
ClH$_4$N	Ammonium chloride	cry	−314.4	−202.9	94.6	84.1
ClK	Potassium chloride (KCl)	cry	−436.5	−408.5	82.6	51.3
ClKO$_3$	Potassium chlorate (KClO$_3$)	cry	−397.7	−296.3	143.1	100.3
ClKO$_4$	Potassium perchlorate (KClO$_4$)	cry	−432.8	−303.1	151.0	112.4
ClLi	Lithium chloride (LiCl)	cry	−408.6	−384.4	59.3	48.0
ClNa	Sodium chloride (NaCl)	cry	−411.2	−384.1	72.1	50.5
ClNaO$_2$	Sodium chloride (NaClO$_2$)	cry	−307.0			
ClNaO$_3$	Sodium chlorate (NaClO$_3$)	cry	−365.8	−262.3	123.4	
Cl$_2$	Chlorine (Cl$_2$)	gas	0.0	0.0	223.1	33.9
Cl$_2$Cu	Copper chloride (CuCl$_2$)	cry	−220.1	−175.7	108.1	71.9
Cl$_2$Mn	Manganese chloride (MnCl$_2$)	cry	−481.3	−440.5	118.2	72.9
Cl$_3$U	Uranium chloride (UCl$_3$)	cry	−866.5	−799.1	159.0	102.5
Cl$_4$Si	Silicon tetrachloride	liq	−687.0	−619.8	239.7	145.3
Co	Cobalt	cry	0.0	0.0	30.0	24.8
CoH$_2$O$_2$	Cobalt hydroxide (Co(OH)$_2$)	cry	−539.7	−454.3	79.0	
CoO	Cobalt oxide (CoO)	cry	−237.9	−214.2	53.0	55.2
Co$_3$O$_4$	Cobalt oxide (Co$_3$O$_4$)	cry	−891.0	−774.0	102.5	123.4
Cr	Chromium	cry	0.0	0.0	23.8	23.4
CrF$_3$	Chromium fluoride (CrF$_3$)	cry	−1159.0	−1088.0	93.9	78.7
Cr$_2$FeO$_4$	Chromium iron oxide (FeCr$_2$O$_4$)	cry	−1444.7	−1343.8	146.0	133.6
Cr$_2$O$_3$	Chromium oxide (Cr$_2$O$_3$)	cry	−1139.7	−1058.1	81.2	118.7
Cs	Cesium	cry	0.0	0.0	85.2	32.2
CsF	Cesium fluoride	cry	−553.5	−525.5	92.8	51.1
Cs$_2$O	Cesium oxide (Cs$_2$O)	cry	−345.8	−308.1	146.9	76.0
Cu	Copper	cry	0.0	0.0	33.2	24.4
CuO	Copper oxide (CuO)	cry	−157.3	−129.7	42.6	42.3
CuO$_4$S	Copper sulfate (CuSO$_4$)	cry	−771.4	−662.2	109.2	
CuS	Copper sulfide (CuS)	cry	−53.1	−53.6	66.5	47.8
Cu$_2$	Copper (Cu$_2$)	gas	484.2	431.9	241.6	36.6
Cu$_2$O	Copper oxide (Cu$_2$O)	cry	−168.6	−146.0	93.1	63.6
Cu$_2$S	Copper sulfide (Cu$_2$S)	cry	−79.5	−86.2	120.9	76.3
F$_2$	Fluorine (F$_2$)	gas	0.0	0.0	202.8	31.3
F	Fluorine	gas	79.4	62.3	158.8	22.7
FH	Hydrogen fluoride	gas	−273.3	−275.4	173.8	
FK	Potassium fluoride (KF)	cry	−567.3	−537.8	66.6	49.0
FLi	Lithium fluoride (LiF)	cry	−616.0	−587.7	35.7	41.6
FNa	Sodium fluoride (NaF)	cry	−576.6	−546.3	51.1	46.9
F$_2$HK	Potassium hydrogen fluoride (KHF$_2$)	cry	−927.7	−859.7	104.3	76.9
F$_2$HNa	Sodium hydrogen fluoride (NaHF$_2$)	cry	−920.3	−852.2	90.9	75.0
F$_2$Mg	Magnesium fluoride	cry	−1124.2	−1071.1	57.2	61.6
F$_2$O$_2$U	Uranyl fluoride	cry	−1648.1	−1551.8	135.6	103.2
F$_2$Si	Difluorosilylene (SiF$_2$)	gas	−619.0	−628.0	252.7	43.9
F$_2$Zn	Zinc fluoride	cry	−764.4	−713.3	73.7	65.7
F$_3$OP	Phosphoryl fluoride	gas	−1254.3	−1205.8	285.4	68.8
F$_3$P	Phosphorus trifluoride	gas	−958.4	−936.9	273.1	58.7
F$_4$S	Sulfur fluoride (SF$_4$)	gas	−763.2	−722.0	299.6	77.6
F$_6$S	Sulfur fluoride (SF$_6$)	gas	−1220.5	−1116.5	291.5	97.0
F$_6$U	Uranium fluoride (UF$_6$)	cry	−2197.0	−2068.5	227.6	166.8
Fe	Iron	cry	0.0	0.0	27.3	25.1

Molecular formula	Name	State	$\Delta_f H°$ kJ/mol	$\Delta_f G°$ kJ/mol	$S°$ J/mol K	C_p J/mol K
FeO$_4$S	Iron sulfate (FeSO$_4$)	cry	-928.4	-820.8	107.5	100.6
FeS	Iron sulfide (FeS)	cry	-100.0	-100.4	60.3	50.5
FeS$_2$	Iron sulfide (FeS$_2$)	cry	-178.2	-166.9	52.9	62.2
Fe$_2$O$_3$	Iron oxide (Fe$_2$O$_3$)	cry	-824.2	-742.2	87.4	103.9
Fe$_3$O$_4$	Iron oxide (Fe$_3$O$_4$)	cry	-1118.4	-1015.4	146.4	143.4
H$_2$	Hydrogen (H$_2$)	gas	0.0	0.0	130.7	28.8
H	Hydrogen	gas	218.0	203.3	114.7	20.8
HI	Hydrogen iodide	gas	26.5	1.7	206.6	29.2
HKO	Potassium hydroxide (KOH)	cry	-424.8	-379.1	78.9	64.9
HLi	Lithium hydride (LiH)	cry	-90.5	-68.3	20.0	27.9
HNO$_2$	Nitrous acid (HONO)	gas	-79.5	-46.0	254.1	45.6
HNO$_3$	Nitric acid	liq	-174.1	-80.7	155.6	109.9
HNa	Sodium hydride	cry	-56.3	-33.5	40.0	36.4
HNaO	Sodium hydroxide (NaOH)	cry	-425.6	-379.5	64.5	59.5
HO	Hydroxyl (OH)	gas	39.0	34.2	183.7	29.9
HO$_2$	Hydroperoxy (HOO)	gas	10.5	22.6	229.0	34.9
H$_2$Mg	Magnesium hydride	cry	-75.3	-35.9	31.1	35.4
H$_2$MgO$_2$	Magnesium hydroxide	cry	-924.5	-833.5	63.2	77.0
H$_2$O	Water	liq	-285.8	-237.1	70.0	75.3
H$_2$O$_2$	Hydrogen peroxide	liq	-187.8	-120.4	109.6	89.1
H$_2$O$_2$Sn	Tin hydroxide (Sn(OH)$_2$)	cry	-561.1	-491.6	155.0	
H$_2$O$_2$Zn	Zinc hydroxide	cry	-641.9	-553.5	81.2	
H$_2$O$_4$S	Sulfuric acid	liq	-814.0	-690.0	156.9	138.9
H$_2$S	Hydrogen sulfide	gas	-20.6	-33.4	205.8	34.2
H$_3$N	Ammonia (NH$_3$)	gas	-45.9	-16.4	192.8	35.1
H$_3$O$_4$P	Phosphoric acid	cry	-1284.4	-1124.3	110.5	106.1
		liq	-1271.7	-1123.6	150.8	145.0
H$_3$P	Phosphine	gas	5.4	13.4	210.2	37.1
H$_4$IN	Ammonium iodide	cry	-201.4	-112.5	117.0	
H$_4$N$_2$	Hydrazine	liq	50.6	149.3	121.2	98.9
H$_4$N$_2$O$_3$	Ammonium nitrate	cry	-365.6	-183.9	151.1	139.3
H$_4$Si	Silane	gas	34.3	56.9	204.6	42.8
H$_8$N$_2$O$_4$S	Ammonium sulfate	cry	-1180.9	-901.7	220.1	187.5
He	Helium	gas	0.0		126.2	20.8
HgI$_2$	Mercury iodide (HgI$_2$) (red)	cry	-105.4	-101.7	180.0	
HgO	Mercury oxide (HgO) (red)	cry	-90.8	-58.5	70.3	44.1
HgS	Mercury sulfide (HgS)	cry	-58.2	-50.6	82.4	48.4
Hg$_2$	Mercury (Hg$_2$)	gas	108.8	68.2	288.1	37.4
Hg$_2$O$_4$S	Mercury sulfate (Hg$_2$SO$_4$)	cry	-743.1	-625.8	200.7	132.0
I	Iodine	gas	106.8	70.2	180.8	20.8
IK	Potassium iodide	cry	-327.9	-324.9	106.3	52.9
IKO$_3$	Potassium iodate	cry	-501.4	-418.4	151.5	106.5
ILi	Lithium iodide	cry	-270.4	-270.3	86.8	51.0
INa	Sodium iodide	cry	-287.8	-286.1	98.5	52.1
INaO$_3$	Sodium iodate	cry	-481.8			92.0
K	Potassium	cry	0.0	0.0	64.7	29.6
KMnO$_4$	Potassium permanganate	cry	-837.2	-737.6	171.7	117.6
KNO$_2$	Potassium nitrite	cry	-369.8	-306.6	152.1	107.4
KNO$_3$	Potassium nitrate	cry	-494.6	-394.9	133.1	96.4
K$_2$O$_4$S	Potassium sulfate	cry	-1437.8	-1321.4	175.6	131.5
K$_2$S	Potassium sulfide (K$_2$S)	cry	-380.7	-364.0	105.0	

Molecular formula	Name	State	$\Delta_f H°$ kJ/mol	$\Delta_f G°$ kJ/mol	$S°$ J/mol K	C_p J/mol K
Li	Lithium	cry	0.0	0.0	29.1	24.8
Li_2	Lithium (Li_2)	gas	215.9	174.4	197.0	36.1
Li_2O	Lithium oxide (Li_2O)	cry	−597.9	−561.2	37.6	54.1
Li_2O_3Si	Lithium metasilicate	cry	−1648.1	−1557.2	79.8	99.1
Li_2O_4S	Lithium sulfate	cry	−1436.5	−1321.7	115.1	117.6
Mg	Magnesium	cry	0.0	0.0	32.7	24.9
MgN_2O_6	Magnesium nitrate	cry	−790.7	−589.4	164.0	141.9
MgO	Magnesium oxide	cry	−601.6	−569.3	27.0	37.2
MgO_4S	Magnesium sulfate	cry	−1284.9	−1170.6	91.6	96.5
MgS	Magnesium sulfide	cry	−346.0	−341.8	50.3	45.6
Mn	Manganese	cry	0.0	0.0	32.0	26.3
$MgNa_2O_4$	Sodium permanganate	cry	−1156.0			
MnO	Manganese oxide (MnO)	cry	−385.2	−362.9	59.7	45.4
MnS	Manganese sulfide (MnS)	cry	−214.2	−218.4	78.2	50.0
Mn_2O_3	Manganese oxide (Mn_2O_3)	cry	−959.0	−881.1	110.5	107.7
Mn_2O_4Si	Manganese silicate (Mn_2SiO_4)	cry	−1730.5	−1632.1	163.2	129.9
N_2	Nitrogen (N_2)	gas	0.0	0.0	191.6	29.1
N	Nitrogen	gas	472.7	455.5	153.3	20.8
$NNaO_2$	Sodium nitrite	cry	−358.7	−284.6	103.8	
$NNaO_3$	Sodium nitrate	cry	−467.9	−367.0	116.5	92.9
NO	Nitrogen oxide	gas	90.25	86.57	210.8	29.84
NO_2	Nitrogen dioxide	gas	33.2	51.3	240.1	37.2
N_2O	Nitrous oxide	gas	82.1	104.2	219.9	38.5
N_2O_3	Nitrogen trioxide	liq	50.3			
N_2O_4	Dinitrogen tetroxide	gas	9.16	97.89	304.29	77.28
N_2O_5	Nitrogen pentoxide	cry	−43.1	113.9	178.2	143.1
Na	Sodium	cry	0.0	0.0	51.3	28.2
NaO_2	Sodium superoxide (NaO_2)	cry	−260.2	−218.4	115.9	72.1
Na_2	Sodium (Na_2)	gas	142.1	103.9	230.2	37.6
Na_2O	Sodium oxide (Na_2O)	cry	−414.2	−375.5	75.1	69.1
Na_2O_2	Sodium peroxide (Na_2O_2)	cry	−510.9	−447.7	95.0	89.2
Na_2O_4S	Sodium sulfate	cry	−1387.1	−1270.2	149.6	128.2
Ne	Neon	gas	0.0		146.3	20.8
Ni	Nickel	cry	0.0	0.0	29.9	26.1
NiO_4S	Nickel sulfate ($NiSO_4$)	cry	−872.9	−759.7	92.0	138.0
NiS	Nickel sulfide (NiS)	cry	−82.0	−79.5	53.0	47.1
O	Oxygen	gas	249.2	231.7	161.1	21.9
OP	Phosphorus oxide (PO)	gas	−28.5	−51.9	222.8	31.8
O_2Pb	Lead oxide (PO_2)	cry	−277.4	−217.3	68.6	64.6
O_2S	Sulfur dioxide	gas	−296.8	−300.1	248.2	39.9
O_2Si	Silicon dioxide (α-quartz)	cry	−910.7	−856.3	41.5	44.4
O_2U	Uranium oxide (UO_2)	cry	−1085.0	−1031.8	77.0	63.6
O_3	Ozone	gas	142.7	163.2	238.9	39.2
O_3PbSi	Lead metasilicate ($PbSiO_3$)	cry	−1145.7	−1062.1	109.6	90.0
O_3S	Sulfur trioxide	gas	−395.7	−371.1	256.8	50.7
O_4SZn	Zinc sulfate	cry	−982.8	−871.5	110.5	99.2
P	Phosphorus (white)	cry	0.0	0.0	41.1	23.8
	Phosphorus (red)	cry	−17.6		22.8	21.2
Pb	Lead	cry	0.0	0.0	64.8	26.4
PbS	Lead sulfide (PbS)	cry	−100.4	−98.7	91.2	49.5
Pt	Platinum	cry	0.0	0.0	41.6	25.9

Molecular formula	Name	State	$\Delta_f H°$ kJ/mol	$\Delta_f G°$ kJ/mol	$S°$ J/mol K	C_p J/mol K
PtS	Platinum sulfide (PtS)	cry	−81.6	−76.1	55.1	43.4
PtS$_2$	Platinum sulfide (PtS$_2$)	cry	−108.8	−99.6	74.7	65.9
S	Sulfur	cry (rhombic)	0.0	0.0	32.1	22.6
	Sulfur	cry (monoclinic)	0.3			
S$_2$	Sulfur (S$_2$)	gas	128.6	79.7	228.2	32.5
Si	Silicon	cry	0.0	0.0	18.8	20.0
Sn	Tin (white)	cry	0.0		51.2	27.0
	Tin (gray)	cry	−2.1	0.1	44.1	25.8
Zn	Zinc	cry	0.0	0.0	41.6	25.4
		gas	130.4	94.8	161.0	20.8

Compounds containing carbon:

Molecular formula	Name	State	$\Delta_f H°$ kJ/mol	$\Delta_f G°$ kJ/mol	$S°$ J/mol K	C_p J/mol K
C	Carbon (graphite)	cry	0.0	0.0	5.7	8.5
	Carbon (diamond)	cry	1.9	2.9	2.4	6.1
CAgN	Silver cyanide (AgCN)	cry	146.0	156.9	107.2	66.7
CBaO$_3$	Barium carbonate (BaCO$_3$)	cry	−1216.3	−1137.6	112.1	85.3
CBrN	Cyanogen bromide	cry	140.5			
CCaO$_3$	Calcium carbonate (clacite)	cry	−1207.6	−1129.1	91.7	83.5
	Calcium carbonate (aragonite)	cry	−1207.8	−1128.2	88.0	82.3
CCl$_2$F$_2$	Dichlorodifluoromethane	gas	−477.4	−439.4	300.8	72.3
CCl$_3$F	Trichlorofluoromethane	liq	−301.3	−236.8	225.4	121.6
CCuN	Copper cyanide (CuCN)	cry	96.2	111.3	84.5	
CFe$_3$	Iron carbide (Fe$_3$C)	cry	25.1	20.1	104.6	105.9
CFeO$_3$	Iron carbonate (FeCO$_3$)	cry	−740.6	−666.7	92.9	82.1
CKN	Potassium cyanide (KCN)	cry	−113.0	−101.9	128.5	66.3
CKNS	Potassium thiocyanate (KSCN)	cry	−200.2	−178.3	124.3	88.5
CK$_2$O$_3$	Potassium carbonate (KCO$_3$)	cry	−1151.0	−1063.5	155.5	114.4
CMgO$_3$	Magnesium carbonate (MgCO$_3$)	cry	−1095.8	−1012.1	65.7	75.5
CNNa	Sodium cyanide (NaCN)	cry	−87.5	−76.4	115.6	70.4
CNNaO	Sodium cyanate	cry	−405.4	−358.1	96.7	86.6
CNa$_2$O$_3$	Sodium carbonate (NaCO$_3$)	cry	−1130.7	−1044.4	135.0	112.3
CO	Carbon monoxide	gas	−110.5	−137.2	197.7	29.1
CO$_2$	Carbon dioxide	gas	−393.5	−394.4	213.8	37.1
CO$_3$Zn	Zinc carbonate (ZnCO$_3$)	cry	−812.8	−731.5	82.4	79.7
CS$_2$	Carbon disulfide	liq	89.0	64.6	151.3	76.4
CSi	Silicon carbide (cubic)	cry	−65.3	−62.8	16.6	26.9
CHBr$_3$	Tribromomethane	liq	−28.5	−5.0	220.9	130.7
CHClF$_2$	Chlorodifluoromethane	gas	−482.6		280.9	55.9
CHCl$_3$	Trichloromethane	liq	−134.5	−73.7	201.7	114.2
CHN	Hydrogen cyanide	liq	108.9	125.0	112.8	70.6
CH$_2$	Methylene	gas	390.4	372.9	194.9	33.8
CH$_2$I$_2$	Diiodomethane	liq	66.9	90.4	174.1	134.0
CH$_2$O	Formaldehyde	gas	−108.6	−102.5	218.8	35.4
CH$_2$O$_2$	Formic acid	liq	−424.7	−361.4	129.0	99.0
CH$_3$	Methyl	gas	145.7	147.9	194.2	38.7
CH$_3$Cl	Chloromethane	gas	−81.9		234.6	40.8
CH$_3$NO$_2$	Nitromethane	liq	−113.1	−14.4	171.8	106.6
CH$_4$	Methane	gas	−74.4	−50.3	186.3	35.3
CH$_4$N$_2$O	Urea	cry	−333.6			
CH$_4$O	Methanol	liq	−239.1	−166.6	126.8	81.1

Molecular formula	Name	State	$\Delta_f H°$ kJ/mol	$\Delta_f G°$ kJ/mol	$S°$ J/mol K	C_p J/mol K
C_2	Carbon (C_2)	gas	831.9	775.9	199.4	43.2
C_2Ca	Calcium carbide	cry	−59.8	−64.9	70.0	62.7
C_2ClF_3	Chlorotrifluoroethylene	gas	−555.2	−523.8	322.1	83.9
C_2Cl_4	Tetrachloroethylene	liq	−50.6	3.0	266.9	143.4
$C_2Cl_4F_2$	1,1,1-Tetrachloro-2,2-difluoroethane	gas	−489.9	−407.0	382.9	123.4
C_2H_2	Acetylene	gas	228.2	210.7	200.9	43.9
$C_2H_2Cl_2$	1,1-Dichloroethylene	liq	−23.9	24.1	201.5	111.3
C_2H_2O	Ketene	gas	−47.5	−48.3	247.6	51.8
$C_2H_2O_4$	Oxalic acid	cry	−821.7		109.8	91.0
$C_2H_3Cl_3$	1,1,1-Trichloroethane	liq	−177.4		227.4	144.3
		gas	−144.6		323.1	93.3
C_2H_3N	Acetonitrile	liq	31.4	77.2	149.6	91.4
$C_2H_3NaO_2$	Sodium acetate	cry	−708.8	−607.2	123.0	79.9
C_2H_4	Ethylene	gas	52.5	68.4	219.6	43.6
$C_2H_4Cl_2$	1,1-Dichloroethane	liq	−158.4	−73.8	211.8	126.3
		gas	−127.7	−70.8	305.1	76.2
$C_2H_4O_2$	Acetic acid	liq	−484.5	−389.9	159.8	123.3
		gas	−432.8	−374.5	282.5	66.5
C_2H_5I	Iodoethane	liq	−40.2	14.7	211.7	115.1
C_2H_6	Ethane	gas	−83.8	−31.9	229.6	52.6
C_2H_6O	Dimethyl ether	gas	−184.1	−112.6	266.4	64.4
C_2H_6O	Ethanol	liq	−277.7	−174.8	160.7	112.3
C_2H_6S	Ethanethiol	liq	−73.6	−5.5	207.0	117.9
C_2H_7N	Dimethylamine	gas	−18.5	68.5	273.1	70.7
C_3H_7N	Cyclopropylamine	liq	45.8		187.7	147.1
C_3H_8	Propane	gas	−104.7			
C_3H_8O	1-Propanol	liq	−302.6		193.6	143.9
$C_3H_8O_3$	Glycerol	liq	−668.5		206.3	218.9
C_4H_4O	Furan	liq	−62.3		177.0	115.3
$C_4H_4O_4$	Fumaric acid	cry	−811.7		168.0	142.0
C_4H_6	1,3-Butadiene	liq	87.9		199.0	123.6
$C_4H_6O_2$	Methyl acrylate	liq	−362.2		239.5	158.8
C_4H_8	Isobutene	liq	−37.5			
C_4H_8	Cyclobutane	liq	3.7			
C_4H_8O	Butanal	liq	−239.2		246.6	163.7
C_4H_8O	Isobutanal	liq	−247.4			
$C_4H_8O_2$	1,4-Dioxane	liq	−353.9		270.2	152.1
$C_4H_8O_2$	Ethyl acetate	liq	−479.3		257.7	170.7
$C_4H_{10}O$	1-Butanol	liq	−327.3		225.8	177.2
$C_4H_{10}O$	2-Butanol	liq	−342.6		214.9	196.9
$C_4H_{12}Si$	Tetramethylsilane	liq	−264.0	−100.0	277.3	204.1
C_5H_8	Cyclopentene	liq	4.4		201.2	122.4
C_5H_{10}	1-Pentene	liq	−46.9		262.6	154.0
C_5H_{10}	Cyclopentane	liq	−105.1		204.5	128.8
C_5H_{12}	Isopentane	liq	−178.5		260.4	164.8
C_5H_{12}	Neopentane	gas	−168.1			
$C_5H_{12}O$	Butyl methyl ether	liq	−290.6		295.3	192.7
C_6H_6	Benzene	liq	49.0			136.3
C_6H_6O	Phenol	cry	−165.1		144.0	127.4
$C_6H_{12}O_6$	α-D-Glucose	cry	−1274.4	−910.52	212.1	
C_7H_8	Toluene	liq	12.4			157.3

Molecular formula	Name	State	$\Delta_f H°$ kJ/mol	$\Delta_f G°$ kJ/mol	$S°$ J/mol K	C_p J/mol K
C_7H_8O	Benzyl alcohol	liq	−160.7		216.7	217.9
C_7H_{14}	Cycloheptane	liq	−156.6			
C_7H_{14}	Ethylcyclopentane	liq	−163.4		279.9	
C_7H_{14}	1-Heptene	liq	−97.9		327.6	211.8
C_7H_{16}	Heptane	liq	−224.2			
C_8H_{16}	Cyclooctane	liq	−167.7			
C_8H_{18}	Octane	liq	−250.1			254.6
		gas	−208.6			
C_9H_{20}	Nonane	liq	−274.7			284.4
$C_9H_{20}O$	1-Nonanol	liq	−456.5			
$C_{10}H_8$	Naphthalene	cry	77.9		167.4	165.7
$C_{10}H_{22}$	Decane	liq	−300.9			314.4
$C_{12}H_{10}$	Biphenyl	cry	99.4		209.4	198.4
$C_{12}H_{22}O_{11}$	Sucrose	cry	−2222.1	−1544	360.2	
$C_{12}H_{26}$	Dodecane	liq	−350.9			375.8

PHYSICAL CONSTANTS AND DATA

Avogadro constant	$N_A = 6.022137 \times 10^{23} \text{ mol}^{-1}$
Boltzmann constant	$k_B = 1.38066 \times 10^{-23} \text{ J K}^{-1}$
Gas constant	$R = 8.314 \text{ J mol}^{-1} \text{ K}^{-1}$
	$= 0.082058 \text{ atm L mol}^{-1} \text{ K}^{-1}$
	$= 1.9872 \text{ cal mol}^{-1} \text{ K}^{-1}$
Faraday constant	$F = 9.6485 \times 10^{4} \text{ C mol}^{-1}$
Stefan-Boltzmann constant	$\sigma = (c\beta/4) = 5.6705 \times 10^{-8} \text{ J m}^{-2} \text{ K}^{-4} \text{ s}^{-1}$
Triple point of water*	$T_{tp}(H_2O) = 273.16 \text{ K}$
Zero of Celsius scale*	$T(0°C) = 273.15 \text{ K}$
Molar volume of ideal gas at 1 bar and 273.15 K	$V_0 = 22.711 \text{ L mol}^{-1}$
Permittivity of vacuum*	$\varepsilon_o = 8.854187816 \times 10^{-12} \text{ C}^2 \text{ N}^{-1} \text{ m}^{-2}$
Permeability of vacuum*	$\mu_o = 4\pi \times 10^{-7} \text{ N A}^{-2}$
Speed of light in vacuum*	$c = 2.99792458 \times 10^{8} \text{ m s}^{-1}$
Planck constant	$h = 6.62607 \times 10^{-34} \text{ J s}$
Elementary charge	$e = 1.60218 \times 10^{-19} \text{ C}$
Electron rest mass	$m_e = 9.10939 \times 10^{-31} \text{ kg}$
	$= 5.486 \times 10^{-4} \text{ u}$
Proton rest mass	$m_p = 1.67262 \times 10^{-27} \text{ kg}$
	$= 1.00728 \text{ u}$
Neutron rest mass	$m_n = 1.67493 \times 10^{-27} \text{ kg}$
	$= 1.00867 \text{ u}$
Gravitational constant	$G = 6.6726 \times 10^{-11} \text{ N m}^2 \text{ kg}^{-2}$
Standard gravitational acceleration*	$g = 9.80665 \text{ m s}^{-2} = 32.17 \text{ ft s}^{-2}$
Mass of the earth	$5.98 \times 10^{24} \text{ kg}$
Average radius of the earth	$6.37 \times 10^{3} \text{ km}$
Average earth-sun distance	$1.495 \times 10^{8} \text{ km}$
Radius of the sun	$7 \times 10^{5} \text{ km}$
Sun's energy incident at the top of the atmosphere About 31% of this energy is reflected back into space.	$1340 \text{ J m}^{-2} \text{ s}^{-1} = 0.032 \text{ cal cm}^{-2} \text{ s}^{-1}$

*Exact values by definition

Conversion Factors

Bar	1 bar $= 10^5$ Pa
Standard atmosphere*	1 atm $= 1.01325 \times 10^5$ Pa
Torr (mm Hg)	1 Torr $= 133.322$ Pa
Calorie*	1 cal $= 4.184$ J
Erg	1 erg $= 10^{-7}$ J
Gauss	1 G $= 10^{-4}$ T
Debye	1 D $= 3.33564 \times 10^{-30}$ C m
Atomic mass unit	1 u $= 1.66054 \times 10^{-27}$ kg
Electron volt	1 eV $= 1.60218 \times 10^{-19}$ J
	$= 96.4853$ kJ mol^{-1}
Metric ton	1 metric ton $= 1000$ kg
Pound	1 lb $= 16$ oz $= 0.45359$ kg
Gallon (U.S.)	1 gal $= 4$ quarts $= 3.78541$ L
Gallon (British imperial)	1 gal $= 4$ quarts $= 4.545$ L

---

* Exact values by definition

INDEX

Index compiled by Geoffrey C. Jones